EMS Series of Lectures in Mathematics

Edited by Andrew Ranicki (University of Edinburgh, U.K.)

EMS Series of Lectures in Mathematics is a book series aimed at students, professional mathematicians and scientists. It publishes polished notes arising from seminars or lecture series in all fields of pure and applied mathematics, including the reissue of classic texts of continuing interest. The individual volumes are intended to give a rapid and accessible introduction into their particular subject, guiding the audience to topics of current research and the more advanced and specialized literature.

Previously published in this series:

Katrin Wehrheim, *Uhlenbeck Compactness*
Torsten Ekedahl, *One Semester of Elliptic Curves*
Sergey V. Matveev, *Lectures on Algebraic Topology*
Joseph C. Várilly, *An Introduction to Noncommutative Geometry*
Reto Müller, *Differential Harnack Inequalities and the Ricci Flow*
Eustasio del Barrio, Paul Deheuvels and Sara van de Geer, *Lectures on Empirical Processes*
Iskander A. Taimanov, *Lectures on Differential Geometry*
Martin J. Mohlenkamp, María Cristina Pereyra, *Wavelets, Their Friends, and What They Can Do for You*
Stanley E. Payne and Joseph A. Thas, *Finite Generalized Quadrangles*
Masoud Khalkhali, *Basic Noncommutative Geometry*

Helge Holden
Kenneth H. Karlsen
Knut-Andreas Lie
Nils Henrik Risebro

Splitting Methods for Partial Differential Equations with Rough Solutions

Analysis and MATLAB programs

Authors:

Prof. Helge Holden
Department of Mathematical Sciences
Norwegian University of Science and Technology
NO-7491 TRONDHEIM
NORWAY

and

Centre of Mathematics for Applications
University of Oslo
P.O. Box 1053 Blindern
NO-0316 OSLO
NORWAY

Prof. Kenneth H. Karlsen
Prof. Nils Henrik Risebro
Centre of Mathematics for Applications
Department of Mathematics
University of Oslo
P.O. Box 1053 Blindern
NO-0316 OSLO
NORWAY

Prof. Knut-Andreas Lie
Centre for Mathematics of Applications
SINTEF
Department of Applied Mathematics
P.O.Box 124 Blindern
NO-0314 OSLO
NORWAY

Helge Holden har mottatt støtte fra Det faglitterære fond

2010 Mathematics Subject Classification (primary; secondary): 35L65, 35K65, 65M99, 65M15, 65M12; 35L67, 35D30, 47N40, 35A35

Key words: Operator splitting, nonlinear partial differential equations, evolution equations, hyperbolic conservation laws, degenerate convection-diffusion equations, finite difference methods, finite volume methods, front tracking, Matlab codes

ISBN 978-3-03719-078-4

The Swiss National Library lists this publication in The Swiss Book, the Swiss national bibliography, and the detailed bibliographic data are available on the Internet at http://www.helveticat.ch.

Contact address:
European Mathematical Society Publishing House
Seminar for Applied Mathematics
ETH-Zentrum FLI C4
CH-8092 Zürich
Switzerland

Phone: +41 (0)44 632 34 36
Email: info@ems-ph.org
Homepage: www.ems-ph.org

Printed on acid-free paper produced from chlorine-free pulp. TCF (unendlich)
Printed in Germany

9 8 7 6 5 4 3 2 1

Preface

The book has grown out of a concerted research effort over the last decade. We have enjoyed collaboration with many good friends and colleagues on these problems, in particular, Raimund Bürger, Giuseppe Coclite, Helge Dahle, Magne Espedal, Steinar Evje, Harald Hanche-Olsen, Runar Holdahl, Trygve Karper, Vegard Kippe, Siddhartha Mishra, Xavier Raynaud, and John Towers, and we use this opportunity to thank them for the joy of collaboration.

Our research has been supported in part by the Research Council of Norway. K.-A. Lie gratefully acknowledges support from Simula Research Laboratory for part of his work on this project. K. H. Karlsen has also been supported by an Outstanding Young Investigators Award from the Research Council of Norway. Part of the book was written as part of the international research program on Nonlinear Partial Differential Equations at the Centre for Advanced Study at the Norwegian Academy of Science and Letters in Oslo during the academic year 2008–09.

We have established a web site, www.math.ntnu.no/operatorsplitting, where we will post computer codes in MATLAB[1] for the examples as well as a list of errata. Please let us know if you find errors.

Trondheim and Oslo, April 7, 2010

Helge Holden   Kenneth H. Karlsen   Knut-Andreas Lie   Nils Henrik Risebro

[1]MATLAB is a registered trademark of the MathWorks, Inc.

Contents

1

Introduction

Partial differential equations (PDEs) have become enormously successful as models of physical phenomena. With the rapid increase in computing power in recent years, such models have permeated virtually every physical and engineering problem. The phenomena modeled by partial differential equations become increasingly complicated, and so do the partial differential equations themselves. Often, one wishes a model to capture different aspects of a situation, for instance both convective transport and dispersive oscillations on a small scale. These different aspects of the model are then reflected in a partial differential equation, which may contain terms (operators) that are mathematically very different, making these models hard to analyze, both theoretically and numerically.

A computational scientist is therefore often faced with new and complex equations for which an efficient solution method must be developed. If one is lucky, the equation is of a well-known type, and it is fairly easy to find efficient methods that are simple to implement. In most cases, however, one is not so lucky; good methods may be hard to find, and even good methods may be hard to implement.

A strategy to deal with complicated problems is to "divide and conquer". In the context of equations of evolution type, a rather successful approach in this spirit has been *operator splitting.*

The idea behind this type of approach is that the overall evolution operator is formally written as a sum of evolution operators for each term (operator) in the model. In other words, one splits the model into a set of sub-equations, where each sub-equation is of a type for which simpler and more practical algorithms are available. The overall numerical method is then formed by picking an appropriate numerical scheme for each sub-equation and piecing the schemes together by operator splitting.

In an abstract way one can formulate the method as follows: We want to solve the Cauchy problem

$$\frac{dU}{dt} + \mathcal{A}(U) = 0, \qquad U(0) = U_0, \tag{1.1}$$

where $\mathcal{A}$ is some unspecified operator. Formally (but not very helpful from a computational point of view) the solution reads[1]

$$U(t) = e^{-t\mathcal{A}} U_0. \tag{1.2}$$

[1] Inspired by the case when $\mathcal{A}(U) = AU$, where A is a finite matrix, we formally write $e^{-t\mathcal{A}}$ for the solution operator.

Assume that we can write $\mathcal{A} = \mathcal{A}_1 + \mathcal{A}_2$ in some "natural" way, and that one can solve the sub-problems

$$\frac{dU}{dt} + \mathcal{A}_j(U) = 0, \qquad U(0) = U_0, \quad j = 1, 2, \tag{1.3}$$

more easily with formal solutions

$$U_j(t) = e^{-t\mathcal{A}_j} U_0, \quad j = 1, 2. \tag{1.4}$$

In its simplest form, operator splitting reads as follows: Let $t_n = n\Delta t$ (with Δt small and positive). Approximately, we hope that

$$U(t_{n+1}) \approx e^{-\Delta t \mathcal{A}_2} e^{-\Delta t \mathcal{A}_1} U(t_n). \tag{1.5}$$

For commuting operators we have that $e^{-t\mathcal{A}_2} e^{-t\mathcal{A}_1} = e^{-t\mathcal{A}}$ and the method would be exact. Taking it one step further, one could hope that

$$U(t) = e^{t\mathcal{A}} U_0 = \lim_{\Delta t \downarrow 0,\, t = n\Delta t} \big(e^{-\Delta t \mathcal{A}_2} e^{-\Delta t \mathcal{A}_1}\big)^n U_0, \tag{1.6}$$

(with a limit to be determined) which indeed is the celebrated *Lie–Trotter–Kato formula.* A numerical method is obtained if one replaces the exact solution operators $e^{t\mathcal{A}_j}$ by numerical approximations. All splitting methods are refinements of this basic set-up.

This approach may seem a bit primitive at first glance, but in fact operator splitting has several advantages. Since the operators in the new submodels may be very different, they may also require very different numerical and analytical techniques. Operator splitting allows one to exploit this, and the resulting numerical method may be both simpler to implement and more efficient. By operator splitting, one can combine specialized numerical methods that have been developed to solve a particular class of evolutionary problems (i.e., developed especially for one of the elementary operators) in a fairly straightforward manner. This way, one can choose from a toolbox of highly efficient and well-tested numerical methods for elementary operators that can be combined to solve complicated problems. Indeed, the operator-splitting framework offers great flexibility in replacing one scheme for an elementary operator with another scheme for the same operator. Moreover, the use of operator splitting may also reduce memory requirements, increase the stability range, and even provide methods that are unconditionally stable. For very high dimensional problems this may be the only feasible method. Finally, by resorting to operator splitting, it is also easy to add increasing complexity to a numerical model, since each new term can be an independent numerical module.

The idea of splitting sums of complicated operators into simpler operators that are treated separately, is both easy and fundamental, and as such has appeared under various names in different contexts. We will here indicate some of the historical development, with no ambition of providing a complete survey. One of the first rigorous results is associated with the name of Trotter [268]. The fundamental question he asked was: Given two continuous semi-groups with corresponding generators, how can one define the semigroup corresponding to the sum of the two generators? This corresponds to the equations (1.1)–(1.6) above. The result in the case of finite-dimensional matrices goes back to Sophus Lie and important extensions were provided by Kato [152]; the result is often denoted as the *Lie–Trotter–Kato formula* or simply the *Trotter formula.* Applications by Trotter and Kato were to quantum mechanics. Several refinements of this method exist, for instance, the *Baker–Campbell–Hausdorff formula* expresses the operator $\tilde{A}$ with the property that $e^{-tA_2}e^{-tA_1} = e^{-t\tilde{A}}$.

In a more concrete setting (and prior to Trotter and Kato), Douglas, Peaceman, and Rachford [90, 220] introduced a method called the *alternating direction implicit* (ADI) scheme, where multi-dimensional problems were successfully reduced to repeated one-dimensional problems. The ADI method was soon applied to petroleum reservoir simulation. In the late 1960s, increased computer power made other methods viable for reservoir simulations. Starting in the late 1950s and early 60s, there was an extensive development in the Soviet Union, using what was coined *splitting methods* or the *fractional steps method* as a general method to study a large variety of problems in mathematical physics and several applications. Key advances were made by Yanenko, Samarskii, Marchuk and others. It is impossible here to survey the results obtained; rather we refer to Yanenko's monograph [278], and the comprehensive survey by Marchuk [203]. Related to these methods is the method of *locally one-dimensional* (LOD) methods, where a dimensional splitting in effect reduces the original problem to a series of one-dimensional problems. For a general survey of splitting methods we refer to Hundsdorfer and Verwer [129, Ch. IV]. For matrix-related methods we refer to [208]. Observe that one often finds the same method denoted by different names, and the same name used for different methods. This is due to wide applicability of the method, but it complicates an accurate historical description of the development.

Most of the refinements depend on further knowledge of properties of the underlying sub-problems. Detailed knowledge of the behavior of solutions can make rather powerful methods. Here we will analyze operator splitting for a class of nonlinear partial differential equations, see Section 1.2, with the property that the solutions are *rough*, i.e., the solutions are functions of limited regularity and may even contain jump discontinuities, called shocks, so that the equations have to be interpreted in the sense of distributions.

Operator splitting may not always be the right answer. The extent to which operator splitting will give an effective overall method depends on the coupling of different elementary operators and the dynamics of the evolution problem. If the elementary operators are weakly coupled—that is, if the interaction of the different physical phenomena has a long time scale—an operator-splitting scheme will be efficient over a wide range of sizes for the splitting steps. Furthermore, for higher-dimensional problems it may be the only feasible method. On the other hand, if the operators interact significantly over a short time scale, operator splitting may be subject to severe restrictions on the splitting step. For nonlinear operators, interaction between elementary operators is often nonlinear, and splitting them into separate steps may result in large and unwanted errors. To prevent or remedy such splitting errors requires a thorough understanding of the underlying error mechanisms.

1.1 Purpose of the book

The purpose of this book is to give an introduction to various types of operator-splitting methods for constructing discontinuous, but physically relevant, solutions of nonlinear mixed hyperbolic-parabolic partial differential equations. The theory is illustrated by several examples and MATLAB code for most of the examples is posted on the web site

www.math.ntnu.no/operatorsplitting

The class of equations is very rich, and contains, for instance, hyperbolic conservation laws, heat (diffusion) equations, porous medium equations, two-phase reservoir flow equations, as well as (strongly) degenerate convection-diffusion equations with applications to sedimentation. These equations are frequently also referred to as degenerate (or degenerate parabolic) convection-diffusion equations. A significant part of this book is devoted to reporting the results of applying operator-splitting methods to a variety of convection dominated problems, including problems coming from flow in porous media, shallow water waves, and gas dynamics. Along the way we make an effort to provide enough (algorithmic) details so as to enable the readers themselves to implement the presented methods without too much effort. Another significant part of this book aims at introducing the reader to the basic parts of a theoretical foundation of operator-splitting methods for convection-dominated problems possessing solutions with limited regularity or even discontinuous solutions. Although the theory is restricted to problems consisting of scalar and weakly coupled systems of equations, it nevertheless provides guiding principles for designing accurate and efficient operator-splitting methods for systems of equations. A novelty of this book is that it develops a theoretical framework for operator-splitting methods based on recent 'hyperbolic' techniques. This enables us to treat the whole spectrum of equations in a unified manner, ranging from purely hyperbolic equations possessing shock wave (discontinuous) solutions, via degenerate parabolic equations admitting solutions with limited regularity or even shock wave solutions in the case of degeneracy on intervals, to uniformly parabolic convection-diffusion equations possessing smooth solutions. Furthermore, since it turns out that the hyperbolic arguments also apply to many weakly coupled systems of partial differential equations, we will in fact develop a convergence theory for a general class of weakly coupled systems of equations.

Hence this theory not only covers scalar equations but also many physically interesting weakly coupled systems of hyperbolic and parabolic equations. For some of the operator-splitting methods we identify intrinsic splitting-error mechanisms as well as present procedures for reducing the errors originating from these mechanisms.

1.2 The class of PDEs discussed in the book

We will in the following develop a theoretical framework for operator-splitting methods in the setting of systems of weakly coupled nonlinear partial differential equations of the type

$$\begin{aligned} u_t^\kappa + \sum_i F_i^\kappa(u^\kappa)_{x_i} &= \Delta A^\kappa(u^\kappa) + g^\kappa(U), \qquad (x,t) \in \mathbb{R}^d \times [0,T], \\ u^\kappa\big|_{t=0} &= u_0^\kappa, \quad \kappa = 1, \dots, K, \end{aligned} \tag{1.7}$$

with $U(x,t) = (u^1(x,t), \dots, u^K(x,t))$. The term *weakly coupled* means that the equations are coupled only through the source term $g^\kappa(U)$. The diffusive term is assumed to satisfy

$$\frac{dA^\kappa}{du}(u) \geq 0, \qquad \kappa = 1, \dots, K, \quad A^\kappa(0) = 0,$$

where the essential condition is the first one, under which (1.7) is referred to as *degenerate* or sometimes *degenerate parabolic.* A mild form of degeneracy occurs if for some κ we have $\frac{dA^\kappa}{du}(u) = 0$ for one or several values of u, in which case one often speaks of point degeneracy. A more severe form of degeneracy occurs if for some κ we have $\frac{dA^\kappa}{du}(u) = 0$ for u in some interval. In this case one often says that (1.7) is strongly degenerate. In other words, (1.7) is strongly degenerate if A^κ is constant on intervals. In general, the system (1.7) possesses solutions with limited regularity, i.e., weak solutions in the sense of distributions. Despite the restriction "weakly coupled", partial differential equations like (1.7) include several important model equations.

When $g^\kappa \equiv 0$ for all κ, the system (1.7) becomes a set of independent scalar partial differential equations. In particular, the scalar conservation law

$$u_t + \nabla \cdot f(u) = 0 \tag{1.8}$$

is a simple special case of (1.7) for $K = 1$. The regularized conservation law

$$u_t + \nabla \cdot f(u) = \Delta u \tag{1.9}$$

is another equation within the class analyzed here. Included is also the heat equation

$$u_t = \Delta u, \tag{1.10}$$

the one-point degenerate porous medium equation

$$u_t = \Delta u^m, \qquad m \geq 1, \tag{1.11}$$

the two-point degenerate two-phase reservoir flow equation

$$u_t + \Big(\frac{u^2}{u^2 + (1-u)^2}\Big)_x = \big(u(1-u)\big)_{xx}, \tag{1.12}$$

as well as the nonlinear, possibly strongly degenerate, convection-diffusion equation

$$u_t + \nabla \cdot f(u) = \Delta A(u), \qquad A' \geq 0. \tag{1.13}$$

An example of a strongly degenerate convection-diffusion equation is provided by the theory of sedimentation-consolidation processes [47]. In this theory a typical choice of A satisfies

$$A'(u) \begin{cases} = 0, & u \in [0, u_c], \\ > 0, & u \notin [0, u_c], \end{cases}$$

where $u_c > 0$ is a given constant, i.e., A is flat (constant) on the interval $[0, u_c]$. On $[0, u_c]$, equation (1.13) reduces to a hyperbolic equation (1.8). Consequently, degenerate convection-diffusion equations will in general possess all the features of hyperbolic conservation laws, including the existence of shock wave (discontinuous) solutions, the necessity of using weak solutions, the loss of uniqueness of weak solutions, the need for additional selection criteria (entropy conditions) to restore uniqueness, and so forth.

1.3 Operator splitting for initial-value problems

Let us revisit the abstract approach (1.1)–(1.6). Writing the system (1.7) as an abstract Cauchy problem

$$\frac{dU}{dt} + \mathcal{A}(U) = 0, \qquad U(0) = U_0, \tag{1.14}$$

with solution $U(t) = \mathcal{S}_t U_0$, the operator $\mathcal{A}$ can often be decomposed as a sum of elementary (simpler) operators in a natural way. As an example assume that $\mathcal{A} = \mathcal{A}_1 + \mathcal{A}_2$. Using the semigroup notation $U^j = \mathcal{S}_t^j U_0$ for the solution of

$$\frac{dU^j}{dt} + \mathcal{A}_j(U^j) = 0, \qquad U^j(0) = U_0, \qquad j = 1, 2, \tag{1.15}$$

we approximate the solution of (1.14) by

$$U(n\Delta t) \approx \big[\mathcal{S}^2_{\Delta t}\mathcal{S}^1_{\Delta t}\big]^n U_0. \tag{1.16}$$

An alternative splitting formula is obtained by reversing the order of the operators, but this will in general give a different approximation.

The aim is to prove a Trotter formula like

$$U(t) = \mathcal{S}_t U_0 = \lim_{n\to\infty} \left[\mathcal{S}^2_{\Delta t}\mathcal{S}^1_{\Delta t}\right]^n U_0 = \lim_{n\to\infty} \left[\mathcal{S}^2_{t/n}\mathcal{S}^1_{t/n}\right]^n U_0.$$

To obtain a numerical solution, we replace the exact solutions operators $\mathcal{S}_j$ by approximations, with the goal of proving that the Trotter formula still holds.

The operator splitting in (1.16) is only first-order accurate. As an alternative, one can use the so-called *Strang splitting*,

$$U(n\Delta t) \approx \left[\mathcal{S}^1_{\Delta t/2}\mathcal{S}^2_{\Delta t}\mathcal{S}^1_{\Delta t/2}\right]^n U_0. \tag{1.17}$$

which is formally second-order accurate for sufficiently smooth solutions. The two operator splittings (1.16) and (1.17) are examples of so-called *multiplicative* operator splittings, which will be the main focus in this book.

Multiplicative operator splitting is closely related to the ADI method. To explain the idea behind ADI, we replace the exact evolution operators in (1.15) by standard forward/backward Euler approximations. Writing $U^n = U(n\Delta t)$, the classical ADI method reads

$$\begin{aligned} U^{n+\frac{1}{2}} + \mathcal{A}_1\big(U^{n+\frac{1}{2}}\big) &= -\mathcal{A}_2\big(U^n\big), \\ U^{n+1} + \mathcal{A}_2\big(U^{n+1}\big) &= -\mathcal{A}_1\big(U^{n+\frac{1}{2}}\big). \end{aligned} \tag{1.18}$$

Another class of operator splitting is the so-called *additive operator splitting* (AOS). The first-order equivalent of (1.16) reads

$$U(n\Delta t) \approx \left[\frac{1}{2}(\mathcal{S}^2_{2\Delta t} + \mathcal{S}^1_{2\Delta t})\right]^n U_0, \tag{1.19}$$

whereas the second-order equivalent of the Strang splitting reads

$$U(n\Delta t) \approx \left[\tfrac{1}{2}(\mathcal{S}^1_{\Delta t}\mathcal{S}^2_{\Delta t} + \mathcal{S}^2_{\Delta t}\mathcal{S}^1_{\Delta t})\right]^n U_0. \tag{1.20}$$

There are two main motivations for these operator splittings. First of all, the result of an additive operator splitting is independent of the order in which the operators are applied. For the multiplicative operator splitting, the operators will generally not commute in the nonlinear case, which means that the result depends on the order in which the operators are applied. This is the main reason why AOS methods are gaining popularity within image processing, even though they generally are less accurate than multiplicative splittings. The second advantage of AOS methods is that, since the operators are applied independently, they can be computed in parallel. AOS methods are therefore often used in combination with parallel processing.

1.4 Operator splitting for convection-diffusion equations

As already mentioned, we will in this book study nonlinear evolutionary PDEs of mixed hyperbolic-parabolic type. By advocating operator splitting, the numerical solution of the abstract problem (1.14) is reduced to the numerical solution of simplified problems of the type (1.15), for which one may utilize highly efficient methods which are tailor-made for each simplified subproblem. In recent years, we have witnessed an immense activity in developing sophisticated numerical methods for hyperbolic partial differential equations. We refer to [15, 67, 101, 104, 107, 108, 115, 159, 175, 176, 260, 262] for an introduction to modern numerical methods for hyperbolic equations. It is a reasonable strategy to attempt to utilize some of these hyperbolic solvers as building blocks in numerical methods for convection-diffusion problems. Indeed, in this book we make use of a diversity of hyperbolic solvers, including monotone schemes such as the upwind and Godunov schemes, quasi-monotone schemes, front tracking, large-time-step Godunov or Glimm methods, characteristic Galerkin methods, second-order MUSCL schemes, and high-order nonoscillatory central schemes. It is well-known that an accurate numerical approximation of convective and diffusive processes is a very difficult matter. This is especially true if convection dominates diffusion, which is the quintessential case. Accurate numerical simulations in such cases are often complicated by excessive amounts of unphysical oscillations or numerical diffusion. Often numerical methods based operator splitting and modern hyperbolic solvers avoid undue amounts of oscillations and diffusion.

A typical splitting approach for convection-diffusion equations involves not only hyperbolic equations modeling convection effects, but also (possibly degenerate) parabolic equations imitating diffusion effects. In this book, we will rely on very simple finite-difference schemes to approximate these parabolic equations. However, there exists a diversity of numerical methods that have been developed over the last fifty years—including finite-difference, finite-volume, and finite-element methods. To learn about numerical methods for elliptic and parabolic equations we invite the reader to take a closer look at one or several of the references [40, 101, 104, 112, 129, 136, 154, 209, 230, 241, 247, 259–261, 269].

1.5 Rigorous analysis of operator-splitting methods

A key focus of this book is the analysis of what happens when the exact solution operators $\mathcal{S}_j$ are replaced by approximate solvers. To this end, we provide a general theoretical framework by which it follows that if the approximate solvers for $\mathcal{S}_j$, cf. (1.15), satisfy certain properties, then the corresponding operator-splitting method will converge to the exact solution of the underlying partial differential equation. This framework includes scalar and weakly coupled systems of nonlinear partial differential equations containing various combinations of hyperbolic and parabolic effects, cf. (1.7). As a pedagogic device to convey to novice readers the fundamental parts of this framework, we will frequently illustrate the main methodological concepts and results on simplified problems.

Let us recall that operator-splitting methods can always be analyzed in terms of accuracy by straightforward Taylor expansions, at least formally. For recent work on such analysis of operator splitting from the point of view of the Lie operator formalism, see [171]. However, in terms of rigorous analysis, this approach is not satisfactory, since nonlinear partial differential equations in general will possess solutions that exhibit complex behavior in small regions in space and time, i.e., sharp transitions or even singularities like shock waves (discontinuities). Moreover, even if the underlying exact solution is smooth, it can be that the operator splitting is composed of solution operators $\mathcal{S}_j$ that may produce nonsmooth solutions. An example of this case is provided by viscous operator splitting of nonlinear convection-diffusion equations, in which the nonlinear convection operator may introduce discontinuities into an otherwise smooth solution.

The general convergence framework developed in this book, in the context of fully discrete operator-splitting methods for weakly coupled systems of equations containing a synthesis of hyperbolic and parabolic effects, is based on the so-called Kružkov L^1-entropy solution theory. This pioneering theory was originally developed by Kružkov [161] for first-order quasilinear hyperbolic equations and only recently extended by Carrillo [50] to second-order quasilinear degenerate parabolic equations. Our convergence theory includes and extends previous (L^1) convergence results for problem-specific operator-splitting methods. For weakly coupled systems of hyperbolic conservation laws we also provide abstract L^1-error estimates for dimensional-splitting methods. Hence, in order to verify convergence (or convergence rates), one only has to check whether *each* method satisfies certain assumptions, whereupon convergence follows. When applied in a specific situation, these abstract error estimates avoid Kružkov's usual doubling of variables. We consider a variety of semi-discrete and fully discrete operator-splitting methods, including dimensional splitting, viscous splitting, flux splitting, and source-term splitting, and verify that the conditions needed to apply the abstract convergence results hold. The main advantage of the L^1-approach is that it makes it possible to have a unifying convergence theory for hyperbolic, parabolic, and mixed hyperbolic-parabolic problems. In the parabolic setup, where alternative approaches are possible, the L^1-approach has the advantage that it yields results that are independent of the Peclet number, i.e., the ratio of convection forces to diffusion forces. At this point it should be stressed that the development of a unifying theoretical framework is possible only due to the recent forming of a mathematical theory for discontinuous solutions of strongly degenerate convection-diffusion equations [20, 45, 48–50, 55–58, 102, 147, 148, 178, 202, 205, 207, 222, 231, 232, 251, 252, 270].

The idea behind operator splitting is certainly an old one and has been comprehensively described, for example, in [63, 129, 203, 206, 246]. The new, and to a certain extent original, aspect of our presentation lies in the systematic use of numerical schemes and mathematical theory associated with hyperbolic equations. We use this hyperbolic approach to construct splitting methods and a corresponding unifying convergence theory for degenerate convection-diffusion problems. In doing so we are building on and extending our previous (unified) analysis of operator splitting methods [119, 120], which again is based on ideas that have evolved over several years [34–37, 43, 93, 97, 100, 114, 121, 124–126, 131–133, 139–143, 145, 146, 182]. Consult [93] for a review of this activity. Another unconventional facet of our presentation is the focus on splitting methods that allow for large time-steps. The use of large time-step methods for the convection step, like front tracking, has some advantages. For example, in the setting of a nonlinear convection-diffusion equations and an implicit diffusion solver, the resulting operator-splitting methods become unconditionally stable in the sense that there is no CFL condition (named for its originators Courant, Friedrichs, and Lewy) restricting the time step. Indeed, it has always been our firm belief that the time-step in a numerical method should be dictated by the dynamics of the equation and not by the spatial discretization. For convection-diffusion equations, it turns out that a practicable time-step is highly dictated by the degree of (nonlinear) interaction between convective and diffusive forces. Unfortunately, large time-steps can lead to fronts (sharp transitions in the solution) that are too wide: A recurrent theme in this book is that it is possible to identify and reduce this kind of splitting errors, thereby yielding accurate large time-step methods, along the lines of the approach initiated in [146] and further developed and analyzed in [100, 140–143]. The approach in [146] was motivated by an idea introduced in [92] and further expanded on in a series of papers [74–78].

Besides viscous splitting methods for convection-diffusion problems, we will devote considerable attention to methods for hyperbolic problems based on dimensional splitting [71, 125, 139, 180, 182] as well as source splitting [170, 177, 223, 253, 254].

1.6 Topics not treated in the book

Before we end this introductory chapter, let us list some important topics that are not treated in this book. First of all, there are of course many numerical approaches that do not rely on operator splitting, cf. the lists of references given above for hyperbolic and parabolic problems and cf. [6, 32, 44, 46, 51, 60, 96, 98, 99, 102, 111, 147, 164, 207, 216, 217] for mixed hyperbolic-parabolic problems. Regarding convection-diffusion problems, we omit the Godunov-mixed operator splitting methods [81–84, 274] and the recent fast explicit operator-splitting methods [61, 62]. Next, we do not address the numerical solution of so-called elliptic-parabolic problems, see, e.g., [3, 5, 24, 130, 137, 138, 197, 219, 236, 237]. Moreover, we do not discuss hyperbolic and mixed hyperbolic-parabolic problems with discontinuous flux, see, for example, [149–151, 153, 266, 267]. Nor do we go into operator splitting for reaction-diffusion equations, as in for example [86–88, 235, 245]. Nonlinear convection-diffusion equations can be seen as toy models for the fundamental equations of fluid flow—the Navier–Stokes equations—and viscous operator splitting for the Navier–Stokes equations has been analyzed and applied in a great number of papers, see [16–18, 200, 279–287] and, for example, [192, 193, 272] for parallel splitting methods. None of these papers will be examined herein. For operator splitting applied to the Boltzmann equation, which describes the statistical distribution of particles in a fluid, see, for example, [69]. Another important class of equations that is omitted in this book is that of the Hamilton–Jacobi and Hamilton–Jacobi–Bellman equations, see [14, 103, 131–133, 188, 189, 210, 244, 248, 263]. operator-splitting methods for such equations arising in the context of finance have been used and analyzed in several papers, see, for example, [12, 13, 42, 264, 265]. Let us also mention that a convergence theory for splitting methods in the setting of maximal monotone operators on Hilbert spaces has been developed in [188], see also [210]. For operator splitting of the KdV (Korteweg–de Vries) equation, which models waves on shallow water surfaces, we refer to [121, 123, 255] and the references therein. Operator splitting from the point of view of semigroups has been a topic of study in for example [89, 155–157]. Variants of the Schrödinger equation, which is the fundamental equation of nonrelativistic quantum mechanics, have been approximated by operator splitting in [10, 11, 27, 273], see also [127] for the Maxwell–Dirac system.

1.7 Organization of the book

The book is organized as follows: In Chapter 2 we give the reader a taste of the content in terms of some simple examples of elementary operator splittings. Moreover, we discuss the convergence of splitting approximations and briefly touch upon errors common to them. The splittings will all be semi-discrete in the sense that there will be analytical solutions available for the split-operators. In Chapter 3 we present central elements of the mathematical framework in which to analyze, both from a mathematical and numerical point of view, second-order quasilinear degenerate parabolic equations. This theory, which can be viewed as a generalization of the well-known Kružkov theory [161] of entropy solutions for first-order hyperbolic conservation laws, provides the foundation for the convergence theory for operator splitting developed in the last section of the chapter. The theory is demonstrated in Chapter 4 by applying it to one-dimensional convection-diffusion problems. We consider a variety of semi-discrete and fully discrete splitting methods and verify that the conditions needed to apply the abstract convergence theory hold. Moreover, we present several numerical examples to highlight the use of operator splitting as a basis for developing efficient numerical schemes for convection-diffusion equations. In particular, we discuss underlying error mechanisms, and in some cases suggest strategies to reduce the splitting errors. In Chapter 5 we extend the approach originating in Chapter 3 to yield not only convergence of operator-splitting methods, but also precise error estimates, at least in the context of hyperbolic problems. We also present many numerical examples using dimensional splitting, discuss error mechanisms, and go into how to choose the splitting step to optimize runtime versus numerical errors. Chapter 6 is devoted to numerical examples for systems of equations; these systems are not covered by the rigorous analysis in the previous chapters. In particular, we discuss applications from porous media flow and for two systems of conservation laws: the Euler equations of gas dynamics and the shallow-water equations. Finally, the purpose of Appendix A is to provide the novice reader with a brief introduction to numerical methods for hyperbolic problems, many of which will be used as building blocks in the splitting algorithms discussed in the following chapters.

The theory is illustrated by many examples throughout the text. For many of the examples in the book, runtimes are given. Note that the examples were developed over a period of several years and on various computers. Thus the runtimes will be considerably lower today, however, the relative times should remain unchanged.

The purpose of computing
is insight, not numbers.
— R. W. HAMMING

1.8 MATLAB programs

The theory described in this book is applicable in many different settings. One of the attractions is that it is reasonably easy to develop computer codes that can be used for the computation of (approximate) solutions which can be used for theoretical study as well as numerical results. To make the transition from theory to computer code easier, we offer computer codes in MATLAB (version 7.7, R2008b) for almost all examples in the book. Note that the front tracking code is considerable slower when programmed in MATLAB compared with codes in C or C++. MATLAB codes are posted on the web site

www.math.ntnu.no/operatorsplitting

Feel free to use and modify them. Please let us know if you find bugs or possible improvements which we can post on the web site. If the computer codes are used for scientific work, we ask that you refer to the present book. We have decided to keep the computer codes on the web rather than include them in the book. The reasons for this are threefold: (i) It is easier to correct and update on the web than in a printed book; (ii) The book becomes smaller and can be used independently of the computer codes; (iii) Computer codes, and in particular their syntax and structure, change more rapidly than mathematical theory.

1.9 A guide to the reader

'Begin at the beginning,' the King said gravely,
'and go on till you come to the end: then stop.'
— Alice's Adventures in Wonderland

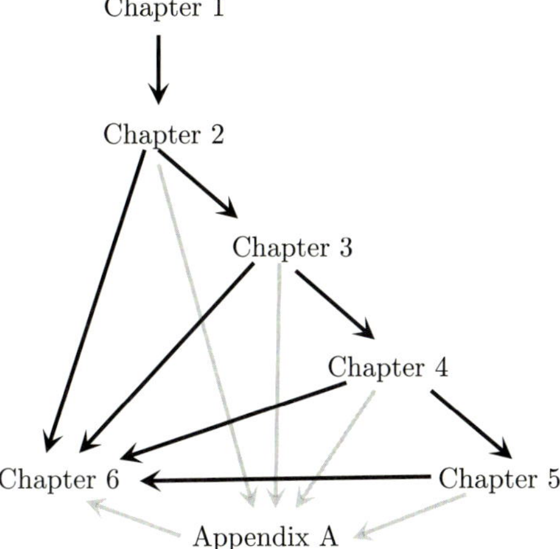

Chapter 6 focuses on applications of operator splitting in various contexts, and can be read independently of Chapters 3–5. The thin arrows mark a path through Appendix A for those unfamiliar with numerical methods for conservation laws.

1.10 Notation

For the d-dimensional ball with radius $r > 0$, we use the notation $\mathcal{B}_r$, i.e.,

$$\mathcal{B}_r = \left\{x \in \mathbb{R}^d \;\middle|\; |x| \le r\right\}.$$

For partial derivatives we use all the different standard notation, e.g.,

$$\frac{\partial f}{\partial x_j}(x) = f_{x_j}(x) = \partial_{x_j} f(x), \quad x = (x_1, \dots, x_d) \in \mathbb{R}^d,$$
$$\frac{\partial^{|\alpha|} f}{\partial x^\alpha}(x) = D^\alpha f(x), \quad \alpha \text{ multi-index.}$$

Special differential operators are as usual given by

$$\Delta f(x) = \sum_j \frac{\partial^2 f}{\partial x_j^2}(x),$$
$$\nabla f(x) = (f_{x_1}(x), \dots, f_{x_d}(x)),$$
$$\nabla \cdot F(x) = \sum_j \partial_{x_j} F_j(x), \quad F = (F_1, \dots, F_d).$$

We will frequently be working in Lebesgue spaces, and we use standard notation: If $\Omega \subseteq \mathbb{R}^d$, we have for functions $f\colon \Omega \to \mathbb{R}$ that

$$\|f\|_{L^p(\Omega)} = \begin{cases} \left(\int_\Omega |f(x)|^p \, dx\right)^{1/p}, & \text{for } p \in [1, \infty), \\ \operatorname{ess\,sup}_{x \in \Omega} |f(x)|, & \text{for } p = \infty, \end{cases}$$
$$L^p(\Omega) = \{f\colon \Omega \to \mathbb{R} \mid \|f\|_{L^p(\Omega)} < \infty\}.$$

Local versions of the same spaces are defined by

$$L^p_{\mathrm{loc}}(\Omega) = \left\{f\colon \Omega \to \mathbb{R} \;\middle|\; f\chi_K \in L^p(\Omega) \text{ for all compact sets } K\right\},$$

where we use the notation χ_K for the characteristic function of the set K. Let $C^p = C^p(\Omega)$, $p = 1, \dots, \infty$, denote the space of functions $f\colon \Omega \to \mathbb{R}$ possessing continuous partial derivatives of order $\le p$. In addition

$$C_0^p = C_0^p(\Omega) = \left\{f \in C^p(\Omega) \;\middle|\; \operatorname{supp} f \text{ compact}\right\}. \tag{1.21}$$

For vector-valued functions $f\colon \Omega \to \mathbb{R}^K$ we write $C^p(\Omega; \mathbb{R}^K) = C^p(\Omega; \mathbb{R})^K$, etc., for the corresponding spaces.

More generally, for $\Pi_T = \mathbb{R}^d \times (0, T]$, we will often need to consider functions $u\colon \Pi_T \to \mathbb{R}^K$ as elements of various Bochner spaces; that is, we consider the functions $t \mapsto u(\,\cdot\,, t)$. For instance, we will employ the space

$$L^\infty(0, T; L^1(\mathbb{R}^d)) = \{u\colon \Pi_T \to \mathbb{R} \mid \operatorname{ess\,sup}_{t \in (0,T]} \|u(\,\cdot\,, t)\|_1 < \infty\}.$$

The space $C(0,T;L^1(\mathbb{R}^d))$ consists of functions $u\colon \Pi_T \to \mathbb{R}$ such that the map $t \mapsto u(\,\cdot\,,t)$ is continuous in the L^1 norm. Finally, the space $C(0,T;L^1(\mathbb{R}^d;\mathbb{R}^K))$ consists of functions $u\colon \Pi_T \to \mathbb{R}^K$ with $t \mapsto u(\,\cdot\,,t) \in \mathbb{R}^K$ continuous in the norm in $L^1(\mathbb{R}^d;\mathbb{R}^K)$.

The Lipschitz constant of a function $f\colon \Omega \to \mathbb{R}^K$ is by definition

$$\|f\|_{\mathrm{Lip}} = \|f\|_{\mathrm{Lip}(\Omega)} = \sup_{\substack{x,y\in\Omega \\ x\neq y}} \frac{|f(x)-f(y)|}{|x-y|}. \tag{1.22}$$

The corresponding space of Lipschitz functions is given by

$$\mathrm{Lip}(\Omega) = \Big\{ f\colon \Omega \to \mathbb{R}^K \;\Big|\; \|f\|_{\mathrm{Lip}(\Omega)} < \infty \Big\}, \tag{1.23}$$

with local version

$$\mathrm{Lip}_{\mathrm{loc}}(\Omega) = \Big\{ f\colon \Omega \to \mathbb{R} \;\Big|\; \|f\|_{\mathrm{Lip}(K)} < \infty \text{ for each compact set } K \subseteq \Omega \Big\}. \tag{1.24}$$

We will need the concept of total variation for a function, which is defined as follows: Consider first the one-dimensional case. For $f\colon [a,b] \to \mathbb{R}$ ($a = -b = -\infty$ permitted) we let

$$\mathrm{T.V.}\,(f)_{[a,b]} = \mathrm{T.V.}\,(f) = \sup_{a<x_0<\cdots<x_n<b} \sum_j |f(x_{j+1}) - f(x_j)|, \tag{1.25}$$

where the supremum is over all finite partitions $x_0 < x_1 < \cdots < x_n$. For functions in L^p spaces a refinement is needed (often called *essential variation*): We still use the definition above, but restrict the points $x_0 < x_1 < \cdots < x_n$ to points of approximate continuity of f, thereby obtaining a definition that is independent of the equivalence classes used in the proper definition of L^p spaces. Functions of bounded variation are defined as follows:

$$\mathrm{BV}([a,b]) = \{ f \in L^1([a,b]) \mid \mathrm{T.V.}\,(f) < \infty \}. \tag{1.26}$$

For functions of several variables we use the following definition: Let $f\colon \mathbb{R}^2 \to \mathbb{R}$. Then we define the Tonelli variation

$$\mathrm{T.V.}\,(f) = \int_{\mathbb{R}} \mathrm{T.V.}\,(f(\,\cdot\,,y))_x \,dy + \int_{\mathbb{R}} \mathrm{T.V.}\,(f(x,\,\cdot\,))_y \,dx, \tag{1.27}$$

where $\mathrm{T.V.}\,(\,\cdot\,)_x$ and $\mathrm{T.V.}\,(\,\cdot\,)_y$ denote total variation with respect to the x and y variables respectively. Extensions to n variables are straightforward.

A function $f \in L^1(\Omega)$ is said to be of bounded total variation if its first-order derivative in the sense of distributions can be represented by a finite Radon measure, more precisely,

$$-\int_\Omega f \frac{\partial \phi}{\partial y_j}\,dy = \int_\Omega \phi \,d\mu_j, \qquad \phi \in C_0^\infty(\Omega),\ j = 1,\dots,d \tag{1.28}$$

with $|\mu_j|(\Omega) < \infty$. The set of all functions of bounded total variation is denoted by $\mathrm{BV}(\Omega)$. We denote the total variation of f by $|Df|_\Omega$ and define it by

$$|Df|_\Omega = \sup\left\{ \int_\Omega f\, \nabla\cdot\phi\, dy \mid \phi \in C_0^\infty(\Omega;\mathbb{R}^n), \|\phi\|_\infty \le 1 \right\}. \tag{1.29}$$

The local version is defined as follows: We say that a function $f \in L^1_{\mathrm{loc}}(\Omega)$ is in $f \in \mathrm{BV}_{\mathrm{loc}}(\Omega)$ if for each open set V, whose closure is contained in Ω, we have $|Df|_V < \infty$. We equip $\mathrm{BV}(\Omega)$ with the norm

$$\|f\|_{\mathrm{BV}} = \|f\|_{L^1(\Omega)} + |Df|_\Omega, \tag{1.30}$$

which makes $\mathrm{BV}(\Omega)$ into a Banach space [4, p. 121]. Then using Riesz's theorem on functionals in the space of continuous functions, we obtain that $\mathrm{BV}(\Omega)$ can equivalently be defined as

$$\mathrm{BV} = \mathrm{BV}(\Omega) = \left\{ f \in L^1(\Omega) \mid \|f\|_{\mathrm{BV}\Omega} < \infty \right\}.$$

Note that for a function $f \in L^1_{\mathrm{loc}}(\Omega)$, we have that $\mathrm{BV}_{\mathrm{loc}}(\Omega)$ if and only if

$$\int_{R^{d-1}} \mathrm{T.V.}\,(f(\tilde{x}))_{\hat{x}_i}\, d\tilde{x} < \infty \tag{1.31}$$

for all compact rectangles $R^{d-1} \subset \mathbb{R}^{d-1}$ where we have for each i that $x = (x_1, \dots, x_d) = (x_1, \dots, x_{i-1}, \hat{x}_i, x_{i+1}, \dots, x_d)$, $\tilde{x} = (x_1, \dots, x_{i-1}, x_{i+1}, \dots, x_d)$ (see [288, Thm. 5.3.5]). It is well known that the following inclusions hold:

$$\mathrm{BV}(\Omega) \subset L^{\frac{d}{d-1}}(\Omega) \text{ for } d > 1 \text{ and } \mathrm{BV}(\Omega) \subset L^\infty(\Omega) \text{ for } d = 1.$$

Furthermore,

$$\mathrm{BV}(\Omega) \text{ is compactly imbedded into } L^p(\Omega) \text{ for } 1 \le p < \frac{d}{d-1}.$$

See, e.g., [4, 95, 288] for an extensive discussion about BV functions.

When discussing difference schemes, we shall also be needing discrete versions of these norms. For a sequence $U = \{U_i\}_{i\in\mathbb{Z}}$, we define

$$\|U\|_p = \begin{cases} \left(\sum_{i\in\mathbb{Z}} |U_i|^p\right)^{1/p} & \text{if } p < \infty, \\ \sup_{i\in\mathbb{Z}} |U_i| & \text{for } p = \infty. \end{cases}$$

This is extended in the natural way to several dimensions, if we let $i \in \mathbb{Z}^d$ denote a multiindex $i = (i_1, \dots, i_d)$, and let the above sum and supremum be taken over $\mathbb{Z}^d$.

Throughout this book, by the notation Const_X we shall mean a "constant" depending on X only.

2

Simple Examples of Semi-Discrete Operator Splitting

The purpose of this short chapter is to provide a prologue to operator splitting for some simplified cases. Through several examples we give a demonstration of how evolutionary equations can be given a natural decomposition in terms of simpler equations and how this decomposition can be used to solve the equation approximately. Moreover, we discuss the convergence of the approximations and briefly touch upon errors committed by the operator splittings. The splittings will all be semi-discrete in the sense that we will use exact solution operators for all the splitting steps. In many of the examples there will be analytical solutions available for the solution operators associated with the simplified subproblems.

As we saw in Chapter 1, evolutionary equations can be described by the abstract Cauchy problem

$$\frac{dU}{dt} + \mathcal{A}(U) = 0, \qquad U(0) = U_0, \tag{2.1}$$

where $\mathcal{A}$ is a suitable operator. We will write $\mathcal{A}$ as a sum of more elementary operators, say

$$\mathcal{A} = \mathcal{A}_1 + \cdots + \mathcal{A}_\ell. \tag{2.2}$$

In many cases this decomposition arises naturally. Each operator $\mathcal{A}_j$ may represent a differentiation in a spatial direction, or $\mathcal{A}$ model a composition of different physical phenomena like convection, reaction, or diffusion. But the decomposition may also be a result of some deeper understanding of the equation.

The idea of operator splitting is to choose a decomposition of the operator $\mathcal{A}$ so that each of the sub-operators $\mathcal{A}_j$ give equations that are simpler to solve, e.g.,

$$\frac{dU^j}{dt} + \mathcal{A}_j(U^j) = 0, \qquad U^j(0) = U_0, \qquad j = 1, \dots, \ell. \tag{2.3}$$

In the following, $U^j(t) = \mathcal{S}_t^j U_0$ will denote the solution of (2.3). Once the solution is known for all sub-operators, we can choose a small time-step and apply the sub-operators sequentially to construct an approximate solution of (2.1). Mathematically, this process can be written as

$$U(n\Delta t) \approx \left[\mathcal{S}_{\Delta t}^\ell \cdots \mathcal{S}_{\Delta t}^1\right]^n U_0. \tag{2.4}$$

In doing so, we will generally make some error, but our hope is that this error will decrease as we increase the number of steps in the construction. In the limit, we expect that the approximation will converge to the true solution

$$U(t) = \lim_{\substack{\Delta t \to 0, n \to \infty \\ t = n\Delta t}} \left[\mathcal{S}^{\ell}_{\Delta t} \cdots \mathcal{S}^{1}_{\Delta t}\right]^n U_0. \tag{2.5}$$

In the rest of this section we will give a few examples of elementary operator splittings.

Example 2.1. (Finite-dimensional matrices) *Consider the system of ordinary differential equations*

$$u_t + Cu = 0, \quad u|_{t=0} = u_0 \tag{2.6}$$

where C is an $m \times m$ matrix. Assume now that C can be decomposed as the sum of two $m \times m$ matrices A and B. Thus the elementary equations become

$$\begin{aligned} v_t + Av &= 0, \quad v_{t=0} = v_0, \\ w_t + Bw &= 0, \quad w|_{t=0} = w_0. \end{aligned} \tag{2.7}$$

The corresponding elementary solutions are given by

$$\begin{aligned} v(t) &= \mathcal{S}^1_t v_0 = \exp(-tA)v_0 = \sum_{m=0}^{\infty} \frac{(-t)^m}{m!} A^m v_0, \\ w(t) &= \mathcal{S}^2_t w_0 = \exp(-tB)w_0 = \sum_{m=0}^{\infty} \frac{(-t)^m}{m!} B^m w_0, \end{aligned} \tag{2.8}$$

respectively. Introduce matrices[1]

$$U_{\Delta t} = \exp(-\Delta t(A+B)), \quad V_{\Delta t} = \exp(-\Delta t A)\exp(-\Delta t B). \tag{2.9}$$

We see that

$$U^n_{\Delta t} - V^n_{\Delta t} = \sum_{m=0}^{n-1} U^m_{\Delta t}(U_{\Delta t} - V_{\Delta t})V^{n-1-m}_{\Delta t},$$

which implies

$$\begin{aligned} \left\|U^n_{\Delta t} - V^n_{\Delta t}\right\| &= \left\| \sum_{m=0}^{n-1} U^m_{\Delta t}(U_{\Delta t} - V_{\Delta t})V^{n-1-m}_{\Delta t} \right\| \\ &\le \sum_{m=0}^{n-1} \left\|U^m_{\Delta t}(U_{\Delta t} - V_{\Delta t})V^{n-1-m}_{\Delta t}\right\| \\ &\le \sum_{m=0}^{n-1} \left\|U_{\Delta t}\right\|^m \left\|U_{\Delta t} - V_{\Delta t}\right\| \left\|V_{\Delta t}\right\|^{n-1-m} \\ &\le n \max\{\left\|U_{\Delta t}\right\|, \left\|V_{\Delta t}\right\|\}^{n-1} \left\|U_{\Delta t} - V_{\Delta t}\right\| \\ &\le n \left\|U_{\Delta t} - V_{\Delta t}\right\| \exp(t(\|A\| + \|B\|)), \end{aligned} \tag{2.10}$$

[1] The derivation is based on [225, Theorem VIII.29].

where $t = n\Delta t$. *Since*

$$U_{\Delta t} = I - \Delta t C + \frac{\Delta t^2}{2} C^2 + \mathcal{O}(\Delta t^3),$$

we see that

$$\begin{aligned}
U_{\Delta t} - V_{\Delta t} &= I - \Delta t C + \frac{\Delta t^2}{2} C^2 + \mathcal{O}(\Delta t^3) \\
&\quad - \big(I - \Delta t A + \frac{\Delta t^2}{2} A^2 + \mathcal{O}(\Delta t^3)\big)\big(I - \Delta t B + \frac{\Delta t^2}{2} B^2 + \mathcal{O}(\Delta t^3)\big) \\
&= -\Delta t(C - (A+B)) + \frac{\Delta t^2}{2}(AB - BA) + \mathcal{O}(\Delta t^3)) \\
&= \frac{\Delta t^2}{2}[A, B] + \mathcal{O}(\Delta t^3)
\end{aligned}$$

where we have introduced the commutator $[A, B] = AB - BA$. *Thus*

$$\|U_{\Delta t}^n - V_{\Delta t}^n\| \le \frac{t\Delta t}{2} \exp(t(\|A\| + \|B\|))\, \|[A, B]\| + \mathcal{O}(\Delta t^2),$$

which in particular shows that

$$\lim_{\substack{\Delta t \to 0, n \to \infty \\ t = n\Delta t}} [\exp(-\Delta t A)\exp(-\Delta t B)]^n = \exp(-t(A+B)). \tag{2.11}$$

We conclude that the limit

$$\big[\mathcal{S}^2_{\Delta t}\mathcal{S}^1_{\Delta t}\big]^n u_0 = \big[\exp(-\Delta t A)\exp(-\Delta t B)\big]^n u_0 \underset{\substack{\Delta t \to 0, n \to \infty \\ t = n\Delta t}}{\longrightarrow} \exp(-tC)u_0 = \mathcal{S}_t u_0 \tag{2.12}$$

is indeed the exact solution of (2.6).

Example 2.2. (Multi-dimensional heat equation) *Consider first the linear heat equation in two space dimensions*

$$u_t = \Delta u, \quad u|_{t=0} = u_0. \tag{2.13}$$

The natural decomposition of the operator is in terms of the spatial coordinates; that is, $\mathcal{A} = \Delta = \partial_x^2 + \partial_y^2$. *It is well-known that*

$$\begin{aligned}
v_t &= v_{xx}, \quad v|_{t=0} = v_0, \\
w_t &= w_{yy}, \quad w|_{t=0} = w_0,
\end{aligned} \tag{2.14}$$

have solutions

$$\begin{aligned}
v(x, y, t) &= \mathcal{S}^1_t v_0 = \int_{\mathbb{R}} G_t(x - \xi) v_0(\xi, y)\, d\xi, \\
w(x, y, t) &= \mathcal{S}^2_t w_0 = \int_{\mathbb{R}} G_t(y - \eta) w_0(x, \eta)\, d\eta,
\end{aligned} \tag{2.15}$$

respectively, where G is the heat kernel

$$G_t(x) = (4\pi t)^{-1/2} \exp(-x^2/4t). \tag{2.16}$$

The heat kernel satisfies the property

$$\int_{\mathbb{R}} G_t(x-y) G_s(y-z)\, dy = G_{t+s}(x-z). \tag{2.17}$$

Thus

$$\begin{aligned}
&\left[S^2_{\Delta t} S^1_{\Delta t}\right]^n u_0 \qquad\qquad (2.18)\\
&= \int_{\mathbb{R}^{2n}} G_{\Delta t}(x-\xi_n) G_{\Delta t}(\xi_n - \xi_{n-1}) \cdots G_{\Delta t}(\xi_2 - \xi_1) \\
&\quad \times G_{\Delta t}(y-\eta_n) G_{\Delta t}(\eta_n - \eta_{n-1}) \cdots G_{\Delta t}(\eta_2 - \eta_1) u_0(\xi_1, \eta_1)\, d\xi_n \cdots d\xi_1 d\eta_n \cdots d\eta_1 \\
&= \int_{\mathbb{R}^2} G_{n\Delta t}(x-\xi) G_{n\Delta t}(y-\eta) u_0(\xi,\eta)\, d\xi d\eta \\
&= \int_{\mathbb{R}^2} G_t(x-\xi) G_t(y-\eta) u_0(\xi,\eta)\, d\xi d\eta, \quad t = n\Delta t. \qquad\qquad (2.19)
\end{aligned}$$

Observe that in this case the limit $\Delta t \to 0$, $n \to \infty$ with $t = n\Delta t$ is trivial. However, applying the product structure of the heat kernel (cf. the method of separation of variables) we see that

$$G_t(x-\xi) G_t(y-\eta) = \frac{1}{4\pi t} \exp\Big(-\frac{(x-\xi)^2 + (y-\eta)^2}{4t}\Big), \tag{2.20}$$

which is the two-dimensional heat kernel. Thus we find, as in the previous example, that the right-hand side of (2.19) is indeed the exact solution (2.13). Physically, the operator splitting can be interpreted as follows: first we allow heat to diffuse in the x-direction a time Δt, then heat diffuses in the y-direction for a time Δt, and so on.

As seen in the introduction, (1.16) is not the only possible operator splitting for (2.13). Indeed, for this simple equation it turns out that all the operator splittings discussed in the introduction (multiplicative, additive and ADI) are appropriate and will produce reasonable results, at least in the limit $\Delta t \to 0$. As we will see in the next example, this may not always be true.

Example 2.3. (Linear transport) *Consider next the linear hyperbolic equation*

$$u_t + a u_x + b u_y = 0, \quad u|_{t=0} = u_0, \tag{2.21}$$

where a, b are constants. The equation describes the passive advection of a conserved quantity u in a constant velocity field (a, b) and the exact solution is given by

$$u(x, y, t) = u_0(x - at, y - bt).$$

The natural way to split the corresponding solution operator is, of course, in terms of spatial coordinates; in fact, this has been done in (2.21), that is, $\mathcal{A} = (a,b)\cdot\nabla = a\partial_x + b\partial_y$. The solution of each of the two sub-problems

$$\begin{aligned} v_t + av_x &= 0, \quad v|_{t=0} = v_0, \\ w_t + bw_y &= 0, \quad w|_{t=0} = w_0, \end{aligned} \tag{2.22}$$

is given by

$$\begin{aligned} v(x,y,t) &= \mathcal{S}_t^1 v_0 = v_0(x-at,y), \\ w(x,y,t) &= \mathcal{S}_t^2 w_0 = w_0(x,y-bt), \end{aligned} \tag{2.23}$$

respectively. Thus we find that the operator splitting solution becomes

$$\left[\mathcal{S}_{\Delta t}^2 \mathcal{S}_{\Delta t}^1\right]^n u_0 = u_0(x - na\Delta t, y - nb\Delta t). \tag{2.24}$$

Physically, the operator splitting can be interpreted as follows: first we advect the quantity u_0 linearly in the x-direction a time Δt, then we advect the quantity a time Δt in the y-direction, and so on. Since the process is linear and the two sub-operators commute, it should come as no surprise that we obtain the exact solution if $n\Delta t = t$.

A similar argument can be used to show that we obtain the exact solution also for the alternative Strang formula (1.17):

$$\left[\mathcal{S}_{\Delta t/2}^1 \mathcal{S}_{\Delta t}^2 \mathcal{S}_{\Delta t/2}^1\right]^n u_0 = u_0(x - na\Delta t, y - nb\Delta t). \tag{2.25}$$

However, the additive operator splitting defined by (1.19) will not reproduce the exact solution unless u_0 is separable. This follows from:

$$\left[\tfrac{1}{2}(\mathcal{S}_{2\Delta t}^2 + \mathcal{S}_{2\Delta t}^1)\right] u_0 = \tfrac{1}{2}\big(u_0(x - 2a\Delta t, y) + u_0(x, y - 2b\Delta t)\big),$$

which is different from the exact solution $u_0(x - a\Delta t, y - b\Delta t)$. On the other hand, the second-order additive splitting given by (1.20) will reproduce the correct solution:

$$\begin{aligned} \left[\tfrac{1}{2}(\mathcal{S}_{\Delta t}^1 \mathcal{S}_{\Delta t}^2 + \mathcal{S}_{\Delta t}^2 \mathcal{S}_{\Delta t}^1)\right] u_0 &= \tfrac{1}{2}\big(u_0(x - a\Delta t, y - b\Delta t) + u_0(x - a\Delta t, y - b\Delta t)\big) \\ &= u_0(x - a\Delta t, y - b\Delta t). \end{aligned}$$

In the rest of this book, we will only consider multiplicative operator splittings and in the following examples we focus on the simple splitting (1.16). We will return in Chapter 6 to the issue of choosing good operator splittings, where we discuss various fully discrete operator splittings in which the operators $\mathcal{S}$ are replaced by numerical methods.

Example 2.4. (Linear transport and diffusion) *We can easily combine the two previous examples by analysing*

$$u_t + au_x = u_{xx}, \quad u|_{t=0} = u_0. \tag{2.26}$$

In this case the evolution operator contains two different physical mechanisms: linear advection and diffusion. It is therefore natural to use the corresponding elementary equations

$$\begin{aligned} v_t + av_x &= 0, \quad v|_{t=0} = v_0, \\ w_t &= w_{xx}, \quad w|_{t=0} = w_0, \end{aligned} \tag{2.27}$$

as building blocks in the operator splitting algorithm. We have seen above that the convection and the diffusion equation have solutions

$$\begin{aligned} u(x,t) &= \mathcal{S}_t^1 u_0 = u_0(x - at), \\ u(x,t) &= \mathcal{S}_t^2 u_0 = \int_{\mathbb{R}} G_t(x-\xi) u_0(\xi)\, d\xi. \end{aligned} \tag{2.28}$$

Thus

$$\begin{aligned} \left[\mathcal{S}_{\Delta t}^2 \mathcal{S}_{\Delta t}^1\right]^n u_0 &= \int_{\mathbb{R}} G_{n\Delta t}(x - an\Delta t - \xi) u_0(\xi)\, d\xi \\ &\underset{\substack{\Delta t \to 0, n\to\infty \\ t = n\Delta t}}{\longrightarrow} \int_{\mathbb{R}} G_t(x - at - \xi) u_0(\xi)\, d\xi = u(x,t), \end{aligned} \tag{2.29}$$

which is easily seen to be the solution of (2.26), *either by direct computation, or by introducing new variables* $(t,x) \mapsto (s = t, y = x - at)$. *Physically, the operator splitting first transports matter, then it allows matter to diffuse out, and so on.*

We have so far only considered linear equations, where we have been able to show convergence of the operator splitting by fairly simple means. Once we leave the safe realm of linear operators, the situation becomes more involved as the next example shows.

Example 2.5. (Viscous conservation laws) *Let us now consider*

$$u_t + f(u)_x = u_{xx}, \quad u|_{t=0} = u_0. \tag{2.30}$$

As in Example 2.4 *we separate the convective and the diffusive parts of the operator. Let* $\mathcal{S}_t^1$ *and* $\mathcal{S}_t^2$ *denote the exact solution operators of the corresponding convective and diffusive subproblems*

$$\begin{aligned} v_t + f(v)_x &= 0, \quad v|_{t=0} = v_0, \\ w_t &= w_{xx}, \quad w|_{t=0} = w_0. \end{aligned} \tag{2.31}$$

The lack of explicit expressions for the solution of scalar conservation laws generally necessitates the use of a numerical method to approximate the $\mathcal{S}_t^1$ *operator. We will return to this with a much more thorough discussion in the next section.*

For now, we will use abstract arguments to obtain the desired convergence of the semi-discrete splitting. Define

$$U^{n+1} = \left[\mathcal{S}_{\Delta t}^2 \mathcal{S}_{\Delta t}^1\right] U^n, \quad U^0 = u_0 \tag{2.32}$$

From the theory of conservation laws, see Holden and Risebro [126], we know that

$$\left\|\mathcal{S}_t^1 v_0\right\|_{L^\infty(\mathbb{R})} \leq \|v_0\|_{L^\infty(\mathbb{R})}, \qquad \text{T.V.}\left(\mathcal{S}_t^1 v_0\right) \leq \text{T.V.}\left(v_0\right), \tag{2.33}$$

where $\text{T.V.}(u)$ *denotes the total variation of* u *(see [126, App. A]). From the explicit formula for the solution of the heat equation we immediately infer that*

$$\left\|\mathcal{S}_t^2 w_0\right\|_{L^\infty(\mathbb{R})} \leq \|w_0\|_{L^\infty(\mathbb{R})}. \tag{2.34}$$

For the total variation we easily find

$$\begin{aligned} \frac{1}{h}\left|\mathcal{S}_t^2 w_0(x+h) - \mathcal{S}_t^2 w_0(x)\right| &\leq \frac{1}{h}\int_{\mathbb{R}} \left|G_t(x+h-\xi)w_0(\xi) - G_t(x-\xi)w_0(\xi)\right| d\xi \\ &= \frac{1}{h}\int_{\mathbb{R}} G_t(\xi)\left|w_0(x+h-\xi) - w_0(x-\xi)\right| d\xi \\ &\leq \text{T.V.}\left(w_0\right)\int_{\mathbb{R}} G_t(\xi)\, d\xi = \text{T.V.}\left(w_0\right). \end{aligned} \tag{2.35}$$

Since both elementary solutions are bounded in L^∞ *and have bounded variation, we conclude that the same is true for the operator splitting approximations*

$$\|U^n\|_{L^\infty(\mathbb{R})} \leq \left\|U^0\right\|_{L^\infty(\mathbb{R})}, \quad \text{T.V.}\left(U^n\right) \leq \text{T.V.}\left(U^0\right). \tag{2.36}$$

Having obtained boundedness of the sequence of approximations, we can use a compactness argument to prove convergence. Helly's theorem [126, Cor. A.7] yields the existence of a subsequence $\Delta t \to 0$ *such that* $U^n \to u(t)$ *with* $n = t/\Delta t$.

While establishing the desired limit, the analysis in the previous example leaves many questions unanswered: We have not defined what is meant by a solution of $u_t + f(u)_x = 0$, indeed it is well-known that solutions of this equation develop singularities (loss of smoothness) in finite time. Hence one has to study weak solutions and carefully analyze the question of uniqueness. Thus it leaves open the question of whether the limit is the correct solution of the original equation or not. On the more technical side, we have for each t shown the existence of a subsequence $\Delta t \to 0$ for which we have convergence of U^n (with $n = t/\Delta t$). However, one would like to have one subsequence that applies to all t. A thorough discussion of these issues is central in this book.

Example 2.6. (Balance laws) *Consider*

$$u_t + f(u)_x = h(u), \quad u|_{t=0} = u_0. \tag{2.37}$$

In this balance law there are two competing evolutionary mechanisms, nonlinear transport $f(u)_x$ and reaction $h(u)$, which give the natural decomposition of the corresponding nonlinear evolution operator $\mathcal{A}$. The solutions of

$$\begin{aligned} v_t + f(v)_x = 0, &\qquad v|_{t=0} = v_0, \\ w_t = h(w), &\quad w|_{t=0} = w_0 \end{aligned} \tag{2.38}$$

are denoted by $v = \mathcal{S}_t^1 v_0$ and $w = \mathcal{S}_t^2 w_0$, respectively. As in the previous example, we will use a compactness argument to show convergence of the sequence of operator splitting solutions. To do so, we must first show that the two elementary operators $\mathcal{S}_t^1$ and $\mathcal{S}_t^2$ are bounded in L^∞ and have bounded variation.

To avoid blow-up (in finite time) of the solution of the ordinary differential equation, we assume that h is uniformly Lipschitz continuous, that is,

$$|h(u) - h(v)| \le \|h\|_{\text{Lip}} |u - v| .$$

Then

$$|w|_t \le |h(w)| \le |h(0)| + \|h\|_{\text{Lip}} |w| \le C(1 + |w|)$$

for some constant C. Furthermore, Gronwall's inequality [94, p. 624ff] yields

$$|w| \le e^{Ct}(1 + |w_0|).$$

Thus the family of functions U^n defined recursively by

$$U^{n+1} = \left[\mathcal{S}_{\Delta t}^2 \mathcal{S}_{\Delta t}^1\right] U^n, \quad U^0 = u_0 \tag{2.39}$$

satisfies

$$\|U^n\|_{L^\infty(\mathbb{R})} \le e^{CT}(1 + \|U^0\|_{L^\infty(\mathbb{R})}), \tag{2.40}$$

whenever $t \le T$. As for the total variation, we observe that if w is the difference between two solutions, i.e., $w = u - v$, then

$$|w|_t \le \|h\|_{\text{Lip}} |w| ,$$

which implies that

$$|w(t)| \le e^{Ct} |w_0| .$$

Thus, whenever $t \le T$, we have

$$\text{T.V.}\,(U^n) \le e^{CT}\,\text{T.V.}\,(U^0) .$$

Again, Helly's theorem secures the existence of the limit of U^n for a subsequence $\Delta t \to 0$ with $n = t/\Delta t$. For further details, see the next section.

Explicit formulas can easily be obtained in the linear case with $f(u) = au$ and $h(u) = cu$. Also the nonlinear case with $f(u) = au$ and $h(u) = 2\sqrt{u}$ is solvable. In this case we find that operator splitting converges to the solution $(t + \sqrt{u_0(x - at)})^2$ if $u_0 \ge 0$; see Section 5.2 and Remark 5.16.

As for the viscous conservation law, convergence analysis for the balance law leaves many questions open. We have shown that operator splitting converges to a limit, but does this limit satisfy the balance law (2.37), and if so, in what sense? Obviously, the correct solution of (2.37) must be interpreted in a weak sense, since the equation may develop singularities in finite time, similarly to the corresponding homogeneous conservation law, $u_t + f(u)_x = 0$. So, we must somehow establish that the operator splitting limit is a weak solution, but can we generally be assured that this is the only solution to (2.37)?

Example 2.7. (Two-dimensional hyperbolic conservation laws) *Let us now consider*

$$u_t + f(u)_x + g(u)_y = 0, \quad u|_{t=0} = u_0. \tag{2.41}$$

Let $\mathcal{S}_t^1$ and $\mathcal{S}_t^2$ denote the exact solution operators of the corresponding one-dimensional conservation laws in the x and y-directions,

$$\begin{aligned} v_t + f(v)_x &= 0, \quad v|_{t=0} = v_0, \\ w_t + g(w)_y &= 0, \quad w|_{t=0} = w_0. \end{aligned} \tag{2.42}$$

Following the lead of the previous examples we define

$$U^{n+1} = \left[\mathcal{S}_{\Delta t}^2 \mathcal{S}_{\Delta t}^1\right] U^n, \quad U^0 = u_0. \tag{2.43}$$

As before we see that

$$\|U^n\|_{L^\infty(\mathbb{R})} \leq \left\|U^0\right\|_{L^\infty(\mathbb{R})}. \tag{2.44}$$

However, the question of total variation is considerably more difficult in two dimensions than in one. Nevertheless, we obtain boundedness of the total variation also in this case, and by a more refined compactness argument we establish the existence of the limit, which is a weak solution of (2.41)*. We will return to this example later; see Section* 5.1 *and Remark* 5.2.

The reader will probably have observed a common feature in all the above examples. The proof of convergence is obtained by careful analysis of the simplified operators. The main idea behind the overall convergence theory presented in the next chapter is to identify necessary conditions for the individual operators that secure convergence of operator splitting, without having to prove that the approximate solution U^n converges in each case.

Whereas a rigorous convergence analysis is important to give numerical methods a sound mathematical foundation, a computational scientist is often more interested in the qualitative and quantitative properties of a numerical method. In Chapters 4, 5, and 6 we therefore make a more thorough study of operator splitting as a numerical method for nonlinear mixed hyperbolic-parabolic equations. In particular, we will discuss how to choose the splitting step Δt and investigate error mechanisms for nonlinear equations.

3

General Convergence Theory

The main purpose of this book is to describe various operator splittings and to formulate a rigorous and general convergence theory for a wide class of nonlinear partial differential equations that involve a convection operator combined with a possibly strongly degenerate and nonlinear second-order diffusion operator. The diffusion operator may be zero for some discrete values of the solution ('pointwise degenerate'), it can be zero on intervals in the solution space ('strongly degenerate'), and fairly large for other values of the solution ('uniformly parabolic'). Consequences of partial loss of parabolicity in degenerate parabolic equations are manifested in the exhibition of 'hyperbolic phenomena' like finite speed of propagation or appearance of interfaces. Strongly degenerate parabolic equations exhibit even more novel hyperbolic features such as the appearance of shock waves, loss of uniqueness, and the need of entropy conditions. Recall that a simple example of a strongly degenerate equation is a hyperbolic equation. Hence, strongly degenerate parabolic equations will in general possess discontinuous (weak) solutions. Moreover, discontinuous solutions are not uniquely determined by their initial data. In fact, an additional condition — the entropy condition — is needed to single out the physically relevant weak solution of the problem.

A mathematical framework in which to treat, from a mathematical and numerical point of view, such nonlinear partial differential equations is provided by the L^1 theory of entropy solutions of second-order quasilinear degenerate parabolic equations. This theory can be viewed as a generalization of the well-known Kružkov theory of entropy solutions to first-order hyperbolic conservation laws [161]. For a primer on the mathematical theory of hyperbolic problems, see [41, 73, 126, 201, 221, 239, 240, 243]. However, while the first-order theory is classical, the second-order counterpart has advanced significantly only in recent years. The study of entropy solutions to multi-dimensional degenerate parabolic equations was advanced significantly with the important work [50], which in turn generated a large amount of subsequent activity [20, 45, 48–50, 55–58, 102, 147, 148, 178, 202, 205, 207, 222, 231, 232, 251, 252, 270], see also [271, 275, 276] for some previous approaches (in the one-dimensional context).

In this chapter we develop a rigorous convergence theory for operator splitting methods in the context of weakly coupled systems of strongly degenerate (mixed hyperbolic-parabolic) convection-diffusion equations. To verify convergence of a specific operator-splitting method one only has to check whether each subsolver satisfies certain assumptions, whereupon convergence (to the physically relevant entropy solution) follows. The convergence theory is developed within the discontinuous solution framework of entropy solutions. Central elements of this theory are presented in Sections 3.1 to 3.3, before we present the final theory in Section 3.4. In Chapters 4 and 5, we consider a variety of semi-discrete and fully discrete product formulas, including dimensional splitting, viscous splitting, and flux splitting, and verify that the conditions needed to apply the abstract convergence theory developed in this chapter hold. These two chapters also contain a thorough, qualitative and quantitative discussion of various operator splittings in terms of a few carefully selected and representative numerical examples.

3.1 Mathematical preliminaries

We start by reminding the reader of a very general compactness result, known as Kolmogorov's compactness criterion, which we will be using repeatedly later on when proving convergence of various approximate solutions (see, e.g., [113, 126] for a proof).

In the following, by a modulus of continuity we will mean a nondecreasing continuous function $\nu\colon [0,\infty) \to [0,\infty)$ such that $\nu(0) = 0$.

Lemma 3.1 (Kolmogorov's compactness criterion). *Let $\Omega \subset \mathbb{R}^d$ be an open set. A family of functions $\{u_h\}_{h>0} \subset L^p(\Omega;\mathbb{R})$ with $p \in [1,\infty)$, is compact in $L^p(\Omega;\mathbb{R})$ if and only if:*

1. *there exists a constant $C > 0$ which depends on Ω and $\{u_h\}_{h>0}$ but is independent of h such that*
$$\int_\Omega |u_h(x)|^p \, dx \le C;$$
2. *the family $\{u_h\}_{h>0}$ possesses a common spatial modulus of continuity ν not depending on h, i.e.,*
$$\int_\Omega |u_h(x+y) - u_h(x)|^p \, dx \le \nu_\Omega(|y|)$$
for all h (here u_h is defined to vanish outside Ω);
3. *the family $\{u_h\}_{h>0}$ satisfies*
$$\lim_{\alpha\to\infty} \int_{\{x\in\Omega \,|\, |x|\ge\alpha\}} |u_h(x)|^p \, dx = 0,$$
uniformly in h.

Remark 3.2. Condition 3 is clearly superfluous when Ω is bounded.

The next two lemmas are consequences of Kolmogorov's compactness criterion, Lemma 3.1. These results will be used in the convergence analysis.

Lemma 3.3 (L^p_{loc} compactness lemma). *Let $\{u_h = u_h(x,t)\}_{h>0}$ be a family of functions defined on $\mathbb{R}^d \times (0,T)$, $T>0$ that satisfies:*

1. *there exists a constant $C_1 > 0$ that is independent of h such that*
$$\|u_h(\,\cdot\,,t)\|_{L^p(\mathbb{R}^d)} \le C_1, \qquad \|u_h(\,\cdot\,,t)\|_{L^\infty(\mathbb{R}^d)} \le C_1, \quad t \in (0,T);$$
2. *there exist two moduli of continuity ν_j such that*
$$\|u_h(\,\cdot\, + y, t) - u_h(\,\cdot\,,t)\|_{L^p(\mathbb{R}^d)} \le \nu_1(|y|) + \nu_2(h), \quad t \in (0,T);$$
3. *there exist two moduli of continuity ω_j such that*
$$\|u_h(\,\cdot\,,t+\tau) - u_h(\,\cdot\,,t)\|_{L^p(\mathbb{R}^d)} \le \omega_1(\tau) + \omega_2(h), \quad t \in (0,T-\tau)$$
whenever $\tau \in (0,T)$.

Then $\{u_h\}_{h>0}$ is compact in the strong topology of $L^p_{\mathrm{loc}}(\mathbb{R}^d \times (0,T))$. Moreover, any limit point of $\{u_h\}_{h>0}$ belongs to $L^p(\mathbb{R}^d \times (0,T)) \cap L^\infty(\mathbb{R}^d \times (0,T)) \cap C(0,T;L^p(\mathbb{R}^d))$.

Proof. Consider a sequence $\{h_j\}_{j=1}^\infty$ such that $h_j \to 0$. We are going to apply the Kolmogorov compactness criterion, Lemma 3.1, to show convergence in $L^p(\mathbb{R}^{d+1})$. Denote $X = (x,y)$ and $Y = (y,\tau)$, and

$$\theta(Y,h) = \Big(\int |u_h(X+Y) - u_h(X)|^p \, dX \Big)^{1/p},$$
$$\mu_1(Y) = \nu_1(y) + \omega_1(\tau), \ \mu_2(h) = \nu_2(h) + \omega_2(h).$$

Fix $\epsilon > 0$. Then such choose δ and N so that $\mu_1(|Y|) < \epsilon/2$ for $|Y| < \delta$ and $\mu_2(h_j) < \epsilon/2$ for $j > N$. Next choose

$$\tilde{\delta} = \sup_{j \le N} \{|Y| \mid \theta(Y,h_j) < \epsilon\}.$$

For $|Y| < \min\left\{\delta, \tilde{\delta}\right\}$ we find

$$\theta(Y,h_j) < \epsilon, \text{ for all } j.$$

Now we can apply the Kolmogorov compactness criterion, Lemma 3.1, to determine the existence of a convergent subsequence in $L^p(\mathbb{R}^{d+1})$. However, to obtain convergence in $C(0,T;L^p(\mathbb{R}^d))$ we need a more refined argument. For each $t \in [0,T]$ we can find a sequence $h_j \to 0$ such that $u_{h_j}(\,\cdot\,,t)$ converges to $u(\,\cdot\,,t)$ in $L^p_{\mathrm{loc}}(\mathbb{R}^d)$. Let $E \subset [0,T]$ be a countable dense subset and let $K \subset \mathbb{R}^d$ be a compact set. By using a diagonal argument we can find a subsequence (not relabeled) $h_j \to 0$ such that

$$\int_K \left|u_{h_j}(x,t) - u(x,t)\right|^p \, dx \to 0, \quad h_j \to 0, \, t \in E. \tag{3.1}$$

Given a positive ϵ there exists a δ such that $\omega_1(\delta) < \epsilon$ and $\omega_2(\delta) < \epsilon$. Fix $t \in [0,T]$. Then we can find an $s \in E$ such that $|s-t| < \delta$. Thus

$$\left\|u_{h_j}(\,\cdot\,,t) - u_{h_j}(\,\cdot\,,s)\right\|_{L^p(K)} \le \omega_1(|t-s|) + \omega_2(h_j) < 2\epsilon \tag{3.2}$$

for all $h_j < \delta$. In addition,

$$\left\|u_{h_\ell}(\,\cdot\,,s) - u_{h_j}(\,\cdot\,,s)\right\|_{L^p(K)} \le \epsilon \tag{3.3}$$

for all $h_\ell, h_j < \delta$. From the triangle inequality

$$\begin{aligned}\left\|u_{h_\ell}(\,\cdot\,,t) - u_{h_j}(\,\cdot\,,t)\right\|_{L^p(K)} &\le \left\|u_{h_\ell}(\,\cdot\,,t) - u_{h_\ell}(\,\cdot\,,s)\right\|_{L^p(K)} \\ &\quad + \left\|u_{h_\ell}(\,\cdot\,,s) - u_{h_j}(\,\cdot\,,s)\right\|_{L^p(K)} \\ &\quad + \left\|u_{h_j}(\,\cdot\,,s) - u_{h_j}(\,\cdot\,,t)\right\|_{L^p(K)} \le 5\epsilon,\end{aligned}$$

it follows that $u_{h_j}(\,\cdot\,,t)$ converges to $u(\,\cdot\,,t)$ in $L^p_{\mathrm{loc}}(\mathbb{R}^d)$ for each $t \in [0,T]$. The bounded convergence theorem then shows that

$$\sup_{t\in[0,T]} \left\|u_{h_j}(x,t) - u(x,t)\right\|_{L^p(K)} \to 0, \quad h_j \to 0, \tag{3.4}$$

thereby proving that any limit point is in $L^p(\mathbb{R}^d \times (0,T))$. The fact that the limit point is in $L^\infty(\mathbb{R}^d \times (0,T))$ follows from assumption (1). Finally, taking $h \to 0$ in assumption (3) shows that the limit is in $C(0,T; L^p(\mathbb{R}^d))$. □

To prove that the approximate solutions possess some L^1 time continuity, given that they possess some L^1 space continuity, we shall need the following version of a celebrated interpolation lemma due to Kružkov [160] (see, e.g., Lemma 4.10 in [126] for a proof):

Lemma 3.4 (Kružkov interpolation lemma). *Let $z(x,t)$ be a bounded measurable function defined in the cylinder $\mathcal{B}_{r+\hat{r}} \times [0,T]$, $\hat{r} \ge 0$. For $t \in [0,T]$ and $\rho \le \hat{r}$, assume that u possesses a spatial modulus of continuity*

$$\sup_{|y|\le\rho} \int_{\mathcal{B}_r} |z(x+y,t) - z(x,t)|\, dx \le \nu_{r,T,\hat{r}}(\rho; z), \tag{3.5}$$

where $\nu_{r,T,\hat{r}}$ does not depend on t. Suppose that for any $\phi \in C_0^\infty(\mathcal{B}_r)$ and any $t_1, t_2 \in [0,T]$ and for some integer $m \ge 0$,

$$\left|\int_{\mathcal{B}_r} (z(x,t_2) - z(x,t_1))\,\phi(x)\, dx\right| \le \mathrm{Const}_{r,T} \left(\sum_{|\alpha|\le m} \|D^\alpha \phi\|_{L^\infty(\mathcal{B}_r)}\right) |t_2 - t_1|, \tag{3.6}$$

where α denotes a multiindex. Then for t and $t+\tau \in [0,T]$ and for all $\varepsilon \in (0,\hat{r}]$,

$$\int_{\mathcal{B}_r} |z(x,t+\tau) - z(x,t)|\, dx \le \mathrm{Const}_{r,T} \left(\varepsilon + \nu_{r,T,\hat{r}}(\varepsilon; z) + \frac{|\tau|}{\varepsilon^m}\right). \tag{3.7}$$

Remark 3.5. If we choose $\varepsilon = |\tau|^{1/(m+1)}$, we see that z also possesses a temporal modulus of continuity given by

$$\omega_{r,T,\hat{r}}(\tau; z) := \text{Const}_{r,T} \left(|\tau|^{1/(m+1)} + \nu_{r,T,\hat{r}}(|\tau|^{1/(m+1)}; z)\right).$$

If the total variation of $z(\cdot\,, t)$ on $\mathbb{R}^d$ is uniformly bounded, then we can choose $\nu_{r,T,\hat{r}}$ as

$$\nu_{r,T,\hat{r}}(\rho; u) = \text{Const}_{r,T}\, \rho,$$

and hence (3.7) can be replaced by

$$\omega_{r,T,\hat{r}}(\tau; u) = \text{Const}_{r,T}\, |\tau|^{1/(m+1)}.$$

Note that this L^1 Hölder estimate is optimal for solutions of second-order partial differential equations with merely BV data. Finally, it is also possible to use Lemma 3.4 to recover the well-known L^1 Lipschitz continuity in time of solutions of first-order equations with BV data.

We will also need some general results regarding nonlinear mappings between Banach spaces. If X is a Banach space and X^* its dual, a *duality mapping* J is a map $J\colon X \to X^*$ with the properties that for all $x \in X$,

$$\|J(x)\|_{X^*} = \|x\|_X, \quad \text{and} \quad \langle J(x), x\rangle = \|x\|_X^2,$$

where $\langle \cdot\,, \cdot\rangle$ denotes the pairing between X^* and X. If $X = L^1(\Omega)$ where $(\Omega, d\mu)$ is some measure space, then every duality mapping can be written as an integral

$$\langle J(u), v\rangle = \int_\Omega j(u)(x)v(x)\, d\mu(x),$$

with

$$j(u)\,(x) = \begin{cases} \text{sign}\big(u(x)\big), & \text{if } u(x) \neq 0, \\ a(x), & \text{if } u(x) = 0, \end{cases} \tag{3.8}$$

where a is any measurable function with $|a(x)| \leq 1$ almost everywhere w.r.t. $d\mu$.

A mapping $\mathcal{A}\colon D(\mathcal{A}) \subset X \to X$ is called *accretive* if for all pairs $(u, \mathcal{A}(u))$ and $(v, \mathcal{A}(v))$ in the graph of $\mathcal{A}$, and for all duality mappings J we have that

$$\langle J(u - v), \mathcal{A}(u) - \mathcal{A}(v)\rangle \geq 0.$$

If, in addition, $I + \lambda\mathcal{A}$ is surjective, then $\mathcal{A}$ is called m-accretive. By [85, Theorem 13.1] it is sufficient that $\mathcal{A}$ is Lipschitz continuous and accretive for it to be m-accretive.

If X is a function space $L^p(\mathbb{R}^d)$, then a translation $\sigma_y u$ is the function $u(\cdot + y)$ for $y \in \mathbb{R}^d$.

Theorem 3.6. *Let $(\Omega, d\mu)$ be a measure space, and suppose that the (nonlinear, and possibly multivalued) operator $\mathcal{A}\colon L^1(\Omega) \to L^1(\Omega)$ is Lipschitz continuous and accretive. Then for any positive λ and any $u \in L^1(\Omega)$, the equation*

$$\mathcal{T}(u) + \lambda \mathcal{A}(\mathcal{T}(u)) = u$$

has a unique solution $\mathcal{T}(u)$. Furthermore, if

$$\int_\Omega \mathcal{A}(u)\, d\mu = 0, \quad u \in L^1(\Omega),$$

and $\mathcal{A}$ commutes with translations, i.e., $\sigma_y \mathcal{A}(u) = \mathcal{A}(\sigma_y u)$, then we find that the solution operator $\mathcal{T}\colon L^1(\Omega) \to L^1(\Omega)$ has the following properties:

$$\int_\Omega \mathcal{T}(u)\, d\mu = \int_\Omega u\, d\mu, \tag{3.9a}$$

$$\|\mathcal{T}(u) - \mathcal{T}(v)\|_{L^1(\Omega)} \le \|u - v\|_{L^1(\Omega)}, \tag{3.9b}$$

$$\text{T.V.}\,(\mathcal{T}(u))_\Omega \le \text{T.V.}\,(u)_\Omega, \tag{3.9c}$$

$$u \le v \Rightarrow \mathcal{T}(u) \le \mathcal{T}(v), \tag{3.9d}$$

$$\|\mathcal{T}(u)\|_{L^\infty(\Omega)} \le \|u\|_{L^\infty(\Omega)}. \tag{3.9e}$$

See [53] for a proof of this theorem.

3.2 Degenerate parabolic equations

Let us now turn to degenerate parabolic equations, which we will assume to be in the form

$$\partial_t u + \nabla \cdot f(u) = \Delta A(u), \qquad u(x,0) = u_0(x), \tag{3.10}$$

where $(x,t) \in \Pi_T := \mathbb{R}^d \times (0,T)$, $T > 0$ is fixed, and $u = u(t,x)$ is the scalar unknown function that is sought. The initial function $u_0(x)$ satisfies

$$u_0 \in L^1(\mathbb{R}^d) \cap L^\infty(\mathbb{R}^d). \tag{3.11}$$

The diffusion function $A(u)$ is a scalar function that satisfies

$$A(u) \in \text{Lip}_{\text{loc}}(\mathbb{R}), \qquad A(\,\cdot\,) \text{ nondecreasing with } A(0) = 0. \tag{3.12}$$

Finally, the convection flux $f(u)$ is a vector-valued function that satisfies

$$f(u) = (f_1(u), \dots, f_d(u)) \in \text{Lip}_{\text{loc}}(\mathbb{R}; \mathbb{R}^d), \quad f(0) = 0. \tag{3.13}$$

When (3.26) is *non-degenerate* (uniformly parabolic), i.e., $A'(u)$ is larger than a positive constant for all u, it is well-known that (3.26) admits a unique classical solution [168]. This is at variance with the degenerate case in which $A'(u)$ may vanish for some values of u. A simple example of a degenerate equation is the porous medium equation defined for $u \in [0, 1]$ by

$$\partial_t u = \partial_x^2(u^m), \qquad m > 1,$$

which is degenerate at $u = 0$. In general, a manifestation of the degeneracy is in the finite speed of propagation of disturbances: if at some fixed time the solution u has compact support, then it will continue to have compact support for all later times. The function u may have a singularity if it approaches the value zero, and it becomes necessary to explore (continuous) weak solutions rather than classical solutions. We refer to the book by Samarskiĭ *et al.* [233] for a summary of the subject of degenerate equations.

A vital condition for the uniqueness of weak solutions to (3.10) is that $A(\cdot)$ is strictly increasing, that is, A is without plateaus (see discussion below). On the other hand, if there exists at least one interval $[\alpha, \beta]$ on which $A(u)$ is flat, then we say that (3.10) is *strongly degenerate.* Going into particulars, the purely hyperbolic equation

$$\partial_t u + \nabla \cdot f(u) = 0$$

is an example of a strongly degenerate equation. Hence, we conclude that strongly degenerate equations in general will possess discontinuous solutions. Accordingly it is necessary to work with weak solutions. However, due to neglected physical mechanisms, weak solutions are not uniquely determined by their initial data. An additional condition is therefore needed to single out a unique weak solution. This additional condition is known as the entropy condition, and the corresponding weak solutions are called entropy weak solutions.

Let $\eta\colon \mathbb{R} \to \mathbb{R}$ be a convex C^2 function, and introduce the corresponding functions $q = (q_1, \dots, q_d)\colon \mathbb{R} \to \mathbb{R}^d$ and $r\colon \mathbb{R} \to \mathbb{R}$ defined by

$$q'(u) = \eta'(u) f'(u), \qquad r'(u) = \eta'(u) A'(u), \qquad u \in \mathbb{R}.$$

We call η an *entropy function,* q and r *entropy fluxes,* and (η, q, r) an *entropy, entropy flux triple.*

Definition 3.7. A function $u(x, t)$ is an *entropy weak solution* of the Cauchy problem (3.10) if the following conditions hold:

1. $u \in L^\infty(\Pi_T) \cap L^\infty(0, T; L^1(\mathbb{R}^d))$;
2. $\nabla A(u) \in L^2(\Pi_T; \mathbb{R}^d)$;
3. for any entropy, entropy flux triple (η, q, r) and for all nonnegative $\phi \in C_0^\infty(\mathbb{R}^d \times [0, T))$,

$$\iint_{\Pi_T} \left(\eta(u)\partial_t\phi + q(u) \cdot \nabla\phi + r(u)\Delta\phi\right) dt\, dx + \int_{\mathbb{R}^d} \eta(u_0)\phi(x, 0)\, dx \geq 0. \quad (3.14)$$

By an approximation argument (see, e.g., [126]) we can replace (3.14) by

$$\iint_{\Pi_T} \left(|u-c|\,\partial_t\phi + \operatorname{sign}(u-c)\,(f(u)-f(c))\cdot\nabla\phi + |A(u)-A(c)|\,\Delta\phi\right) dt\,dx \\ + \int_{\mathbb{R}^d} |u_0-c|\,\phi(x,0)\,dx \geq 0, \quad 0\leq\phi\in C_0^\infty(\mathbb{R}^d\times[0,T)) \tag{3.15}$$

for all real constants c. We refer to $u\mapsto|u-c|$ as the Kružkov entropy function and $u\mapsto\operatorname{sign}(u-c)\,(f(u)-f(c))$ and $u\mapsto|A(u)-A(c)|$ as the corresponding Kružkov entropy flux functions, indexed by $c\in\mathbb{R}$. When $A'\equiv 0$, (3.15) coincides with Kružkov's entropy condition for hyperbolic equations [161].

If we first take $c>\|u\|_{L^\infty(\Pi_T)}$ and subsequently $c<-\|u\|_{L^\infty(\Pi_T)}$ in the entropy condition (3.15), we can deduce that u satisfies the weak formulation of (3.10) for $\phi\in C_0^\infty(\mathbb{R}^d\times[0,T))$,

$$\iint_{\Pi_T}\Big(u\partial_t\phi + f(u)\cdot\nabla\phi + A(u)\Delta\phi\Big)\,dx\,dt + \int u_0(x)\phi(x,0)\,dx = 0. \tag{3.16}$$

If $A(\,\cdot\,)$ is strictly increasing, one can replace the entropy formulations (3.14) or (3.15) by the weak formulation (3.16) and maintain well-posedness [50].

We have the following uniqueness theorem [148].

Theorem 3.8. *Assume that* (3.11), (3.12), *and* (3.13) *hold. Let u, v be two entropy weak solutions of the Cauchy problem* (3.10) *with initial data v_0 and u_0, respectively. Then*

$$\int_{\mathbb{R}^d}\big|v(x,t)-v(x,t)\big|\,dx \leq \int_{\mathbb{R}^d}\big|v_0(x)-u_0(x)\big|\,dx, \quad t>0. \tag{3.17}$$

In particular, there exists at most one entropy weak solution of (3.10).

Regarding existence of entropy weak solutions we refer to [122], but also to later chapters in which existence will be a consequence of the convergence of an operator-splitting method.

The entropy condition for degenerate parabolic equations was first proposed by Vol′pert and Hudjaev [271], who also established the existence of an entropy solution by passing to the limit in a parabolic regularization. In the one-dimensional case, the L^1 contraction property (implying uniqueness) of entropy weak solutions was proved in [276], see also [275] and [21–23]. The L^1 contraction property for multi-dimensional equations was obtained more than a decade later by Carrillo [50] for the homogeneous Dirichlet boundary value problem. Using the pioneering approach by Carrillo [50], the Cauchy problem (3.10) was treated in [144, 148]. Unbounded entropy weak solutions were analyzed in [55]. The nonhomogeneous Dirichlet boundary value problem is treated in [205] and also in [207]. Some other initial-boundary value problems arising in the theory of sedimentation-consolidation processes were studied in [45].

Let us now attempt to motivate the notion of entropy weak solutions specified in Definition 3.7. According to, e.g., [271], an entropy weak solution u to (3.10) can be constructed by the vanishing viscosity method. More precisely, let u_ρ be the unique classical solution to the uniformly parabolic equation

$$\partial_t u_\rho + \operatorname{div} f(u_\rho) = \Delta A(u_\rho) + \rho \Delta u_\rho, \qquad \rho > 0. \tag{3.18}$$

In view of the a priori $(L^1, L^\infty, \mathrm{BV})$ estimates found in [271], there exists a limit function u such that (along a subsequence)

$$u_\rho \to u \text{ a.e. and in } C(0,T; L^1(\mathbb{R}^d)) \text{ as } \rho \to 0, \tag{3.19}$$

and $u \in L^\infty(\Pi_T) \cap L^\infty(0,T; L^1(\mathbb{R}^d))$. Let (η, q, r) be a convex entropy, entropy flux triple. Multiplying (3.18) by $\eta'(u_\rho)$ yields

$$\partial_t \eta(u_\rho) + \nabla \cdot q(u_\rho) - \Delta r(u_\rho) - \rho \Delta \eta(u_\rho) = -\left(n_\rho^{\eta''} + m_\rho^{\eta''}\right), \tag{3.20}$$

where the parabolic dissipation and the entropy dissipation measures $n_\rho^{\eta''}(t,x)$ and $m_\rho^{\eta''}(t,x)$ are defined respectively by

$$n_\rho^{\eta''} = \eta''(u_\rho) A'(u_\rho) \left|\nabla u_\rho\right|^2 \geq 0, \qquad m_\rho^{\eta''} = \rho \eta''(u_\rho) \left|\nabla u_\rho\right|^2 \geq 0.$$

Integrating (3.20) over Π_T yields

$$\int_{\mathbb{R}^n} \eta(u(T,x))\, dx + \iint_{\Pi_T} \left(n_\rho^{\eta''} + m_\rho^{\eta''}\right)(x,t)\, dx\, dt \leq \int_{\mathbb{R}^n} \eta(u_0(x))\, dx.$$

Specifically, the choice $\eta = u^2/2$ discloses that $\sqrt{A'(u_\rho)} \nabla u_\rho$ is bounded in $L^2(\Pi_T)$ independently of ρ, and because of that, so is $\nabla A(u_\rho)$. Consequently, the limit u in (3.19) satisfies $\nabla A(u) \in L^2(\Pi_T)$. Multiplying (3.20) by a nonnegative test function $\phi \in C_0^\infty(\mathbb{R}^d \times [0,T))$, integrating by parts over Π_T, and finally sending $\rho \to 0$ in the resulting inequality, we obtain that the limit u defined in (3.19) satisfies the entropy condition (3.14).

Remark 3.9. Regarding the first part of Definition 3.7, we should point out that solutions constructed, as outlined above by the vanishing viscosity method or through the convergence of operator-splitting methods (as will be detailed later), will belong to $C(0,T; L^1(\mathbb{R}^d))$. As a consequence, the initial condition can be set apart from (3.14). Concretely, this means that (3.14) can be replaced by

$$\iint_{\Pi_T} \left(\eta(u) \partial_t \phi + q(u) \cdot \nabla \phi + r(u) \Delta \phi\right) dt\, dx \geq 0, \tag{3.21}$$

for $0 \leq \phi \in C_0^\infty(\mathbb{R}^d \times [0,T))$ and

$$\lim_{t \to 0} \int_{\mathbb{R}^d} |u(x,t) - u_0(x)|\, dx = 0. \tag{3.22}$$

Although we will not deal with these equations in this book, let us mention that recently [58] the entropy solution theory was extended to cover more general equations than (3.10), namely, quasilinear *anisotropic* degenerate parabolic equations of the form

$$\partial_t u + \operatorname{div} f(x,t,u) = \operatorname{div}\left(a(x,t,u)\nabla u\right), \tag{3.23}$$

where we could have added a source term as well. In (3.23), $f = (f_1, \dots, f_d)$ is a vector–valued flux function and $a = (a_{ij})$ is a symmetric matrix-valued diffusion function of the form

$$\begin{cases} a(x,t,u) = \sigma^a(x,t,u)\sigma^a(x,t,u)^\top \geq 0, \\ \sigma^a(x,t,u) \in \mathbb{R}^{d\times K}, \quad 1 \leq K \leq d. \end{cases} \tag{3.24}$$

More explicitly, the components of a read

$$a_{ij}(x,t,u) = \sum_{k=1}^{K} \sigma^a_{ik}(x,t,u)\sigma^a_{jk}(x,t,u), \qquad i,j = 1,\dots,d.$$

Nonnegativity of the matrix $a(u,t,x)$ is interpreted in the usual sense, meaning that for each fixed $(x,t,u) \in \Pi_T \times \mathbb{R}$ there holds

$$\sum_{i,j=1}^{d} a_{ij}(x,t,u)\lambda_i\lambda_j \geq 0, \qquad \lambda = (\lambda_1,\dots,\lambda_d) \in \mathbb{R}^d.$$

The anisotropic diffusion case in (3.23) with coefficients independent of x and t was treated by Chen and Perthame [58], who developed a notion of kinetic/entropy solutions containing an explicit parabolic dissipation measure and also a certain chain-rule property (not needed when the diffusion coefficient a is a scalar function). The L^1 contraction property of entropy solutions was proved in [58] by developing a kinetic formulation and using the regularization by convolution. An alternative theory was developed in [19] based on Kružkov's device of doubling variables and a notion of renormalized entropy solutions. There are also some other recent papers dealing with the anisotropic diffusion case. In [222], the relation between dissipative solutions and entropy solutions was studied and the convergence of certain relaxation approximations was established. In [57], a kinetic framework was introduced for deriving explicit continuous dependence estimates and convergence rates for approximate entropy solutions. In the context of semigroup solutions for the isotropic case, continuous dependence estimates had been derived earlier in [66]. A theory of existence, uniqueness, and continuous dependence of entropy solutions for the case with (x,t)-dependent coefficients, cf. (3.23), was developed in [56].

3.3 Weakly coupled systems of degenerate parabolic equations

We have seen that a fairly well-developed L^1 theory for degenerate parabolic equations has emerged in recent years. On the other hand, to date there exists no general theory for degenerate parabolic systems, although the general theory of *uniformly-parabolic* systems is by now a mature subject [91, 158, 256], see also [243] for some special uniformly-parabolic systems and [79, 128, 226] for some results on parabolic systems with weaker parabolicity conditions. Strongly degenerate parabolic systems occur in applications such as sedimentation and consolidation of polydisperse suspensions [26]. Moreover, the theory of hyperbolic systems is rather well developed (at least in one spatial dimension) [41, 59, 73, 126, 201, 239, 240]. Although no general theory exists for systems of degenerate parabolic equations with strong coupling, it is rather straightforward to extend the mathematical theory for scalar, strongly degenerate, parabolic equations to weakly coupled systems of such equations [122], which we now discuss briefly. Applications include combustion [198, 215], hydrology [224], biology [30], relaxation [135, 211], and resonance phenomena [199].

We consider weakly coupled systems of nonlinear degenerate parabolic equations of the form

$$u_t^\kappa + \operatorname{div} F^\kappa(u^\kappa) = \Delta A^\kappa(u^\kappa) + g^\kappa(U), \quad (x,t) \in \Pi_T, \quad \kappa = 1, \dots, K. \tag{3.25}$$

Here $U = (u^1, \dots, u^K)$, $F^\kappa(u^\kappa) = (F_1^\kappa(u^\kappa), \dots, F_d^\kappa(u^\kappa))$, and $\Pi_T = \mathbb{R}^d \times (0,T]$ for some T positive. The system (3.25) can be written more compactly as

$$U_t + \sum_i F_i(U)_{x_i} = \Delta A(U) + G(U), \qquad (x,t) \in \Pi_T, \tag{3.26}$$

when we introduce

$$\begin{aligned} F_i(U) &= (F_i^1(u^1), \dots, F_i^K(u^K)), \\ A(U) &= (A^1(u^1), \dots, A^K(u^K)), \\ G(U) &= (g^1(U), \dots, g^K(U)). \end{aligned}$$

Observe that the system is coupled through the source term $G(U)$ only. We will here consider the Cauchy problem for (3.26); i.e., we require that

$$U\big|_{t=0} = U_0 \in L^\infty(\mathbb{R}^d; \mathbb{R}^K) \cap L^1(\mathbb{R}^d; \mathbb{R}^K). \tag{3.27}$$

We assume that the nonlinear (convection and diffusion) flux functions satisfy the general conditions

$$\begin{gathered} F^\kappa \in \mathrm{Lip}_{\mathrm{loc}}(\mathbb{R}; \mathbb{R}^d), \quad F^\kappa(0) = 0, \\ A^\kappa \in \mathrm{Lip}_{\mathrm{loc}}(\mathbb{R}), \quad A^\kappa \text{ is nondecreasing with } A^\kappa(0) = 0, \end{gathered} \tag{3.28}$$

where $\kappa = 1, \dots, K$. In addition, we assume that

$$G \in \mathrm{Lip}_{\mathrm{loc}}(\mathbb{R}^K; \mathbb{R}^K), \quad G(0) = 0. \tag{3.29}$$

Letting $\eta\colon \mathbb{R} \to \mathbb{R}$ denote a convex C^2 entropy function, we introduce the associated entropy-flux vectors

$$Q_i = (q_i^1, \dots, q_i^K), \qquad R = (r^1, \dots, r^K),$$

that satisfy the compatibility conditions

$$\begin{aligned} \frac{dq_i^\kappa}{du}(u) &= \eta'(u)\frac{dF_i^\kappa}{du}(u), \qquad \kappa = 1, \dots, K, \quad i = 1, \dots, d, \\ \frac{dr^\kappa}{du}(u) &= \eta'(u)\frac{dA^\kappa}{du}(u), \qquad \kappa = 1, \dots, K. \end{aligned}$$

Definition 3.10. A vector-valued function

$$U \in L^\infty(\Pi_T^0; \mathbb{R}^K) \cap L^1(\Pi_T^0; \mathbb{R}^K) \cap C([0,T]; L^1(\mathbb{R}^d; \mathbb{R}^K))$$

is called an entropy weak solution of the Cauchy problem (3.26)–(3.27) if:

1. For all convex C^2 entropy functions $\eta\colon \mathbb{R} \to \mathbb{R}$ and corresponding entropy fluxes Q_i and R, the following entropy inequality holds for $\kappa = 1, \dots, K$:

$$\eta(u^\kappa)_t + \sum_i q_i^\kappa(u^\kappa)_{x_i} - \Delta r^\kappa(u^\kappa) \le \eta'(u^\kappa) g^\kappa(U) \quad \text{in } \mathcal{D}'(\Pi_T^0); \tag{3.30}$$

that is, for any nonnegative test function $\phi(x,t) \in C_0^\infty(\Pi_T^0)$,

$$\begin{aligned} \iint_{\Pi_T} \Big(\eta(u^\kappa)\phi_t + \sum_i q_i^\kappa(u^\kappa)\phi_{x_i} + r^\kappa(u^\kappa)\Delta\phi\Big)\, dt\, dx \ge \\ - \iint_{\Pi_T} \eta'(u^\kappa) g^\kappa(U)\phi \, dt\, dx - \int_{\mathbb{R}^d} \eta\left(u_0^\kappa(x)\right)\phi(x,0)\, dx. \end{aligned}$$

2. We have that

$$\nabla_x A^\kappa(u^\kappa) \in L^2\left(\Pi_T\right), \qquad \kappa = 1, \dots, K. \tag{3.31}$$

It is a standard matter to see that it is equivalent to require that (3.30) holds for the Kružkov entropies $\eta = \eta(\xi, k) = |\xi - k|$ and entropy fluxes $q_i^\kappa = q_i^\kappa(\xi, k) = \mathrm{sign}(\xi - k)(f_i^\kappa(\xi) - f_i^\kappa(k))$ and $r^\kappa = r^\kappa(\xi, k) = |A^\kappa(\xi) - A^\kappa(k)|$ for all k, where $k, \xi \in \mathbb{R}$. That is, (3.30) can be replaced by[1]

$$\begin{aligned} |u^\kappa - k|_t + \sum_i \left[\mathrm{sign}(u^\kappa - k)\big(F_i^\kappa(u^\kappa) - F_i^\kappa(k)\big)\right]_{x_i} \\ - \Delta \left|A^\kappa(u^\kappa) - A^\kappa(k)\right| \le \mathrm{sign}(u^\kappa - k) g^\kappa(U) \quad \text{in } \mathcal{D}'\left(\Pi_T^0\right) \end{aligned} \tag{3.32}$$

[1]Observe that $\mathrm{sign}(u^\kappa - k)(A^\kappa(u^\kappa) - A^\kappa(k)) = |A^\kappa(u^\kappa) - A^\kappa(k)|$.

for all k. For the sake of our own convenience, we will switch back and forth between (3.30) and (3.32). If we first take $k > \text{ess sup}\, u^\kappa(x,t)$ and subsequently $k < \text{ess inf}\, u^\kappa(x,t)$ in (3.32), then we deduce that u^κ satisfies the equation

$$u_t^\kappa + \sum_i F_i^\kappa(u^\kappa)_{x_i} = \Delta A^\kappa(u^\kappa) + g^\kappa(U) \quad \text{in } \mathcal{D}'(\Pi_T^0).$$

Hence, an entropy weak solution U satisfies the system (3.26) in the sense of distributions. Furthermore, in the context of conservation laws, (3.32) coincides with Kružkov's celebrated entropy condition [161] in the scalar case as well as the extension of this condition to a weakly coupled system of conservation laws [212, 228, 229].

Finally, we have the following uniqueness theorem [122].

Theorem 3.11. *Assume that* (3.28), (3.29), *and* (3.28) *hold. Let* V, U *be two entropy weak solutions of the Cauchy problem* (3.25)–(3.27) *with initial data* V_0, U_0 *respectively. Then for* $t > 0$,

$$\int_{\mathbb{R}^d} \left|V(x,t) - U(x,t)\right| dx \le \sqrt{K} \exp\Big(K\|G\|_{\text{Lip}}\, t\Big) \int_{\mathbb{R}^d} \left|V_0(x) - U_0(x)\right| dx. \quad (3.33)$$

In particular, there exists at most one entropy weak solution of (3.25)–(3.27).

Regarding existence of entropy weak solutions we refer to [122], but also to later chapters in which existence will be a consequence of the convergence of an operator splitting method.

3.4 A general convergence theory

Having outlined a rigorous mathematical framework for studying weakly coupled systems of strongly degenerate parabolic convection-diffusion equations, we now come to the main novel point in this book: the development of a general convergence theory, which can be seen as a continuation of the work started in [119]. Using this theory, one can then prove the convergence of a specific operator splitting by verifying that each subsolver satisfies certain assumptions with regard to boundedness and entropy production.

To develop the theory, we first write the weakly coupled system (3.26)–(3.27) of convection-diffusion equations as an abstract Cauchy problem of the form

$$\frac{dU}{dt} + \mathcal{A}(U) = 0, \qquad U|_{t=0} = U_0, \quad (3.34)$$

where the nonlinear operator $\mathcal{A}$ is given by

$$\mathcal{A}(U) = \sum_i F_i(U)_{x_i} - \Delta A(U) - G(U). \quad (3.35)$$

Let $\mathcal{A}^1, \dots, \mathcal{A}^\ell$ be a splitting of the differential operator $\mathcal{A}$, i.e.,

$$\mathcal{A} = \mathcal{A}^1 + \cdots + \mathcal{A}^\ell, \qquad \ell \in \mathbb{N}, \tag{3.36}$$

where each $\mathcal{A}^l$ is also assumed to be a differential operator but possibly with a source term. Corresponding to this operator splitting, we assume that there exists a splitting of the entropy fluxes and the source term; that is, we have

$$\begin{aligned} Q_i &= Q_i^1 + \cdots + Q_i^\ell, \qquad i = 1, \dots, d, \\ R &= R^1 + \cdots + R^\ell, \\ G &= G^1 + \cdots + G^\ell, \end{aligned} \tag{3.37}$$

where

$$\begin{aligned} Q_i^l &= (q_i^{l1}, \dots, q_i^{lK}), \qquad l = 1, \dots, \ell, \\ R^l &= (r^{l1}, \dots, r^{lK}), \qquad l = 1, \dots, \ell, \\ G^l &= (g^{l1}, \dots, g^{lK}), \qquad l = 1, \dots, \ell. \end{aligned}$$

For a fixed $l = 1, \dots, \ell$, let us now consider the Cauchy problem

$$\frac{dV^l}{dt} + \mathcal{A}^l(V^l) = 0, \qquad V^l|_{t=0} = V_0^l, \quad V^l = (v^{l1}, \dots, v^{lK}). \tag{3.38}$$

Let $V^l(t) = \mathcal{S}_t^l V_0^l$ denote an exact or approximate entropy weak solution of (3.38). We are interested in building approximate entropy weak solutions of (3.26)–(3.27) using the product formula, thus

$$U(x,t) \approx U^n, \qquad t = n\Delta t,$$

where

$$U^n = \left[\mathcal{S}_{\Delta t}^\ell \circ \cdots \circ \mathcal{S}_{\Delta t}^1\right]^n U^0. \tag{3.39}$$

Here U^0 is equal, or approximately equal, to U_0. The aim is to show that U^n converges to the solution U provided the exact or approximate solution operators $\mathcal{S}^l$ satisfy certain criteria.

To investigate the convergence properties of the product formula (3.39), we need to work with functions that are not only defined for each $t = n\Delta t$, but for all t. To this end, we introduce functions

$$U^{n,l} = \mathcal{S}_{\Delta t}^l \circ \cdots \circ \mathcal{S}_{\Delta t}^1 U^n, \qquad l = 1, \dots, \ell, \qquad U^{n,0} = U^n. \tag{3.40}$$

Thus $U^{n,\ell} = U^{n+1}$. Following [71], we next introduce the auxiliary function $U_{\Delta t} \colon \Pi_T^0 \to \mathbb{R}^K$,

$$U_{\Delta t}(t) = \mathcal{S}_{\ell(t - t_{n,l-1})}^l U^{n,l-1} = (u_{\Delta t}^1, \dots, u_{\Delta t}^K), \quad \text{for} \quad t \in [t_{n,l-1}, t_{n,l}), \tag{3.41}$$

for $n \in \mathbb{N}_0$ and $l = 1, \dots, \ell$. Here

$$t_{n,l} = (n + \frac{l}{\ell})\Delta t, \quad l = 0, \dots, \ell, \tag{3.42}$$

and $\Pi_T^0 = \Pi_T \cup \{\mathbb{R}^d \times \{0\}\}$. Note that we have suppressed the dependence in $U_{\Delta t}$ on the discretization parameters associated with the operators $\mathcal{S}_t^\ell, \dots, \mathcal{S}_t^1$ if these are approximate (numerical) solution operators. This is justified by assuming that these discretization parameters are related to the splitting time step Δt in a manner implying that they tend to zero with Δt. Observe that

$$U_{\Delta t}(x, n\Delta t) = U^n = \left[\mathcal{S}_{\Delta t}^\ell \circ \dots \circ \mathcal{S}_{\Delta t}^1\right]^n U^0(x).$$

In this book we shall always let $\nu(h; U)$ denote the spatial modulus of continuity of the function $U\colon \mathbb{R}^d \to \mathbb{R}^K$. More precisely, we set $\nu(h; U)$ to be a continuous function $\nu(\,\cdot\,; U)\colon [0, \infty) \to [0, \infty)$ such that

$$\nu(0; U) = 0, \qquad \nu(\rho; U) \geq \sup_{|z| \leq \rho} \int_{\mathbb{R}^d} |U(x + z) - U(x)|\, dx. \tag{3.43}$$

If $U \in \mathrm{BV}(\mathbb{R}^d)$, then $\nu(h; U)$ can be chosen to be $h\|U\|_{\mathrm{BV}}$. Similarly, we use $\omega(h; U)$ to denote the temporal modulus of continuity of a function $U\colon \mathbb{R}^d \times \mathbb{R} \to \mathbb{R}^K$, i.e., we let $\omega_t(\tau; U)$ be a continuous function such that

$$\omega_t(0; U) = 0, \qquad \omega_t(\tau; U) \geq \sup_{|\varepsilon| \leq \tau} \int_{\mathbb{R}^d} |U(x, t + \varepsilon) - U(x, t)|\, dx. \tag{3.44}$$

Without loss of generality we may assume that both ν and ω are nondecreasing.

Let us assume that the approximate solution $U_{\Delta t}\colon \Pi_T^0 \to \mathbb{R}^K$ satisfies the following three estimates uniformly in Δt:

1. Uniform boundedness:

$$\operatorname{ess\,sup}_{\Pi_T^0} |U_{\Delta t}| \leq \mathrm{Const}_T \tag{3.45}$$

for some positive constant Const_T not depending on Δt.

2. There exists a spatial modulus of continuity:

$$\sup_{|z| \leq \rho} \int_{\mathcal{B}_r} |U_{\Delta t}(x + z, t) - U_{\Delta t}(x, t)|\, dx \leq \nu_{r,T}(\rho; U_{\Delta t}), \tag{3.46}$$

for $t \in [0, T]$, where $\nu_{r,T}\colon [0, \infty) \to [0, \infty)$ does not depend on Δt, but it is allowed to depend on r and T.

3. Weak local Lipschitz continuity in time:

$$\begin{aligned} &\left| \int_{\mathcal{B}_r} (U_{\Delta t}(x, t) - U_{\Delta t}(x, s))\, \phi(x)\, dx \right| \\ &\qquad \leq \mathrm{Const}_{r,T} \max_{\mathcal{B}_r} \Big(|\phi| + \sum_i |\phi_{x_i}| + \sum_{i,j} |\phi_{x_i x_j}| \Big) |t - s|, \end{aligned} \tag{3.47}$$

for $s, t \in [0, T]$ and some constant $\mathrm{Const}_{r,T} > 0$ not depending on Δt and any vector-valued test function $\phi \in C_0^\infty(\mathcal{B}_r; \mathbb{R}^K)$.

To apply the operator splitting technique we need properties (3.45)–(3.47). Next we formulate sufficient conditions on the simplified operators so that these conditions imply (3.45)–(3.47). In Chapters 4 and 5 we present several splitting methods for which it is possible to demonstrate that (3.45)–(3.47) hold.

Lemma 3.12. *Let $V^l(t) = \mathcal{S}_t^l V_0^l$ denote the solution of*

$$\frac{dV^l}{dt} + \mathcal{A}^l(V^l) = 0, \qquad V^l|_{t=0} = V_0^l, \quad V^l = (v^{l1}, \dots, v^{lK}), \quad l = 1, \dots \ell. \tag{3.48}$$

Assume that each function $V^l = V^l(t)$ satisfies

$$\left\|V^l(t)\right\|_{L^\infty(\mathbb{R}^d)} \le \exp(ct) \left\|V^l(0)\right\|_{L^\infty(\mathbb{R}^d)}, \tag{3.49}$$

$$\int_{\mathcal{B}_r} \left|V^l(x+y,t) - V^l(x,t)\right| dx \le C \exp(ct) \int_{\mathcal{B}_r} \left|V^l(x+y,0) - V^l(x,0)\right| dx, \tag{3.50}$$

$$\left| \int_{\mathcal{B}_r} \left(V^l(x,t) - V^l(x,s)\right) \phi(x)\, dx \right| \le \mathrm{Const}_{r,T} \max_{\mathcal{B}_r} \Big(\sum_{|\alpha| \le 2} |D^\alpha \phi| \Big) |t-s|, \tag{3.51}$$

for s and $t \in [0, T]$, $\phi \in C_0^\infty(\mathcal{B}_r; \mathbb{R}^K)$, and constants c and C that are independent of Δt. Then the function $U_{\Delta t}(t)$ defined by (3.39), (3.40), *and* (3.41) *satisfies the estimates* (3.45)–(3.47), *assuming that the initial approximation has a spatial modulus of continuity.*

Proof. By induction we see that

$$\|U_{\Delta t}(t)\|_{L^\infty(\mathbb{R}^d)} \le \exp(c\,\ell t_n) \|U_{\Delta t}(0)\|_{L^\infty(\mathbb{R}^d)} \le \exp(c\,\ell T) \|U_{\Delta t}(0)\|_{L^\infty(\mathbb{R}^d)},$$

for $t \in [t_{n-1}, t_n)$, which proves uniform boundedness, that is, inequality (3.45). To establish a spatial modulus of continuity, let $t \in [t_{n,l-1}, t_{n,l})$. Then

$$\begin{aligned}
&\int_{\mathcal{B}_r} |U_{\Delta t}(x+z,t) - U_{\Delta t}(x,t)|\, dx \\
&\qquad \le \exp(c\ell(t - t_{n,l-1})) \int_{\mathcal{B}_r} \left|U^{n,l-1}(x+z,t) - U^{n,l-1}(x,t)\right| dx \\
&\qquad \le \exp(c\Delta t) \exp(c(l-1)\Delta t) \int_{\mathcal{B}_r} \left|U^{n-1}(x+z,t) - U^{n-1}(x,t)\right| dx \\
&\qquad \le \exp(c\,\ell(n+1)\Delta t) \int_{\mathcal{B}_r} \left|U^0(x+z,t) - U^0(x,t)\right| dx \\
&\qquad \le \exp(c\,\ell(n+1)\Delta t)\, \nu_{r,T}(\rho; U^0), \qquad |z| \le \rho,
\end{aligned}$$

which proves (3.46). To prove (3.47) it suffices to consider $s = 0$ and $t \in [t_{n,l-1}, t_{n,l})$. Thus

$$\begin{aligned}
&\Big|\int_{\mathcal{B}_r} (U_{\Delta t}(x,t) - U_{\Delta t}(x,0))\phi(x)\,dx\Big| \\
&\quad \le \Big|\int_{\mathcal{B}_r} (U_{\Delta t}(x,t) - U^{n,l-1}(x))\phi(x)\,dx\Big| \\
&\qquad + \sum_{m=1}^{l-1} \Big|\int_{\mathcal{B}_r} (U^{n,m}(x) - U^{n,m-1}(x))\phi(x)\,dx\Big| \\
&\qquad + \sum_{p=0}^{n-1}\sum_{m=1}^{\ell} \Big|\int_{\mathcal{B}_r} (U^{p,m}(x) - U^{p,m-1}(x))\phi(x)\,dx\Big| \\
&\quad \le \mathrm{Const}_{r,T} \max_{\mathcal{B}_r}\Big(|\phi| + \sum_i |\phi_{x_i}| + \sum_{i,j} |\phi_{x_i x_j}|\Big)(l\Delta t + \ell n\Delta t).
\end{aligned}$$

□

Theorem 3.13 (Convergence). *Suppose that the approximate solutions* $\{U_{\Delta t}\}$ *satisfy the three estimates* (3.45), (3.46), *and* (3.47). *Then there exists a subsequence of* $\{U_{\Delta t}\}$ *which converges in* $L^1_{\mathrm{loc}}(\Pi_T; \mathbb{R}^K)$ *to a limit function* $U \in L^\infty(\Pi_T; \mathbb{R}^K)$.

Proof. Fix $r > 0$. In view of (3.45), there exists a constant $\mathrm{Const}_{r,T} > 0$ such that

$$\int_{\mathcal{B}_r}\int_0^T |U_{\Delta t}(x,t)|\,dt\,dx \le \mathrm{Const}_{r,T}.$$

Furthermore, according to (3.46), there exists a spatial modulus of continuity $\nu_{r,T}\colon [0,\infty) \to [0,\infty)$, such that

$$\sup_{|z|\le\rho} \int_{\mathcal{B}_r}\int_0^T |U_{\Delta t}(x+z,t) - U_{\Delta t}(x,t)|\,dt\,dx \le \nu_{r,T}(\rho; U_{\Delta T}). \tag{3.52}$$

For each $t \in [0,T]$ we can apply Lemma 3.1 to find a subsequence $\Delta t \to 0$ such that $u^\kappa_{\Delta t}(\,\cdot\,,t)$ converges in L^1_{loc} to a function $u^\kappa(\,\cdot\,,t)$. Let now $\{t_n\} \subset [0,T]$ be a dense, countable set. By a standard diagonal argument we can find one common subsequence $\Delta t_k \to 0$ such that $u^\kappa_{\Delta t}(\,\cdot\,,t_n) \to u^\kappa(\,\cdot\,,t_n)$ for all n.

To prove time continuity of the splitting approximation $U_{\Delta t}(\,\cdot\,,t)$ we will use a different technique than in, e.g., [71, 125]. That is, we will rely on a weak Lipschitz continuity in time of $u_{\Delta t}(\,\cdot\,,t)$, cf. (3.47), and the Kružkov interpolation Lemma 3.4. In view of (3.47) and (3.52), Remark 3.5 then implies that we can find a temporal modulus of continuity $\omega^\kappa_{r,T}\colon [0,\infty) \to [0,\infty)$ such that

$$\int_{\mathcal{B}_r} |u^\kappa_{\Delta t}(x,t+\tau) - u^\kappa_{\Delta t}(x,t)|\,dx \le \omega_{r,T}(|\tau|; u^\kappa_{\Delta t}).$$

To prove that $\{u^\kappa_{\Delta t_k}(\,\cdot\,,t)\}$ is a Cauchy sequence we let $t \in [0,T]$ and write

$$\begin{aligned}\int_{\mathcal{B}_r} \left|u^\kappa_{\Delta t_k}(x,t) - u^\kappa_{\Delta t_{\tilde{k}}}(x,t)\right| dx &\le \int_{\mathcal{B}_r} \left|u^\kappa_{\Delta t_k}(x,t)\,dx - u^\kappa_{\Delta t_k}(x,t_n)\right| dx \\ + \int_{\mathcal{B}_r} \left|u^\kappa_{\Delta t_k}(x,t_n) - u^\kappa_{\Delta t_{\tilde{k}}}(x,t_n)\right| dx &+ \int_{\mathcal{B}_r} \left|u^\kappa_{\Delta t_{\tilde{k}}}(x,t_n) - u^\kappa_{\Delta t_{\tilde{k}}}(x,t)\right| dx. \qquad (3.53)\end{aligned}$$

The middle term can be made small by assumption, while the first and the last terms are small using the temporal modulus. Hence we have established that there exists a subsequence Δt_k such that $u^\kappa_{\Delta t_k}(\,\cdot\,,t)$ converges in L^1_{loc} to a function $u^\kappa(\,\cdot\,,t)$ for all $t \in [0,T]$. Lebesgue's dominated convergence theorem shows convergence in $L^1_{\mathrm{loc}}(\Pi_T;\mathbb{R}^K)$. □

Remark 3.14. The technique used to establish time continuity was first used in [145] for a nonlinear, scalar convection-diffusion equation. Subsequent applications of this technique can be found in [97, 119, 141, 146].

Having established necessary conditions on the suboperators to ensure convergence of $\{U_{\Delta t}\}$, we next look into conditions necessary to guarantee that the limit function U is an entropy weak solution. To this end, we consider an approximate solution $V^l\colon \Pi^0_T \to \mathbb{R}^K$ of (3.38). Thus V^l will not satisfy (3.30) (with q_i^κ replaced by $q_i^{l\kappa}$, etc.) exactly, but rather (3.30) modified with additional error terms. Inspired by [33], we assume that V^l satisfies the entropy inequalities (3.30) with error terms that are bounded linear functionals on C_0 (in the distributional sense); that is, for $l = 1,\dots,\ell$, $\kappa = 1,\dots,K$ and all $0 \le t_1 < t_2 \le T$,

$$\begin{aligned}\eta\left(v^{l\kappa}\right)_t + \sum_i q_i^{l\kappa}\left(v^{l\kappa}\right)_{x_i} - \Delta r^{l\kappa}\left(v^{l\kappa}\right) \le \eta'\left(v^{l\kappa}\right) g^{l\kappa}\left(V^l\right) \\ + \left(\mathcal{E}_{\mathrm{I}}{}^{l\kappa}\right)_t + \sum_i \left(\mathcal{E}_{\mathrm{II}i}{}^{l\kappa}\right)_{x_i} + \sum_{i,j}\left(\mathcal{E}_{\mathrm{III}ij}{}^{l\kappa}\right)_{x_i x_j} + \left(\mathcal{E}_{\mathrm{IV}i}{}^{l\kappa}\right)\end{aligned} \qquad (3.54)$$

in $\mathcal{D}'([t_1,t_2]\times\mathbb{R}^d)$. Recall that this notation means that for a test function $\phi(x,t)$, the linear functional $(\mathcal{E}_{\mathrm{I}}{}^{l\kappa})_t$ takes as argument ϕ_t, the term $(\mathcal{E}_{\mathrm{II}}{}^{l\kappa})_{x_i}$ acts on ϕ_{x_i}, and so on. If these linear functionals are bounded, by the Riesz–Markov theorem (see, e.g., [185, Thm. 6.22])

$$\left|\mathcal{E}_{\mathrm{I}}{}^{l\kappa}\right| \le d\mu_{\mathrm{I}}{}^{l\kappa}, \quad \left|\mathcal{E}_{\mathrm{II}i}{}^{l\kappa}\right| \le d\mu_{\mathrm{II}i}{}^{l\kappa}, \quad \left|\mathcal{E}_{\mathrm{III}ij}{}^{l\kappa}\right| \le d\mu_{\mathrm{III}ij}{}^{l\kappa}, \quad \left|\mathcal{E}_{\mathrm{IV}}{}^{l\kappa}\right| \le d\mu_{\mathrm{IV}}{}^{l\kappa},$$

for some nonnegative Radon measures $d\mu_{\mathrm{I}}{}^{l\kappa}$, $d\mu_{\mathrm{II}i}{}^{l\kappa}$, $d\mu_{\mathrm{III}ij}{}^{l\kappa}$, $d\mu_{\mathrm{IV}}{}^{l\kappa}$. Furthermore, for almost every $t \in [0,T]$, these measures should be Radon measures on $\mathbb{R}^d$. In practice, when deciding whether (3.54) holds, we check if

$$\int_{t_1}^{t_2}\int_{\mathbb{R}^d} \Big(\eta(v^{l\kappa})\phi_t + \sum_i q_i^{l\kappa}(v^{l\kappa})\phi_{x_i} - r^{l\kappa}(v^{l\kappa})\Delta\phi - g^{l\kappa}(V^l)\phi\Big)\,dx\,dt$$

$$
\begin{aligned}
& - \int_{\mathbb{R}^d} \eta(v^{l\kappa}(x,s))\phi(x,s) \Big|_{s=t_1}^{s=t_2} dx \\
& \geq \int_{t_1}^{t_2} \int_{\mathbb{R}^d} \Big(\phi_t \, d\mu_{\mathrm{I}}{}^{l\kappa} + \sum_i \phi_{x_i} \, d\mu_{\mathrm{II}}{}_i^{l\kappa} + \sum_{i,j} \phi_{x_i x_j} \, d\mu_{\mathrm{III}}{}_{ij}^{l\kappa} + \phi \, d\mu_{\mathrm{IV}}{}^{l\kappa} \Big)
\end{aligned}
$$

for all nonnegative test functions ϕ. Furthermore, we must check if these measures are finite, i.e., that for (almost) all $t \in [t_1, t_2]$,

$$
\begin{aligned}
&\int_{\mathbb{R}_d} d\mu_{\mathrm{I}}{}^{l\kappa}(x,t) \leq \mathfrak{E}_{\mathrm{I}}{}^{l\kappa}, && \int_{\mathbb{R}_d} d\mu_{\mathrm{II}}{}_i^{l\kappa}(x,t) \leq \mathfrak{E}_{\mathrm{II}}{}_i^{l\kappa}, \\
&\int_{\mathbb{R}_d} d\mu_{\mathrm{III}}{}_{i,j}^{l\kappa}(x,t) \leq \mathfrak{E}_{\mathrm{III}}{}_{i,j}^{l\kappa}, && \int_{\mathbb{R}_d} d\mu_{\mathrm{IV}}{}_i^{l\kappa}(x,t) \leq \mathfrak{E}_{\mathrm{IV}}{}_i^{l\kappa}.
\end{aligned} \tag{3.55}
$$

Let the approximation be described by a parameter Δt that typically tends to zero (and which we have suppressed in the notation). We assume that the Radon measures tend to zero (in a suitable manner) as $\Delta t \to 0$. In practice we observe that $\mathfrak{E}^{l\kappa}$ vanish as $\Delta t \to 0$, see Chapters 4 and 5 for details. Of course, when $\mathcal{S}_t^1, \dots, \mathcal{S}_t^\ell$ are exact solution operators, these error terms are absent and we obtain so-called semi-discrete splitting methods. For each concrete application of these methods we will have to show the existence of error terms with the properties described above.

For any fixed test function $\phi \in C_0^\infty(\Pi_T)$, $n \in \mathbb{N}_0 = \mathbb{N} \cup \{0\}$, $l = 1, \dots, \ell$, and $\kappa = 1, \dots, K$, let us define the local error function $\Delta t \mapsto \mathcal{E}^{nl\kappa}(\Delta t; \phi)$ by

$$
\begin{aligned}
\mathcal{E}^{nl\kappa}(\Delta t; \phi) = \iint\limits_{\Pi_{n,l}} \Big(\frac{1}{\ell} |\phi_t| \, d\mu_{\mathrm{I}}{}^{l\kappa} + \sum_i |\phi_{x_i}| \, d\mu_{\mathrm{II}}{}_i^{l\kappa} + \sum_{i,j} |\phi_{x_i x_j}| \, d\mu_{\mathrm{III}}{}_{ij}^{l\kappa} + |\phi| \, d\mu_{\mathrm{IV}}{}^{l\kappa} \Big) \\
+ \frac{1}{\ell} \int_{\mathbb{R}^d} |\phi(\,\cdot\,, t_{n,l})| \, d\mu_{\mathrm{I}}{}^{l\kappa}(\,\cdot\,, t_{n,l})
\end{aligned} \tag{3.56}
$$

where $\Pi_{n,l} = \mathbb{R}^d \times (t_{n,l-1}, t_{n,l})$ and

$$
t_n = n\Delta t, \quad t_{n,l} = \Big(n + \frac{l}{\ell} \Big) \Delta t, \quad l = 0, \dots, \ell.
$$

Clearly $t_n = t_{n,0} = t_{n-1,\ell}$. Let $\chi_{n,l}$ be the characteristic function of the set $\Pi_{n,l}$, thus $\chi_{n,l} = \chi_{\Pi_{n,l}}$, and set

$$
\chi_l^\ell = \chi_{\Pi_l} = \sum_n \chi_{n,l}, \quad \Pi_l = \mathbb{R}^d \times \cup_n (t_{n,l-1}, t_{n,l}). \tag{3.57}
$$

For later use we set

$$
\Pi_n = \mathbb{R}^d \times (n\Delta t, (n+1)\Delta t).
$$

Furthermore, let us define the global error function $\Delta t \mapsto \mathcal{E}^\kappa(\Delta t; \phi)$ by

$$
\mathcal{E}^\kappa(\Delta t; \phi) = \ell \sum_{n=0}^{N-1} \sum_{l=1}^{\ell} \mathcal{E}^{nl\kappa}(\Delta t; \phi) \tag{3.58}
$$

$$
\begin{aligned}
= \sum_{l=1}^{\ell} \Bigg[\iint\limits_{\Pi_T} \chi_l^{\ell} \Big(|\phi_t| \, d{\mu_{\mathrm{I}}}^{l\kappa} \\
&+ \ell \Big(\sum_i |\phi_{x_i}| \, d{\mu_{\mathrm{II}}}_i^{l\kappa} + \sum_{i,j} \big|\phi_{x_i x_j}\big| \, d{\mu_{\mathrm{III}}}_{ij}^{l\kappa} + |\phi| \, d{\mu_{\mathrm{IV}}}^{l\kappa} \Big) \Big) \\
&+ \sum_{n=0}^{N-1} \int_{\mathbb{R}^d} |\phi(\,\cdot\,, t_{n,l})| \, d{\mu_{\mathrm{I}}}^{l\kappa}(\,\cdot\,, t_{n,l}) \Bigg],
\end{aligned}
$$

and denote by $\mathcal{E}$ the vector $(\mathcal{E}^1, \dots, \mathcal{E}^K)$. Our next result provides us with the following important approximate entropy inequality:

Lemma 3.15. *Let $U_{\Delta t}$ be defined by (3.40) and (3.41). For $\kappa = 1, \dots, K$ and any nonnegative test function $\phi \in C_0^\infty(\Pi_T^0)$, we have*

$$
\begin{aligned}
&\iint\limits_{\Pi_T} \Big(\eta(u_{\Delta t}^{\kappa}) \phi_t + \sum_i q_i^{\kappa}(u_{\Delta t}^{\kappa}) \phi_{x_i} + r^{\kappa}(u_{\Delta t}^{\kappa}) \Delta\phi \Big) \, dt \, dx \\
&\quad \geq - \iint\limits_{\Pi_T} \eta'(u_{\Delta t}^{\kappa}) g^{\kappa}(U_{\Delta t}) \phi \, dt \, dx - \int_{\mathbb{R}^d} \eta(u_{\Delta t}^{\kappa}(x,0)) \phi(x,0) \, dx \\
&\quad - \mathcal{E}^{\kappa}(\Delta t; \phi) - I_1^{\kappa} - I_2^{\kappa} - I_3^{\kappa},
\end{aligned}
\tag{3.59}
$$

where

$$
\begin{aligned}
I_1^{\kappa} &= \ell \sum_{l=1}^{\ell} \sum_i \iint\limits_{\Pi_T} (\chi_l^{\ell} - \tfrac{1}{\ell}) q_i^{l\kappa}(u_{\Delta t}^{\kappa}) \phi_{x_i} \, dt \, dx, \\
I_2^{\kappa} &= \ell \sum_{l=1}^{\ell} \iint\limits_{\Pi_T} (\chi_l^{\ell} - \tfrac{1}{\ell}) r^{l\kappa}(u_{\Delta t}^{\kappa}) \Delta\phi \, dt \, dx, \\
I_3^{\kappa} &= \ell \sum_{l=1}^{\ell} \iint\limits_{\Pi_T} (\chi_l^{\ell} - \tfrac{1}{\ell}) \eta'(u_{\Delta t}^{\kappa}) g^{l\kappa}(U_{\Delta t}) \phi \, dt \, dx,
\end{aligned}
$$

where χ_l^{ℓ} is the characteristic function of $\Pi_l = \bigcup_n \Pi_{n,l} = \mathbb{R}^d \times \bigcup_n (t_{n,l-1}, t_{n,l})$.

Proof. Fix $\kappa = 1, \dots, K$, let $\phi \in C_0^\infty(\Pi_T)$ be a nonnegative test function, and define a new test function φ on each strip $\Pi_{n,l}$ by

$$
\varphi(x,t) = \phi\left(x, \frac{t}{\ell} + t_{n,l-1} \right).
$$

For $n \in \mathbb{N}_0$ and $l = 1, \dots, \ell$, let us also introduce

$$
V^{n,l}(t) = \mathcal{S}_t^l U^{n,l-1} = (v^{nl1}, \dots, v^{nlK}), \qquad t \geq 0.
$$

Using (3.54) we find that

$$\iint\limits_{\Pi_{n,l}} \Big(\frac{1}{\ell}\eta(u^\kappa_{\Delta t})\phi_t + \sum_i q_i^{l\kappa}(u^\kappa_{\Delta t})\phi_{x_i} + r^{l\kappa}(u^\kappa_{\Delta t})\Delta\phi\Big)\,dt\,dx$$

$$= \frac{1}{\ell}\int_{\mathbb{R}^d}\int_0^{\Delta t}\Big(\eta\big(v^{nl\kappa}(\tau)\big)\varphi_\tau + \sum_i q_i^{l\kappa}\big(v^{nl\kappa}(\tau)\big)\varphi_{x_i} + r^{l\kappa}\big(v^{nl\kappa}(\tau)\big)\Delta\varphi\Big)\,d\tau\,dx$$

$$\geq -\iint\limits_{\Pi_{n,l}} \eta'(u^\kappa_{\Delta t})g^{l\kappa}(U_{\Delta t})\phi\,dt\,dx$$
$$+\frac{1}{\ell}\int_{\mathbb{R}^d}\Big(\eta\big(u^\kappa_{\Delta t}(x,t_{n,l})\big)\phi\,(x,t_{n,l}) - \eta\big(u^\kappa_{\Delta t}(x,t_{n,l-1})\big)\phi\,(x,t_{n,l-1})\Big)\,dx$$
$$+\iint\limits_{\Pi_{n,l}}\Big(\frac{1}{\ell}\mathcal{E}_{\mathrm{I}}{}^{l\kappa}\phi_t + \sum_i \phi_{x_i}\mathcal{E}_{\mathrm{II}}{}_i^{l\kappa} - \sum_{i,j}\phi_{x_ix_j}\mathcal{E}_{\mathrm{III}}{}_{ij}^{l\kappa} - \phi\mathcal{E}_{\mathrm{IV}}{}^{l\kappa}\Big)\,dt\,dx$$
$$-\frac{1}{\ell}\int_{\mathbb{R}^d}\mathcal{E}_{\mathrm{I}}{}^{l\kappa}\,(x,t_{n,l})\,\phi\,(x,t_{n,l})\,dx$$

$$\geq -\iint\limits_{\Pi_{n,l}} \eta'(u^\kappa_{\Delta t})g^{l\kappa}(U_{\Delta t})\phi\,dt\,dx$$
$$+\frac{1}{\ell}\int_{\mathbb{R}^d}\Big(\eta\big(u^\kappa_{\Delta t}(x,t_{n,l})\big)\phi\,(x,t_{n,l}) - \eta\big(u^\kappa_{\Delta t}(x,t_{n,l-1})\big)\phi\,(x,t_{n,l-1})\Big)\,dx$$
$$-\iint\limits_{\Pi_{n,l}}\Big(\frac{1}{\ell}\,|\phi_t|\,d\mu_{\mathrm{I}}{}^{l\kappa} + \sum_i |\phi_{x_i}|\,d\mu_{\mathrm{II}}{}_i^{l\kappa} + \sum_{i,j}|\phi_{x_ix_j}|\,d\mu_{\mathrm{III}}{}_{ij}^{l\kappa} + |\phi|\,d\mu_{\mathrm{IV}}{}^{l\kappa}\Big)$$
$$-\frac{1}{\ell}\int_{\mathbb{R}^d}|\phi(x,t_{n,l})|\,d\mu_{\mathrm{I}}{}^{l\kappa}(x,t_{n,l}).$$

The first equality follows using the change of variables

$$\tau = \ell\,(t - t_{n,l-1})\,.$$

Furthermore, to deduce the first inequality, we have used that $V^{nl}(t)$ satisfies (3.54) in the sense of distributions on $\mathbb{R}^d \times (0,\Delta t)$, with initial data

$$U_{\Delta t}\Big|_{\tau=0} = U^{n,l-1}$$

taken in the strong L^1_{loc} sense. Furthermore, the error $\mathcal{E}_{\mathrm{I}}{}^{l\kappa}$ vanishes at $\tau = 0$. If we first sum the above inequality over $l = 1,\ldots,\ell$ and then sum over $n = 0,\ldots,N-1$, the resulting inequality takes the form (recall (3.56) and (3.58))

$$\iint\limits_{\Pi_T}\Big(\frac{1}{\ell}\eta\,(u^\kappa_{\Delta t})\,\phi_t + \sum_{l=1}^{\ell}\sum_i \chi_l^\ell\, q_i^{l\kappa}\,(u^\kappa_{\Delta t})\,\phi_{x_i} + \sum_{l=1}^{\ell}\sum_i \chi_l^\ell\, r_i^{l\kappa}\,(u^\kappa_{\Delta t})\,\phi_{x_ix_i}\Big)\,dt\,dx$$

$$\geq - \iint\limits_{\Pi_T} \sum_{l=1}^{\ell} \chi_l^\ell \, \eta' \left(u_{\Delta t}^\kappa\right) g^{l\kappa} \left(U_{\Delta t}\right) \phi \, dt \, dx - \sum_{n=0}^{N-1} \sum_{l=1}^{\ell} \mathcal{E}^{nl\kappa}(\Delta t; \phi).$$

Using that $\mathcal{E}^\kappa = \ell \sum_{n,l} \mathcal{E}^{nl\kappa}$, and rewriting the left-hand side we obtain (3.59). □

Remark 3.16. Suppose $U_{\Delta t} \to U$ in $L^2_{\text{loc}}(\Pi_T; \mathbb{R}^K)$ and $\mathcal{E}(\Delta t; \phi) \to 0$ as $\Delta t \to 0$. Now since[2] $\chi_l^\ell(x,t) \rightharpoonup 1/\ell$ in $L^2(\Pi_T)$, we get

$$\iint\limits_{\Pi_T} \chi_l^\ell q_i^{l\kappa} \left(u_{\Delta t}^\kappa\right) \phi_{x_i} \, dt \, dx \to \frac{1}{\ell} \iint\limits_{\Pi_T} q_i^{l\kappa} \left(u^\kappa\right) \phi_{x_i} \, dt \, dx,$$

$$\iint\limits_{\Pi_T} \chi_l^\ell r_i^{l\kappa} \left(u_{\Delta t}^\kappa\right) \phi_{x_i x_i} \, dt \, dx \to \frac{1}{\ell} \iint\limits_{\Pi_T} r_i^\kappa \left(u^\kappa\right) \phi_{x_i x_i} \, dt \, dx.$$

Furthermore, we easily get

$$\iint\limits_{\Pi_T} \chi_l^\ell \eta' \left(u_{\Delta t}^\kappa\right) g^{l\kappa}(U_{\Delta t}) \phi \, dt \, dx \to \frac{1}{\ell} \iint\limits_{\Pi_T} \eta' \left(u^\kappa\right) g^{l\kappa}(U) \phi \, dt \, dx.$$

Passing to the limit in (3.59), with (3.37) in mind and recalling that $u_{\Delta t}^\kappa(\cdot, 0) \to u_0^\kappa$ strongly in $L^1_{\text{loc}}(\mathbb{R})$, we obtain (3.30) and thus that U satisfies the entropy condition (see also Corollary 3.19 below).

Our next result yields a more precise estimate of the entropy discrepancy associated with $U_{\Delta t}$:

Theorem 3.17 (Entropy Estimate). *Fix $\kappa = 1, \dots, K$. Suppose $t \mapsto u_{\Delta t}^\kappa(t)$ is locally uniformly L^1-continuous, i.e., it has a uniform temporal modulus of continuity $\omega_{r,T}^\kappa$ so that*

$$\int_{\mathcal{B}_r} \left| u_{\Delta t}^\kappa(x, t + \tau) - u_{\Delta t}^\kappa(x,t) \right| \, dx \leq \omega_{r,T}^\kappa(|\tau| \, ; u_{\Delta t}^\kappa),$$

where $\omega_{r,T}^\kappa$ does not depend on Δt or $t \in [0, T]$. Furthermore, suppose $u_{\Delta t}^\kappa$ is locally integrable,

$$\int_{\mathcal{B}_r} \left| u_{\Delta t}^\kappa(x,t) \right| \, dx \leq \text{Const}_{r,T},$$

for some constant not depending on Δt or $t \in [0, T]$. Then the entropy discrepancy associated with $u_{\Delta t}^\kappa$ is bounded as follows:

$$\iint\limits_{\Pi_T} \Big(\eta \left(u_{\Delta t}^\kappa\right) \phi_t + \sum_i q_i^\kappa \left(u_{\Delta t}^\kappa\right) \phi_{x_i} + r^\kappa \left(u_{\Delta t}^\kappa\right) \Delta \phi \Big) \, dt \, dx \tag{3.60}$$

[2]We say that $\phi_n \rightharpoonup \phi$ in L^2 if $\int \phi_n \psi \, dx \to \int \phi \psi \, dx$ for all $\psi \in L^2$.

$$\geq -\iint\limits_{\Pi_T} \eta'\left(u^\kappa_{\Delta t}\right) g^\kappa(U_{\Delta t})\, dt\, dx - \int_{\mathbb{R}^d} \eta\left(u^\kappa_{\Delta t}(x,0)\right) \phi(x,0)\, dx$$
$$- \mathcal{E}^\kappa(\Delta t;\phi) - C\,T\,\omega^\kappa_{R,T}(\Delta t; u^\kappa_{\Delta t}), \qquad \kappa = 1,\dots,K,$$

for some $R > 0$ depending on $\mathrm{supp}_x(\phi)$ and some constant C depending in particular on η'' and the derivatives of ϕ up to third order, but not on Δt.

Remark 3.18. If $U_{\Delta t} = (u^1_{\Delta t},\dots,u^K_{\Delta t})$ satisfies (3.45)–(3.47), then by the Kružkov interpolation lemma, Lemma 3.4, $U_{\Delta t}$ satisfies the assumptions of the theorem.

Proof. We are going to use the inequality (3.59) to estimate the entropy discrepancy in (3.30). We start by estimating the term I^κ_1 in (3.59) for fixed $\kappa = 1,\dots,K$. Writing

$$\left(q^{l\kappa}_i\left(u^\kappa_{\Delta t}\right)\phi_{x_i}\right)(t) = \left(q^{l\kappa}_i\left(u^\kappa_{\Delta t}\right)\phi_{x_i}\right)(t_n) + \left[\left(q^{l\kappa}_i\left(u^\kappa_{\Delta t}\right)\phi_{x_i}\right)(t) - \left(q^{l\kappa}_i\left(u^\kappa_{\Delta t}\right)\phi_{x_i}\right)(t_n)\right],$$

we can write I^κ_1 as sum of two terms. The first term integrates to zero since

$$\sum_{l=1}^{\ell} \int_0^T \left(\chi^\ell_l - \frac{1}{\ell}\right) dt = 0,$$

and thus

$$\begin{aligned}
I^\kappa_1 &= \sum_{n,i,l}\Bigg(\ell \iint\limits_{\Pi_{n,l}} q^{l\kappa}_i\left(u^\kappa_{\Delta t}\right)\phi_{x_i}\, dt\, dx - \iint\limits_{\Pi_n} q^{l\kappa}_i\left(u^\kappa_{\Delta t}\right)\phi_{x_i}\, dt\, dx\Bigg)\\
&= \sum_{n,i,l}\Bigg(\ell \iint\limits_{\Pi_{n,l}} \left[\left(q^{l\kappa}_i\left(u^\kappa_{\Delta t}\right)\phi_{x_i}\right)(t) - \left(q^{l\kappa}_i\left(u^\kappa_{\Delta t}\right)\phi_{x_i}\right)(t_n)\right] dt\, dx\\
&\qquad - \iint\limits_{\Pi_n} \left[\left(q^{l\kappa}_i\left(u^\kappa_{\Delta t}\right)\phi_{x_i}\right)(t) - \left(q^{l\kappa}_i\left(u^\kappa_{\Delta t}\right)\phi_{x_i}\right)(t_n)\right] dt\, dx\Bigg)\\
&=: I^\kappa_{11} - I^\kappa_{12}.
\end{aligned}$$

Let $R > 0$ be the smallest integer such that $\mathrm{supp}(\phi(\,\cdot\,,t)) \subset \mathcal{B}_R$ for all $t \in [0,T]$. We then proceed as follows:

$$\begin{aligned}
|I^\kappa_{11}| &\leq \Bigg|\sum_{n,i,l} \ell \iint\limits_{\Pi_{n,l}} \left[q^{l\kappa}_i\left(u^\kappa_{\Delta t}(t)\right) - q^{l\kappa}_i\left(u^\kappa_{\Delta t}(t_n)\right)\right]\phi_{x_i}(t)\, dt\, dx\Bigg|\\
&\quad + \Bigg|\sum_{n,i,l} \ell \iint\limits_{\Pi_{n,l}} \left[\phi_{x_i}(t) - \phi_{x_i}(t_n)\right] q^{l\kappa}_i\left(u^\kappa_{\Delta t}(t_n)\right) dt\, dx\Bigg|
\end{aligned}$$

$$\leq \sum_{n,i,l} \ell \frac{\Delta t}{\ell} \|q^\kappa\|_{\mathrm{Lip}}\, \omega^\kappa_{R,T}(\Delta t; u^\kappa_{\Delta t})\, \|\phi_{x_i}\|_\infty$$

$$+ \sum_{n,i,l} \ell \frac{\Delta t}{\ell} \|\phi_{x_i t}\|_\infty\, \Delta t \int_{\mathcal{B}_R} \left|q_i^{l\kappa}\left(u^\kappa_{\Delta t}(x,t_n)\right)\right| dx$$

$$\leq C_1\left(\phi_x, \|q^\kappa\|_{\mathrm{Lip}}\right) \omega^\kappa_{R,T}(\Delta t; u^\kappa_{\Delta t})\, T + C_2\left(\phi_{xt}, R, T, \|q^\kappa\|_{\mathrm{Lip}}\right) \Delta t\, T,$$

where we have taken into account that $Q_i^l(0) = 0$ so that

$$\int_{\mathcal{B}_R} \left|q_i^{l\kappa}\left(u^\kappa_{\Delta t}(x,t_n)\right)\right| dx \leq \|q^\kappa\|_{\mathrm{Lip}} \int_{\mathcal{B}_R} |u^\kappa_{\Delta t}(x,t_n)|\, dx \leq \|q^\kappa\|_{\mathrm{Lip}}\, \mathrm{Const}_{R,T}.$$

Summing up, we have proved that

$$|I^\kappa_{11}| \leq \tilde{C}\left(\phi_x, \phi_{xt}, R, T, \|q^\kappa\|_{\mathrm{Lip}}\right) \left(\omega^\kappa_{R,T}(\Delta t; u^\kappa_{\Delta t}) + \Delta t\right) T.$$

Estimating I^κ_{12} similarly, we get the desired estimate on I^κ_1:

$$|I^\kappa_1| \leq C\left(\phi_x, \phi_{xt}, R, T, \|q^\kappa\|_{\mathrm{Lip}}\right) \omega^\kappa_{R,T}(\Delta t; u^\kappa_{\Delta t})\, T,$$

for some constant $C > 0$ not depending on Δt. Similar estimates for I_2 and I_3 yield

$$|I^\kappa_2| \leq C\left(\phi_{xx}, \phi_{xxt}, R, T, \|r^\kappa\|_{\mathrm{Lip}}\right) \omega^\kappa_{R,T}(\Delta t; u^\kappa_{\Delta t})\, T,$$

$$|I^\kappa_3| \leq C\left(\phi, \phi_t, R, T, \|\eta'\|_{\mathrm{Lip}}, \|g^\kappa\|_{\mathrm{Lip}}\right) \omega^\kappa_{R,T}(\Delta t; u^\kappa_{\Delta t})\, T,$$

for constants $C > 0$ not depending on Δt. □

The next corollary follows immediately from the above theorem:

Corollary 3.19. *Suppose that* $U_{\Delta t} \to U \in L^\infty\left(\mathbb{R}^d; \mathbb{R}^K\right)$ *in* $L^1_{\mathrm{loc}}(\Pi_T)$ *as* $\Delta t \to 0$ *and that* $\mathcal{E}^\kappa(\Delta t; \phi) \to 0$, $\kappa = 1, \dots, K$, *as* $\Delta t \to 0$ *for all nonnegative test functions* $\phi \in C_0^\infty(\Pi_T)$. *Assume that* $\nabla_x A^\kappa(u^\kappa) \in L^2(\Pi_T)$, *for* $\kappa = 1, \dots, K$. *Then the limit* U *is an entropy weak solution of the Cauchy problem* (3.26)–(3.27).

Remark 3.20. If we use exact solution operators, the error $\mathcal{E}$ is identically zero and hence the semi-discrete product of the solution operators converges to the unique entropy solution.

Remark 3.21. Corollary 3.19 also holds for systems of equations that possess smooth entropies and entropy-fluxes. Furthermore, independently of existence of entropies and entropy-fluxes, the proof can easily be modified to show that the limit is at least a weak solution. In other words, the above reasoning can be used to prove Lax–Wendroff type theorems. See, e.g., [121] for an example of this. For hyperbolic problems, the entropy consistency of operator-splitting methods has also been a topic of study in [52].

Remark 3.22. Precise entropy estimates for specific operator-splitting methods applied to specific convection-diffusion equations have been obtained earlier in [97, 119, 146]. As we will see later, the generality of Theorem 3.17 allows us to derive the entropy estimates obtained in these works, as well as new ones relevant to this book.

4

Convergence Results for Convection-Diffusion Problems

In this chapter we demonstrate how the theory developed in Chapter 3 works in practice by applying it to the one-dimensional convection-diffusion problem

$$u_t + f(u)_x = A(u)_{xx}, \qquad u|_{t=0} = u_0, \tag{4.1}$$

where $A' \geq 0$, $A(0) = 0$. This is a very general equation that may have nonlinearities in both the convective and the diffusive terms. Moreover, since $A'(u)$ is not bounded away from zero, the parabolic equation may contain hyperbolic regions; the conservation law $u_t + f(u)_x = 0$, for instance, is a special case of (4.1). The interplay between the nonlinearities gives rise to a rich set of intricate phenomena, to which we will return later when we discuss the mathematical theory. For simplicity, we will in all the following examples assume simple outflow boundary conditions, unless stated otherwise.

It is difficult to design a single robust scheme that can handle all possible balances of the convective and diffusive effects in (4.1). For instance, if the equation is dominated by convection, nonlinearities in the flux function may lead to solutions with very steep gradients or even discontinuous solutions if the equation contains hyperbolic regions. As a result, it is natural to try to utilize efficient methods developed for hyperbolic conservation laws. A straightforward way to do this is through operator splitting.

To prove that approximate solutions generated by a certain operator splitting converge to the correct solution of (4.1), we must go through the following steps: First, the three basic estimates (3.45), (3.46), and (3.47) are established, which imply convergence to a limit function by the use of Theorem 3.13. Then we demonstrate that the entropy condition (3.30) holds in the limit by proving that the term $-\mathcal{E}(\Delta t; \phi)$ in the entropy discrepancy (3.60) tends to zero with the discretization parameters. Finally, we verify that the L^2 estimate (3.31) is satisfied in the limit.

In the last section of the chapter we discuss the practical applicability of such a splitting in terms of several examples. Apart from backing up the theoretical analysis by two examples, we will elaborate on the error mechanisms associated with viscous splitting. In particular we show certain deficiencies arising due to local entropy loss introduced in the hyperbolic steps and demonstrate how these shortcomings can be amended to give highly efficient numerical schemes capable of handling different balances of nonlinear convective and diffusive terms.

4.1 A semi-discrete splitting method

In Examples 2.4 and 2.5 in Chapter 2 we computed approximate solutions to two convection-diffusion equations of the form (4.1) by splitting them into a convective and a diffusive part. That is, let $\mathcal{S}_t v_0$ denote the solution operator associated with the conservation law $v_t + f(v)_x = 0$ with initial data $v|_{t=0} = v_0$. Similarly, we denote by $w = \mathcal{H}_t w_0$ the weak solution of the nonlinear heat equation $w_t = A(w)_{xx}$, assuming that A satisfies (3.12). Then the simplest *viscous splitting* reads

$$u(x,t) \approx \left[\mathcal{H}_{\Delta t}\,\mathcal{S}_{\Delta t}\right]^n u_0, \qquad t = n\Delta t. \tag{4.2}$$

This splitting is often referred to as the Godunov splitting and is formally only first order. As an alternative to (4.2), we showed that we can use a Strang splitting,

$$u(x,t) \approx \left[\mathcal{H}_{\Delta t/2}\,\mathcal{S}_{\Delta t}\,\mathcal{H}_{\Delta t/2}\right]^n u_0, \qquad t = n\Delta t, \tag{4.3}$$

which has formal order 2. Conceptually, it does not matter in which order the two operators are applied in the two splittings — in actual computations it does. The hyperbolic operator $\mathcal{S}_t$ tends to create discontinuities and should therefore be applied before the parabolic operator $\mathcal{H}_t$.

To prove convergence of the semi-discrete splittings by the use of Theorem 3.13, we start by observing that the two suboperators $\mathcal{S}_t$ and $\mathcal{H}_t$ satisfy the three assumptions (3.49)–(3.51) in Lemma 3.12. Indeed, assuming v_0 to be bounded, integrable, and of finite total variation, we have that $\mathcal{S}_t$ satisfies the following four estimates (see, e.g., [126, 201])

$$\begin{aligned}
\|\mathcal{S}_t v_0\|_{L^\infty(\mathbb{R})} &\le \|v_0\|_{L^\infty(\mathbb{R})}\,,\\
\|\mathcal{S}_t v_0(\,\cdot\,+y) - \mathcal{S}_t v_0\|_{L^1(\mathbb{R})} &\le |y|\,\mathrm{T.V.}\,(v_0)\,,\\
\mathrm{T.V.}\,(\mathcal{S}_t v_0) &\le \mathrm{T.V.}\,(v_0)\,,\\
\|\mathcal{S}_t v_0 - \mathcal{S}_s v_0\|_{L^1(\mathbb{R})} &\le |t-s|\,\mathrm{T.V.}\,(v_0)\,\|f\|_{\mathrm{Lip}}\,.
\end{aligned} \tag{4.4}$$

Similar estimates can be proved for $\mathcal{H}_t$ (see, e.g., [50, 271])

$$\begin{aligned}
\|\mathcal{H}_t w_0\|_{L^\infty(\mathbb{R})} &\le \|w_0\|_{L^\infty(\mathbb{R})}\,,\\
\|\mathcal{H}_t w_0(\,\cdot\,+y) - \mathcal{H}_t w_0\|_{L^1(\mathbb{R})} &\le |y|\,\mathrm{T.V.}\,(w_0)\,,\\
\mathrm{T.V.}\,(\mathcal{H}_t w_0) &\le \mathrm{T.V.}\,(w_0)\,,\\
\|\mathcal{H}_t w_0 - \mathcal{H}_s w_0\|_{L^1(\mathbb{R})} &\le \mathcal{O}(\sqrt{|t-s|}),
\end{aligned} \tag{4.5}$$

whenever the initial data w_0 is bounded, integrable and of finite total variation. Furthermore, we have (see, e.g., [50, 271])

$$-\int_s^t \int_{\mathbb{R}} (A(w(x,r))_x)^2\,dx\,dr = \int_{\mathbb{R}} B(w(x,t))\,dx - \int_{\mathbb{R}} B(w(x,s))\,dx \tag{4.6}$$

where we have introduced the function

$$B(w) = \int_0^w A(v)\,dv, \tag{4.7}$$

which is nondecreasing since $A'(v)$ is nonnegative and $A(0) = 0$.

We can now state the main result of this section in terms of the following theorem:

Theorem 4.1. *Suppose* $u_0 \in L^1(\mathbb{R}) \cap L^\infty(\mathbb{R}) \cap \mathrm{BV}$ *and* $f, A \in \mathrm{Lip}_{\mathrm{loc}}(\mathbb{R})$ *with* $A' \geq 0$, $A(0) = 0$. *Then the semi-discrete viscous splitting methods* (4.2) *and* (4.3) *converge to an entropy weak solution of* (4.1).

Proof. Estimates (4.4) and (4.5) show that the operators $\mathcal{S}_t$ and $\mathcal{H}_t$ satisfy the conditions of Lemma 3.12. Thus Theorem 3.13 implies that $u_{\Delta t}$ converges in $L^1_{\mathrm{loc}}(\Pi_T)$ to some function u in L^∞. For exact solution operators, $\mathcal{S}_t$ and $\mathcal{H}_t$, the error term $\mathcal{E}$ is absent. If we now can prove that $A(u)_x \in L^2(\Pi_T)$ (i.e., (3.31)), it will follow from Corollary 3.19 that u is the entropy solution. To verify that (3.31) is satisfied, we define a new time interpolant $\tilde{u}_{\Delta t}$ by

$$\tilde{u}_{\Delta t}(x,t) = \left[\mathcal{H}_{t-t_n} \circ \mathcal{S}_{\Delta t}\right] u^n, \qquad t \in [t_n, t_{n+1}).$$

This function is close to $u_{\Delta t}$ in the sense that for $t \in [t_n, t_{n,1})$ (recall that $t_{n,1} = (n + \frac{1}{\ell})\Delta t$) we have

$$\begin{aligned}
\|u_{\Delta t}(t) - \tilde{u}_{\Delta t}(t)\|_{L^1(\mathbb{R})} &= \int \left|\mathcal{S}_{2(t-t_n)} u^n - \mathcal{H}_{t-t_n}\mathcal{S}_{\Delta t} u^n\right| dx \\
&\leq \int \left|\mathcal{S}_{2(t-t_n)} u^n - \mathcal{S}_{\Delta t} u^n\right| dx \\
&\quad + \int \left|\mathcal{H}_{t-t_n}\mathcal{S}_{\Delta t} u^n - \mathcal{S}_{\Delta t} u^n\right| dx \\
&\leq \|f\|_{\mathrm{Lip}}\,\mathrm{T.V.}\,(u)\,\Delta t + \mathcal{O}(\sqrt{\Delta t}).
\end{aligned}$$

Similarly, for $t \in [t_{n,1}, t_{n+1})$ we find

$$\|u_{\Delta t}(t) - \tilde{u}_{\Delta t}(t)\|_{L^1(\mathbb{R})} = \int \left|\mathcal{H}_{2(t-t_{n,1})} u^{n,1} - \mathcal{H}_{t-t_n} u^{n,1}\right| dx \leq \mathcal{O}(\sqrt{\Delta t}),$$

where $u^{n,1} = \mathcal{S}_{\Delta t} u^n$. Thus

$$\|u_{\Delta t}(t) - \tilde{u}_{\Delta t}(t)\|_{L^1(\mathbb{R})} = \mathcal{O}(\sqrt{\Delta t}), \quad \|u_{\Delta t} - \tilde{u}_{\Delta t}\|_{L^1(\Pi_T)} = \mathcal{O}(\sqrt{\Delta t}).$$

Furthermore, using that $u_{\Delta t}$ and $\tilde{u}_{\Delta t}$ are bounded, we see that

$$\|u_{\Delta t} - \tilde{u}_{\Delta t}\|_{L^2(\Pi_T)} = \mathcal{O}(\sqrt{\Delta t}).$$

Next we want to prove that $A(\tilde{u}_{\Delta t})_x \in L^2(\Pi_T)$. From (4.6) we find

$$
\begin{aligned}
-\iint_{\Pi_T} \left(A(\tilde{u}_{\Delta t})_x\right)^2 dt\, dx &= \sum_n \int_{\mathbb{R}} \int_{t_n}^{t_{n+1}} \left(A(\tilde{u}_{\Delta t})_x\right)^2 dt\, dx \\
&= \sum_n \int_{\mathbb{R}} \left(B(u^{n+1}) - B(u^{n,1})\right) dx \\
&= \sum_n \int_{\mathbb{R}} \left[\left(B(u^{n+1}) - B(u^{n})\right) + \left(B(u^{n}) - B(u^{n,1})\right)\right] dx \\
&= \int_{\mathbb{R}} \left(B(u^{N}) - B(u^{0})\right) dx - \sum_n \int_{\mathbb{R}} \left(B(u^{n,1}) - B(u^{n})\right) dx \\
&=: I_1 + I_2.
\end{aligned}
$$

Since $\tilde{u}_{\Delta t}$ satisfies (3.45), we have that

$$
\begin{aligned}
|I_1| &\le \Big(\sup_{|u|\le M} A(u)\Big) \int_{\mathbb{R}} \left(|u(x,T)| + |u(x,0)|\right) dx \le C < \infty, \\
|I_2| &\le \Big(\sup_{|u|\le M} A(u)\Big) \sum_n \int_{\mathbb{R}} |S_{\Delta t} u^n - u^n|\, dx \le C \sum_n \Delta t \le C \cdot T < \infty,
\end{aligned}
\tag{4.8}
$$

where we for the moment have set $M = \|\tilde{u}_{\Delta t}\|_{\infty}$. Thus $A(\tilde{u}_{\Delta t})_x \in L^2(\Pi_T)$ uniformly in the discretization parameter Δt. From this, together with the already known fact that $\tilde{u}_{\Delta t} \to u$ a.e. in Π_T, we conclude that

$$
A(\tilde{u}_{\Delta t})_x \rightharpoonup A(u)_x \quad \text{in } L^2(\Pi_T) \text{ as } \Delta t \to 0.
$$

Hence the limit function u satisfies $A(u)_x \in L^2(\Pi_T)$, which proves that u is an entropy weak solution. □

Convergence of the semi-discrete viscous splitting method was established in [145, 146] for the non-degenerate case where $A'(u) > 0$ and in [97] for the strongly degenerate case where $A'(u)$ may be zero on intervals in solution space.

Although not necessary for proving that u is an entropy weak solution, it is possible to obtain refined estimates and compactness properties of $A(\tilde{u}_{\Delta t})$, which will be the topic of the remaining part of this subsection. We start by investigating the spatial modulus of continuity in L^2 of $A(\tilde{u}_{\Delta t})$. To this end, we calculate that

$$
\begin{aligned}
&\iint_{\Pi_T} \Big(A\big(\tilde{u}_{\Delta t}(x+y,t)\big) - A\big(\tilde{u}_{\Delta t}(x,t)\big)\Big)^2 dx\, dt \\
&\qquad = \iint_{\Pi_T} \Big(\int_x^{x+y} A\big(\tilde{u}_{\Delta t}(z,t)\big)_z\, dz\Big)^2 dx\, dt \\
&\qquad \le |y| \iint_{\Pi_T} \int_x^{x+y} \Big(A\big(\tilde{u}_{\Delta t}(z,t)\big)_z\Big)^2 dz\, dx\, dt
\end{aligned}
$$

$$= |y| \iint_{\Pi_T} \int_z^{z-y} \left(A(\tilde{u}_{\Delta t}(z,t))_z\right)^2 dx\, dt\, dz$$
$$= |y|^2 \iint_{\Pi_T} \left(A(\tilde{u}_{\Delta t}(x,t))_x\right)^2 dx\, dt,$$

which implies the spatial modulus

$$\|A(\tilde{u}_{\Delta t}(\cdot + y)) - A(\tilde{u}_{\Delta t})\|_{L^2(\Pi_T)} \le |y| \, \|A(\tilde{u}_{\Delta t})_x\|_{L^2(\Pi_T)} . \tag{4.9}$$

To determine the temporal modulus of continuity we recall that A is nondecreasing, so that $(A(u) - A(v))(u - v) \ge 0$. Thus

$$\begin{aligned}
&\iint_{\Pi_T} \Big(A\big(\tilde{u}_{\Delta t}(x,t+\tau)\big) - A\big(\tilde{u}_{\Delta t}(x,t)\big)\Big)^2 dx\, dt \\
&\le \|A\|_{\mathrm{Lip}} \iint_{\Pi_T} \big[A\big(\tilde{u}_{\Delta t}(x,t+\tau)\big) - A\big(\tilde{u}_{\Delta t}(x,t)\big)\big]\big[\tilde{u}_{\Delta t}(x,t+\tau) - \tilde{u}_{\Delta t}(x,t)\big] dx\, dt \\
&\le \|A\|_{\mathrm{Lip}} \iint_{\Pi_T} \big[A\big(\tilde{u}_{\Delta t}(x,t+\tau)\big) - A\big(\tilde{u}_{\Delta t}(x,t)\big)\big] \int_t^{t+\tau} \partial_t \tilde{u}_{\Delta t}(x,s)\, ds\, dx\, dt \\
&= \|A\|_{\mathrm{Lip}} \iint_{\Pi_T} \big[A\big(\tilde{u}_{\Delta t}(x,t+\tau)\big) - A\big(\tilde{u}_{\Delta t}(x,t)\big)\big] \int_t^{t+\tau} A\big(\tilde{u}_{\Delta t}(x,s)\big)_{xx}\, ds\, dx\, dt \\
&= \|A\|_{\mathrm{Lip}} \iint_{\Pi_T} \big[A\big(\tilde{u}_{\Delta t}(x,t+\tau)\big) - A\big(\tilde{u}_{\Delta t}(x,t)\big)\big] \int_0^{\tau} A\big(\tilde{u}_{\Delta t}(x,s+t)\big)_{xx}\, ds\, dx\, dt \\
&= -\|A\|_{\mathrm{Lip}} \iint_{\Pi_T} \int_0^{\tau} \Big(A\big(\tilde{u}_{\Delta t}(x,t+\tau)\big)_x A\big(\tilde{u}_{\Delta t}(x,s+t)\big)_x \\
&\qquad\qquad - A\big(\tilde{u}_{\Delta t}(x,t)\big)_x A\big(\tilde{u}_{\Delta t}(x,s+t)\big)_x\Big)\, ds\, dx\, dt \\
&\le \|A\|_{\mathrm{Lip}} \int_0^{\tau} \Big(\big\|A\big(\tilde{u}_{\Delta t}(\cdot + \tau)\big)_x\big\|_{L^2(\Pi_T)} \big\|A\big(\tilde{u}_{\Delta t}(\cdot + s)\big)_x\big\|_{L^2(\Pi_T)} \\
&\qquad\qquad + \big\|A\big(\tilde{u}_{\Delta t}\big)_x\big\|_{L^2(\Pi_T)} \big\|A\big(\tilde{u}_{\Delta t}(\cdot + s)\big)_x\big\|_{L^2(\Pi_T)}\Big)\, ds \\
&\le 2\,|\tau| \, \|A\|_{\mathrm{Lip}} \, \|A(\tilde{u}_{\Delta t})_x\|^2_{L^2(\Pi_T)} .
\end{aligned}$$

Thus

$$\|A(\tilde{u}_{\Delta t}(\cdot + \tau)) - A(\tilde{u}_{\Delta t})\|_{L^2(\Pi_T)} \le 2\sqrt{|\tau|} \, \|A(\tilde{u}_{\Delta t})_x\|_{L^2(\Pi_T)} .$$

Applying the Kolmogorov compactness criterion, Lemma 3.1, to the sequence $\{A(\tilde{u}_{\Delta t})\}$ (indexed by the discretization parameter Δt) in $L^2(\Pi_T)$ we conclude that there exists a subsequence such that $A(\tilde{u}_{\Delta t}) \to \bar{A}$ in $L^2(\Pi_T)$ as $\Delta t \to 0$. The estimate

$$\big\|A(u) - \bar{A}\big\|_{L^2(\Pi_T)} \le \|A\|_{\mathrm{Lip}} \, \|u - \tilde{u}_{\Delta t}\|_{L^2(\Pi_T)} + \big\|A(\tilde{u}_{\Delta t}) - \bar{A}\big\|_{L^2(\Pi_T)}$$

shows that $A(u) \in L^2(\Pi_T)$. Finally,

$$\lim_{\Delta t \to 0} \iint_{\Pi_T} A(\tilde{u}_{\Delta t}) \phi_x \, dx\, dt = \iint_{\Pi_T} A(u) \phi_x \, dx\, dt, \quad \phi \in C_0^\infty(\mathbb{R}^2),$$

shows that indeed $A(u)_x \in L^2(\Pi_T)$. Hence, (3.31) is satisfied and the approximate solution generated by (4.2) converges to an entropy weak solution of (4.1).

This viscous splitting approach can easily be extended to multi-dimensional equations as described in, e.g., [119, 120, 145]. Namely, consider the equation

$$u_t + \sum_i f_i(u)_{x_i} = \Delta A(u), \qquad u|_{t=0} = u_0. \tag{4.10}$$

Let $u = \mathcal{D}_t^l u_0$ denote the solution of the one-dimensional problems

$$u_t + f_l(u)_{x_l} = A(u)_{x_l x_l}, \qquad u|_{t=0} = u_0, \quad l = 1, \dots, d.$$

Then using dimensional splitting as introduced in Examples 2.2, 2.3, and 2.7, gives rise to the splitting scheme

$$u^{n+1} = (\mathcal{D}_{\Delta t}^d \circ \cdots \circ \mathcal{D}_{\Delta t}^1) u^n, \quad u^0 = u_0,$$

where each one-dimensional operator $\mathcal{D}_t^l$ can be approximated using the viscous splitting (4.2). Alternatively, let $u = \mathcal{S}_t^l u_0$ denote the solution of

$$u_t + f_l(u)_{x_l} = 0, \qquad u|_{t=0} = u_0, \quad l = 1, \dots, d,$$

and let $u = \mathcal{H}_t u_0$ denote the solution of

$$u_t = \Delta A(u), \qquad u|_{t=0} = u_0.$$

Another possible splitting method then reads

$$u^{n+1} = (\mathcal{H}_{\Delta t} \circ \mathcal{S}_{\Delta t}^d \circ \cdots \circ \mathcal{S}_{\Delta t}^1) u^n, \quad u^0 = u_0.$$

Here $\mathcal{H}_t$ can either be constructed directly or by using dimensional splitting as in Example 2.2.

Convergence of these semi-discrete splitting methods for (4.10), can be proved along the lines that we used for the one-dimensional algorithms. A similar analysis can be performed for the fully discrete splittings introduced in the next section. Dimensional splitting for hyperbolic problems is discussed and analyzed in detail in Chapter 5.

4.2 A fully discrete splitting method

To obtain a numerical method, we must approximate each of the two suboperators $\mathcal{S}_t$ and $\mathcal{H}_t$ by appropriate numerical schemes. A very efficient splitting method is obtained by combining an unconditionally stable front-tracking method, which will be introduced below and discussed in more detail in Appendix A.7, for the hyperbolic step with a standard implicit finite-difference method for the (strongly degenerate) parabolic operator. The corresponding numerical method will be very time efficient for all problems having an inherent dynamics that allows the use of large time-steps. The front-tracking method gives accurate representation and sharp resolution of possible discontinuities in the hyperbolic step, and thus provides a good initial guess for the implicit method, which will therefore converge much faster in case of nonlinear diffusion. Fully discrete viscous splitting methods based on front tracking have been successfully used in a number of works, see [43, 93, 97, 100, 119, 120, 140–143, 145, 146].

Operator splitting for convection-diffusion equations involves parabolic equations, which often are (strongly) degenerate. In this book we will rely only on very simple finite-difference schemes to approximate degenerate parabolic equations [96, 98, 99, 147]. Numerical methods for uniformly elliptic/parabolic equations is a traditional field and numerous textbooks exist [40, 101, 104, 112, 129, 136, 154, 209, 230, 241, 247, 259–261, 269]. Besides finite-difference schemes, other classes of numerical methods have been developed recently for strongly degenerate parabolic equations, cf. for example [6, 32, 51, 60, 102, 111, 164, 207, 216, 217].

When considering fully discrete methods, we shall view the approximations either as piecewise constant functions, or as members of ℓ^p spaces. By a grid $\{i\Delta x\}$, $i \in \mathbb{Z}$, we mean grid cells that are intervals $I_i = ((i-1/2)\Delta x, (i+1/2)\Delta x]$. We also have the time intervals, $I^n = [t_n, t_{n+1})$ with $t_n = n\Delta t$ for $n = 0, 1, 2, \dots$. A sequence $\{U_i\}_{i\in\mathbb{Z}} \in \ell^p$ can be identified with the piecewise constant function

$$U(x) = \sum_i U_i \chi_{I_i}(x).$$

We use a uniform grid defined by the grid size $\Delta x > 0$, and let π be the usual first-order projection (grid block averaging) operator defined on this grid, i.e., $\pi : L^1(\mathbb{R}) \to \ell^\infty$, given by

$$(\pi v)_i = \frac{1}{\Delta x} \int_{I_i} v(x)\, dx.$$

Let $\mathcal{S}_t^{\Delta x,\delta}$ denote the front-tracking operator $\tilde{\mathcal{S}}_t^\delta$ followed by a projection onto the grid $\{i\Delta x\}$ and $\mathcal{H}_t^{k,\Delta x}$ the operator associated with the finite-difference scheme

$$\begin{aligned} w_i^{n+1} - \mu\theta \left[A(w_{i-1}^{n+1}) - 2A(w_i^{n+1}) + A(w_{i+1}^{n+1})\right] \\ = w_i^n + \mu(1-\theta)\left[A(w_{i-1}^n) - 2A(w_i^n) + A(w_{i+1}^n)\right], \end{aligned} \tag{4.11}$$

for $n \geq 0$, and $w_i^0 = u_i^0$ for $i \in \mathbb{Z}$. Here, $\theta \in [0,1]$ and $\mu = \tau/\Delta x^2$, for $\tau > 0$.

Then the fully discrete splitting reads

$$u^{n+1} = \left[\mathcal{H}_{\Delta t}^{\Delta x} \circ \mathcal{S}_{\Delta t}^{\Delta x,\delta}\right] u^n, \qquad u^0 = \pi u_0. \tag{4.12}$$

In a splitting like this, the two substeps will typically have different restrictions on the time-step due to requirements of stability and accuracy. From a numerical point of view it is therefore feasible to use a variable splitting step, i.e., one could replace Δt by Δt_n. For the numerical examples presented later in this book, we adopt the convention that the splitting step is determined in terms of the CFL number for a convective suboperator. This means that the parabolic substep may consist of several steps of the finite-difference scheme, i.e., $\Delta t = M\tau$ for a certain integer $M \geq 1$, in order to satisfy stability restrictions. In the analysis, however, we only consider fixed splitting steps to simplify the notation.

The front-tracking method for scalar conservation laws is based on first replacing the flux function by a continuous and piecewise linear approximation and the initial data by a step function, i.e., a piecewise constant function. Subsequently, one solves the resulting perturbed problem *exactly*, see [117, 118, 126] for details. In this way rarefaction waves are replaced by jump discontinuities. A key property of this method is that there will only be *finitely* many interactions between all discontinuities globally in time.

Furthermore, let f^δ be the piecewise linear approximations to f, and $\tilde{S}_t^\delta$ the solution operator associated with the corresponding one-dimensional equation

$$u_t + f^\delta(u)_x = 0.$$

We shall assume that δ and Δx are related to Δt so that all three tend to zero together.

For the numerical examples presented in the rest of this chapter, we use the fully explicit scheme corresponding to $\theta = 0$ as $\mathcal{H}_{\Delta t}^{\Delta x}$, unless stated otherwise. This is to emphasize simplicity and accuracy at the cost of a somewhat reduced efficiency. In the analysis in Section 4.2.2, however, we use the general scheme given in (4.11).

4.2.1 Convergence in the discrete L^1 norm. In Section 4.1 (and Example 2.5) we established convergence for the semi-discrete splitting. Before going into a rigorous analysis for the fully discrete splitting (4.12), we will provide numerical data to support the claim that this splitting converges. To this end we apply the splitting in two cases and study the convergence in the discrete L^1 norm.

Example 4.2. *Consider Burgers' equation with viscosity*

$$u_t + \left(\tfrac{1}{2}u^2\right)_x = \varepsilon u_{xx}, \qquad u(x,0) = u_0(x). \tag{4.13}$$

Here ε is a scaling parameter that gives the relative balance between advective and viscous forces. We consider the smooth initial data

$$u_0(x) = -\sin(\pi x)\chi_{[-1,1]}(x).$$

To maximize computational efficiency it is desirable to use as large splitting steps as possible due to the unconditionally stable nature of the front-tracking method.

Table 4.1 shows estimated L^1-errors and convergence rates obtained through a grid-refinement study of the operator-splitting scheme. The error is computed

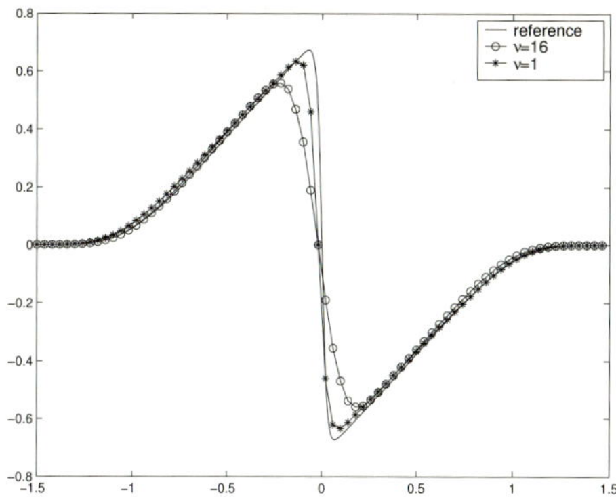

Figure 4.1. Viscous splitting solutions computed on a grid with 75 grid cells with $\varepsilon = 0.01$.

Table 4.1. Estimated L^1-errors and convergence rates at time $t = 1.0$ for Burgers' equation with scaling parameter $\varepsilon = 0.01$ and spatial domain $[-1.5, 1.5]$.

N_x	$\nu = 0.25$		$\nu = 1$		$\nu = 4$		$\nu = 16$	
75	4.86e-02	—	4.29e-02	—	5.85e-02	—	—	—
150	2.38e-02	1.03	2.08e-02	1.04	2.90e-02	1.01	7.50e-02	—
300	1.21e-02	0.98	1.05e-02	0.98	1.46e-02	0.99	4.41e-02	0.77
600	5.93e-03	1.02	5.19e-03	1.02	7.28e-03	1.00	2.37e-02	0.90
1200	2.91e-03	1.03	2.51e-03	1.05	3.59e-03	1.02	1.15e-02	1.04
2400	1.50e-03	0.96	1.19e-03	1.07	1.76e-03	1.03	5.80e-03	0.99

relative to a fine grid solution at time $t = 1.0$ calculated with a scheme using the Engquist–Osher flux function (see (A.31) *in Appendix* A.6.2) *in combination with a standard central-difference approximation of the second-order term. In the experiments, the size of the splitting step Δt has been related to the spatial discretization parameter Δx through the CFL number $\nu = (\Delta t/\Delta x)\max_u |u|$. The experiments indicate convergence over a broad range of CFL numbers. Convergence can of course also be observed for CFL numbers larger than 16 on the finer grids. However, as can be read off from the actual errors, the quality of the splitting solution deteriorates as the relative size of the splitting step increases. This is also illustrated in Figure* 4.1.

Proving that our sequence of approximations converges to the true classical solution $u(x, t = 1)$ is not too difficult since this solution is smooth. Neglecting a few technicalities, all we need to do is to establish that our approximations are bounded and have bounded variation. From (4.4) we have that this is true for the front-tracking subsolution, which is an *exact* solution of a *perturbed* hyperbolic problem, and projecting this onto a regular grid does not alter the fact that the solution is bounded and has bounded variation. Establishing the same bounds on

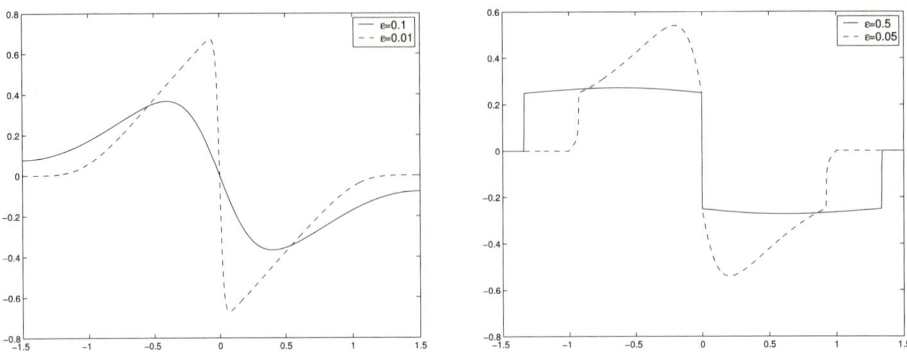

Figure 4.2. Reference solutions at time $t = 1.0$ for Burgers' equation (4.14) for diffusion functions u (left) and $A(u)$ (right).

the finite-difference scheme is a straightforward exercise. Combining the results for the two subsolutions we obtain boundedness for the sequence of approximations as in Section 4.1 and hence convergence to the smooth classical solution; see [145] for a detailed account.

Let us now consider a more difficult example, in which we also allow the diffusion function to be nonlinear.

Example 4.3. *Consider the Burgers' type equation*

$$u_t + \left(\tfrac{1}{2}u^2\right)_x = \varepsilon A(u)_{xx}, \qquad u(x,0) = -\sin(\pi x)\chi_{[-1,1]}(x), \tag{4.14}$$

with diffusion function $A(u)$ given by

$$A'(u) = \begin{cases} 0, & |u| \le 1/4, \\ 1, & \text{otherwise.} \end{cases}$$

Figure 4.2 shows solutions of (4.14) at time $t = 1.0$ compared with solutions of the classical Burgers' equation (4.13) for two different values of ε. Burgers' equation (4.13) is non-degenerate parabolic, and although the solution develops a sharp gradient in the vicinity of the origin, it remains smooth. Equation (4.14), on the other hand, degenerates with a hyperbolic region for $u \in [-1/4, 1/4]$. In the hyperbolic region there are no viscous forces and the solution has developed a stationary shock at the origin. Notice also the nonsmooth transitions between the hyperbolic and the parabolic regions at $x \approx \pm 0.92$ for $\varepsilon = 0.05$.

Table 4.2 shows estimated L^1-errors and convergence rates obtained through a grid-refinement study of the operator-splitting scheme for (4.14). As in Example 4.2, the operator splitting method converges over a broad range of CFL numbers, although the quality of the solution decays as ν increases.

Table 4.2. Estimated L^1-errors and convergence rates at time $t = 1.0$ for (4.14) with scaling parameter $\varepsilon = 0.05$ and spatial domain $[-1.5, 1.5]$.

N_x	$\nu = 0.25$		$\nu = 1$		$\nu = 4$		$\nu = 16$	
75	4.42e-02	—	4.36e-02	—	5.78e-02	—	—	—
150	1.85e-02	1.26	1.80e-02	1.28	2.46e-02	1.23	7.30e-02	—
300	9.59e-03	0.95	9.41e-03	0.93	1.24e-02	0.98	3.73e-02	0.97
600	4.83e-03	0.99	4.80e-03	0.97	6.30e-03	0.98	1.88e-02	0.99
1200	2.30e-03	1.07	2.29e-03	1.07	3.01e-03	1.07	9.07e-03	1.05
2400	8.90e-04	1.37	8.48e-04	1.43	1.27e-03	1.24	4.54e-03	1.00

Once again we have numerical evidence that the operator splitting seems to converge to some function, which can be verified mathematically through a compactness argument. However, whereas (4.13) has classical smooth solutions, (4.14) has discontinuous solutions and hence must take solutions in some weak sense. A weak solution is not necessarily unique and we will therefore need extra conditions to single out the unique physical solution. In the next section we will continue to give a rigorous convergence analysis for the general equation (4.1), which includes both non-degenerate equations like (4.13) and strongly degenerate equations like (4.14).

4.2.2 Convergence analysis. Having shown examples that demonstrate that the fully discrete scheme converges numerically for both smooth and nonsmooth problems, it is now time to go back to the rigorous mathematical analysis. In the notation of Chapter 3 we have two solution operators: $\mathcal{S}^1$ denotes the approximate solution operator for the hyperbolic part and $\mathcal{S}^2$ denotes the approximate solution operator for the (degenerate) parabolic part, which we will also refer to as $\mathcal{H}$. The first operator is defined by letting

$$\mathcal{S}_t^1 \equiv \mathcal{S}_t^{\Delta x,\delta} = \pi \circ \tilde{\mathcal{S}}_t^{\delta}. \tag{4.15}$$

To define the second operator $\mathcal{S}^2$ we view u^0 as a point in ℓ^1. First we define $\mathcal{W}_\tau : \ell^1 \to \ell^1$ by

$$\mathcal{W}_\tau(z) = z + \mu(1-\theta)\mathcal{R}(z),$$

where

$$\big(\mathcal{R}(z)\big)_i = A(z_{i-1}) - 2A(z_i) + A(z_{i+1}).$$

Note that $\mathcal{R}$ and $\mathcal{W}_\tau$ are (Lipschitz) continuous operators since

$$\begin{aligned}\|\mathcal{R}(z) - \mathcal{R}(y)\| &\le 4\,\|A\|_{\mathrm{Lip}}\,\|z-y\|_1\,,\\ \|\mathcal{W}_\tau(z) - \mathcal{W}_\tau(y)\|_1 &\le \Big(1 + 4\mu(1-\theta)\,\|A\|_{\mathrm{Lip}}\Big)\,\|z-y\|_1\,.\end{aligned}$$

For $w \in \ell^1$ we define $\mathcal{T}_\tau(w)$ to be the solution of

$$\mathcal{T}_\tau(w) - \mu\theta\mathcal{R}\left(\mathcal{T}_\tau(w)\right) = w. \tag{4.16}$$

Since this is a nonlinear equation in ℓ^1 we must show that $\mathcal{T}_\tau$ is well defined, and to this end we use Theorem 3.6. First, observe that

$$\sum_i \left(\mathcal{R}(z)\right)_i = 0,$$

and we have already seen that $\mathcal{R}$ is Lipschitz continuous. Therefore, in order to apply Theorem 3.6 it remains to show that $-\mathcal{R}$ is accretive. This means showing that

$$\langle J(z-y), \mathcal{R}(z) - \mathcal{R}(y)\rangle \leq 0,$$

for any duality mapping J and all z and y in ℓ^1. By (3.8), we have that

$$\begin{aligned}
&\langle J(z-y), \mathcal{R}(z) - \mathcal{R}(y)\rangle \\
&\quad = \sum_i \operatorname{sign}\left(z_i - y_i\right) \\
&\qquad\qquad \times \Big(A(z_{i-1}) - A(y_{i-1}) - 2\big(A(z_i) - A(y_i)\big) + A(z_{i+1}) - A(y_{i+1})\Big) \\
&\quad = \sum_i \operatorname{sign}\left(z_i - y_i\right)\Big(A(z_{i-1}) - A(y_{i-1}) + A(z_{i+1}) - A(y_{i+1})\Big) \\
&\qquad - 2\sum_i \left|A(z_i) - A(y_i)\right| \\
&\quad \leq 0,
\end{aligned}$$

and thus $-\mathcal{R}$ is accretive. Hence there exists a unique solution to (4.16). Now we define $\mathcal{H}_t^{\Delta x}$ ($\mathcal{S}^2$ in the notation of Chapter 3) by

$$\mathcal{H}_t^{\Delta x} u^0 = \mathcal{T}_t \circ \mathcal{W}_t u^0. \tag{4.17}$$

If we set $v^{n+1} = \mathcal{H}_{\Delta t}^{\Delta x} v^n$, the componentwise equation for v^n is given by (4.11).

A priori estimates. Next, we show that the a priori estimates (3.45)–(3.47) are satisfied for the approximation defined by

$$u_{\Delta t}(\,\cdot\,,t) \;=\; \mathcal{H}^{\Delta x}_{2(t-\min\{t_{n,1},t\})} \circ \mathcal{S}^{\delta}_{2(\min\{t,t_{n,1}\}-t_n)} \circ \left(\mathcal{H}^{\Delta x}_{\Delta t} \circ \mathcal{S}^{\delta}_{\Delta t}\right)^n u^0 \tag{4.18}$$

for $t \in [t_n, t_{n+1})$. Since $\tilde{\mathcal{S}}_t^\delta$ is an *exact* solution operator for a perturbed problem within the same class of first-order conservation laws but with f replaced by f^δ, the solution operator $\tilde{\mathcal{S}}_t^\delta$ satisfies all the a priori estimates (3.45)–(3.47), and even (4.4) for initial data of finite total variation. This follows from the arguments in Section 4.1. For the projection operator, it is straightforward to show that

$$\left\|\pi v\right\|_{L^\infty(\mathbb{R})} \leq \left\|v\right\|_{L^\infty(\mathbb{R})},$$

$$\|\pi v - v\|_{L^1(\mathbb{R})} \leq \sup_{|\rho|\leq \Delta x} \|v(\,\cdot\, + \rho) - v\|_{L^1(\mathbb{R})} \leq \Delta x \,\mathrm{T.V.}\,(v)\,.$$

In particular, this means that

$$\Delta x \sum_i \left| \left(\mathcal{S}^\delta_{\Delta t}\,(\pi v)\right)_{i+1} - \left(\mathcal{S}^\delta_{\Delta t}\,(\pi v)\right)_i \right| \leq \Delta x \,\mathrm{T.V.}\,(v)\,. \tag{4.19}$$

From this and from (4.12) it follows that $\mathcal{S}^\delta_t$ satisfies the conditions of Lemma 3.12 if $\left\|f^\delta\right\|_{\mathrm{Lip}}$ is bounded.

To show the estimates for $\mathcal{H}^{\Delta x}_{\Delta t}$, we first note that

$$\sum_i \left(\mathcal{W}(z)\right)_i = \sum_i z_i$$

for $z \in \ell^1$. Furthermore, if the CFL-condition

$$(1-\theta)\mu \,\|A\|_{\mathrm{Lip}} \leq \frac{1}{2}, \qquad \mu = \tau/\Delta x^2 \tag{4.20}$$

holds, then

$$\frac{\partial\left(\mathcal{W}_\tau(z)\right)_i}{\partial z_i} \geq 0, \quad \text{and} \quad \frac{\partial\left(\mathcal{W}_\tau(z)\right)_i}{\partial z_{i\pm 1}} \geq 0.$$

This implies that $\mathcal{W}_\tau$ is monotone, i.e.,

$$\|\mathcal{W}_\tau(z)\|_\infty \leq \|z\|_\infty\,, \qquad \|\mathcal{W}_\tau(z)\|_1 \leq \|z\|_1\,, \qquad \mathrm{T.V.}\,(\mathcal{W}_\tau(z)) \leq \mathrm{T.V.}\,(z)\,.$$

Now by Theorem 3.6 we have that

$$\begin{aligned} \|\mathcal{T}_\tau\left(\mathcal{W}_\tau(z)\right)\|_\infty &\leq \|\mathcal{W}_\tau(z)\|_\infty \leq \|z\|_\infty\,, \\ \|\mathcal{T}_\tau\left(\mathcal{W}_\tau(z)\right)\|_1 &\leq \|\mathcal{W}_\tau(z)\|_1 \leq \|z\|_1\,, \\ \mathrm{T.V.}\,(\mathcal{T}_\tau\left(\mathcal{W}_\tau(z)\right)) &\leq \mathrm{T.V.}\,(\mathcal{W}_\tau(z)) \leq \mathrm{T.V.}\,(z)\,. \end{aligned}$$

Thus, if τ and θ are such that the CFL-condition (4.20) holds and the initial data u_0 is of bounded variation, then the approximation $\mathcal{H}^{\Delta x}_\tau u^0$ is of bounded variation. Next we show the weak time estimate (3.47). Let $0 \leq s \leq t \leq \tau$ be such that the CFL-condition (4.20) is satisfied. Then set

$$u(\,\cdot\,,t) = \mathcal{H}^{\Delta x}_{t-s} u(\,\cdot\,,s) = \mathcal{H}^{\Delta x}_{t-s}\mathcal{H}^{\Delta x}_{s} u^0.$$

Set $A_i(t) = A(u_i(t))$, and for $\phi \in C^\infty_0((-r,r))$ define $\phi_i = \frac{1}{\Delta x}\int_{I_i} \phi(x)\,dx$. Then

$$\begin{aligned} \left|\int_{-r}^{r} \left(u(x,t) - u(x,s)\right)\phi(x)\,dx\right| &= \left|\sum_i \int_{I_i} \left(u_i(t) - u_i(s)\right)\phi(x)\,dx\right| \\ &= \left|\Delta x \sum_i \phi_i\left(u_i(t) - u_i(s)\right)\right| \end{aligned}$$

$$
\begin{aligned}
&= \frac{t-s}{\Delta x^2}\left|\Delta x \sum_i \phi_i\Big[(1-\theta)\mathcal{R}\big(u_i(s)\big)+\theta\mathcal{R}\big(u_i(t)\big)\Big]\right| \\
&= \frac{t-s}{\Delta x^2}\Big|\Delta x \sum_i (\phi_{i+1}-\phi_i) \\
&\qquad \times \big[(1-\theta)\big(A_i(s)-A_{i-1}(s)\big)+\theta\big(A_i(t)-A_{i-1}(t)\big)\big]\Big| \\
&\le \frac{t-s}{\Delta x^2}\left\|A\right\|_{\mathrm{Lip}} \sum_i \left|\int_{I_i}\phi(x+\Delta x)-\phi(x)\,dx\right| \\
&\qquad \times \Big[(1-\theta)\left|v_i(t)-v_{i-1}(t)\right|+\theta\left|v_i(s)-v_{i-1}(s)\right|\Big] \\
&\le \frac{t-s}{\Delta x^2}\left\|A\right\|_{\mathrm{Lip}} \sum_i \Big(\int_{I_i}\int_x^{x+\Delta x}\left|\phi'(y)\right|\,dydx\Big) \\
&\qquad \times \Big[(1-\theta)\left|v_i(t)-v_{i-1}(t)\right|+\theta\left|v_i(s)-v_{i-1}(s)\right|\Big] \\
&\le (t-s)\left\|A\right\|_{\mathrm{Lip}}\left\|\phi'\right\|_{L^\infty((-r,r))}\,\mathrm{T.V.}\left(u^0\right).
\end{aligned}
\tag{4.21}
$$

Hence both operators $\mathcal{S}_t^\delta$ and $\mathcal{H}_t^{\Delta x}$ satisfy the conditions of Theorem 3.13, and therefore we can conclude that there exists a function $u \in L^\infty(\Pi_T;\mathbb{R})$ such that $u_{\Delta t} \to u$ in $L^1_{\mathrm{loc}}(\Pi_T;\mathbb{R})$ for a subsequence $\{\Delta t\}$ (which we do not relabel).

To conclude that the limit u is an entropy solution, it remains to show that the global error term $\mathcal{E}$, defined by (3.58) and (3.54), converges to zero as $\Delta t \to 0$, and to establish that $A(u)_x \in L^2(\Pi_T)$.

Entropy error. We start by estimating the error terms. In the notation of Chapter 3, $v^1(\cdot,t) = \mathcal{S}_t^\delta u^0$, and we define $w(\cdot,t)$ to be the entropy solution of the Cauchy problem

$$
w_t + f^\delta(w)_x = 0, \quad w(x,0) = u^0.
$$

That is, $v^1(\cdot,t) = \pi w(\cdot,t)$.

With the notations $\eta(v) = \operatorname{sign}(v-k)$, $q(v) = \operatorname{sign}(v-k)[f(v)-f(k)]$, and $q^\delta(v) = \operatorname{sign}(v-k)[f^\delta(v)-f^\delta(k)]$, we have that for a test function ϕ and for $s>0$,

$$
\begin{aligned}
&\int_{\mathbb{R}}\int_0^s \big(\eta(v^1)\phi_t + q(v^1)\phi_x\big)\,dt\,dx + \int_{\mathbb{R}}\eta(u^0)\phi(x,0)\,dx - \int_{\mathbb{R}}\eta\big(v^1(x,s)\big)\phi(x,s)\,dx \\
&\quad = \int_{\mathbb{R}}\int_0^s \big(\eta(w)\phi_t + q^\delta(w)\phi_x\big)\,dt\,dx \\
&\qquad + \int_{\mathbb{R}}\eta(u^0)\phi(x,0) - \int_{\mathbb{R}}\eta\big(w(x,s)\big)\phi(x,s)\,dx \\
&\qquad - \int_{\mathbb{R}}\int_0^s \big(\eta(w)-\eta(v^1)\big)\phi_t + \big(q^\delta(w)-q(v^1)\big)\phi_x\,dt\,dx \\
&\qquad + \int_{\mathbb{R}}\big[\eta\big(v^1(x,s)\big)-\eta\big(w(x,s)\big)\big]\phi(x,s)\,dx
\end{aligned}
$$

$$\geq -\int_{\mathbb{R}}\int_0^s \big(\eta(w)-\eta(\pi w)\big)\phi_t + \big(q^\delta(w)-q(\pi w)\big)\phi_x\,dt\,dx \\ + \int_{\mathbb{R}} \big[\eta\big(\pi w(x,s)\big)-\eta\big(w(x,s)\big)\big]\phi(x,s)\,dx. \tag{4.22}$$

Using the notation of (3.54) in Chapter 3, we see that

$$\begin{aligned} \mathcal{E}_{\mathrm{I}}{}^1 &= \eta(\pi w)-\eta(w), \\ \mathcal{E}_{\mathrm{II}}{}^1 &= \underbrace{\big(q^\delta(w)-q^\delta(\pi w)\big)}_{\mathcal{E}_{\mathrm{II}}{}^1_a} + \underbrace{\big(q^\delta(v^1)-q(v^1)\big)}_{\mathcal{E}_{\mathrm{II}}{}^1_b}, \\ \mathcal{E}_{\mathrm{IV}}{}^1 &= \left[\eta\left(\pi w\right)-\eta(w)\right]\delta_s(t), \end{aligned}$$

with (possible notational overload) $\delta_s(t)$ denoting the Dirac mass located at $t=s$. All the remaining error terms are zero. The error terms $\mathcal{E}_{\mathrm{I}}{}^1$, $\mathcal{E}_{\mathrm{II}}{}^1_a$, and $\mathcal{E}_{\mathrm{IV}}{}^1$ are similar, and to estimate them we observe that for a Lipschitz continuous function h we have

$$\begin{aligned} \int_{\mathbb{R}} \left|h\big((\pi w)(x,t)\big)-h\big(w(x,t)\big)\right|\,dx &= \sum_i \int_{I_i} \left|h\big(v_i^1(t)\big)-h\big(w(x,t)\big)\right|\,dx \\ &\leq \|h\|_{\mathrm{Lip}} \sum_i \int_{I_i} \left|\frac{1}{\Delta x}\int_{I_i} w(y,t)\,dy - w(x,t)\right|\,dx \\ &\leq \frac{\|h\|_{\mathrm{Lip}}}{\Delta x} \sum_i \int_{I_i}\int_{I_i} \left|w(y,t)-w(x,t)\right|\,dydx \\ &\leq \frac{\|h\|_{\mathrm{Lip}}}{\Delta x} \int_{I_i} \Big(\sup_{|x-z|\leq \Delta x} \int_{\mathbb{R}} \left|w(x,t)-w(z,t)\right|\,dx\Big)\,dy \\ &\leq \Delta x\,\|h\|_{\mathrm{Lip}}\,\mathrm{T.V.}\left(w(\,\cdot\,,t)\right) \leq \Delta x\,\|h\|_{\mathrm{Lip}}\,\mathrm{T.V.}\left(u^0\right). \end{aligned}$$

This means that we can define measures bounding $\mathcal{E}_{\mathrm{I}}{}^1$, $\mathcal{E}_{\mathrm{II}}{}^1_a$ and $\mathcal{E}_{\mathrm{IV}}{}^1$ by

$$\int_{\mathbb{R}} d\mu_{\mathrm{I}}{}^1 = \int_{\mathbb{R}} \left|\eta(\pi w)-\eta(w)\right|\,dx \leq \Delta x\,\mathrm{T.V.}\left(u_0\right), \tag{4.23}$$

$$\int_{\mathbb{R}} d\mu_{\mathrm{II}}{}^1_a = \int_{\mathbb{R}} \left|q^\delta(\pi w)-q^\delta(w)\right|\,dx \leq \Delta x\,\left\|f^\delta\right\|_{\mathrm{Lip}}\,\mathrm{T.V.}\left(u_0\right), \tag{4.24}$$

$$\int_{\mathbb{R}} d\mu_{\mathrm{IV}}{}^1 = \int_{\mathbb{R}} \left|\eta\big((\pi w)(x,s)\big)-\eta\big(w(x,s)\big)\right|\,dx \leq \Delta x\,\mathrm{T.V.}\left(u_0\right). \tag{4.25}$$

Since $\mathcal{E}_{\mathrm{II}}{}^1$ is applied to ϕ_x, we can add an arbitrary constant without changing $\mathcal{E}_{\mathrm{II}}{}^1$, and therefore we can redefine $\mathcal{E}_{\mathrm{II}}{}^1_b$ as

$$\begin{aligned} \mathcal{E}_{\mathrm{II}}{}^1_b = \mathrm{sign}\left(v^1-k\right)&\big(\big(f^\delta-f\big)(v^1)-\big(f^\delta-f\big)(k)\big) \\ &-\underbrace{\mathrm{sign}(-k)\big(\big(f^\delta-f\big)(0)-\big(f^\delta-f\big)(k)\big)}_{\text{a constant}} \end{aligned}$$

$$= \operatorname{sign}\left(v^1 - k\right)\left(\left(f^\delta - f\right)(v^1) - \left(f^\delta - f\right)(0)\right)$$
$$+ \left[\operatorname{sign}(v^1 - k) - \operatorname{sign}(-k)\right]\left(\left(f^\delta - f\right)(v^1) - \left(f^\delta - f\right)(0)\right).$$

With this definition we can choose the measure $d\mu_{\mathrm{II}b}^{\;1}$ by setting

$$\begin{aligned}
\int_{\mathbb{R}} d\mu_{\mathrm{II}b}^{\;1}(\,\cdot\,,t) &= \int_{\mathbb{R}} \left|\operatorname{sign}(v^1 - k)\left((f - f^\delta)(v^1) - (f - f^\delta)(0)\right)\right| dx \\
&\quad + \int_{\mathbb{R}} \left|\operatorname{sign}(v^1 - k) - \operatorname{sign}(-k)\right| \left|(f - f^\delta)(k) - (f - f^\delta)(0)\right| dx \\
&\leq \int_{\mathbb{R}} \left|(f - f^\delta)(v^1) - (f - f^\delta)(0)\right| dx \\
&\quad + 2\int_{\mathbb{R}} \chi_{|k| \leq |v^1|} \left|(f - f^\delta)(0) - (f - f^\delta)(k)\right| dx \\
&\leq \left\|f^\delta - f\right\|_{\mathrm{Lip}} \int_{\mathbb{R}} \left|v^1(x,t)\right| dx \\
&\quad + 2\int_{\mathbb{R}} \max_{|\kappa| \leq |v^1(x,t)|} \left|(f^\delta - f)(\kappa) - (f^\delta - f)(0)\right| dx \\
&\leq 3\left\|f^\delta - f\right\|_{\mathrm{Lip}} \int_{\mathbb{R}} \left|v^1(x,t)\right| dx.
\end{aligned}$$

We can choose the piecewise linear interpolant f^δ such that

$$\left\|f^\delta - f\right\|_{\mathrm{Lip}} \leq \mathrm{Const}\, \|f\|_{\mathrm{Lip}}\, \delta,$$

and with this choice

$$\int_{\mathbb{R}} d\mu_{\mathrm{II}b}^{\;1} \leq \mathrm{Const}\, \|f\|_{\mathrm{Lip}}\, \delta. \tag{4.26}$$

Next we must estimate the entropy discrepancy created by $\mathcal{H}_t^{\Delta x}$, that is, for any constant k, $s \leq \Delta t$ and $v^2(\,\cdot\,,t) = \mathcal{H}_t^{\Delta x} u^0 = \mathcal{T}_t(\mathcal{W}_t(u^0))$ we must estimate

$$A = \int_0^s \int_{\mathbb{R}} \left[\eta(v^2)\phi_t - r\left(v^2\right)\phi_{xx}\right] dx\,dt + \int_{\mathbb{R}} \left[\eta\left(u^0\right)\phi(x,0) - \eta\left(v^2(x,s)\right)\phi(x,s)\right] dx, \tag{4.27}$$

where $\eta(v) = |v - k|$, $r(v) = |A(v) - A(k)|$, and ϕ any nonnegative test function. We start by utilizing the monotonicity properties of $\mathcal{W}_t$ and $\mathcal{T}_t$. For the moment, we use the notation $z \in \ell^\infty = \{z_i\}_{i\in\mathbb{Z}}$, and by $z \leq w$ we mean that $z_i \leq w_i$ for all i. Also, let $a \vee b = \max\{a,b\}$ and $a \wedge b = \min\{a,b\}$, and let $\mathcal{Q}\colon \ell^\infty \to \ell^\infty$ be defined by

$$\left(\mathcal{Q}(z)\right)_i = |A(z_{i-1}) - A(k)| - 2\,|A(z_i) - A(k)| + |A(z_{i+1}) - A(k)|\,.$$

Hoping not to confuse the reader too much, we use the notation $k \in \ell^\infty$ for the constant sequence $\{\dots,k,k,k,k,k,\dots\}$.

Since $\mathcal{W}_t$ is monotone and $\mathcal{W}_t(k) = k$, we have that for $w \in \ell^1$,

$$\mathcal{W}_t(w \vee k) \geq \mathcal{W}_t(w) \vee k,$$
$$\mathcal{W}_t(w \wedge k) \leq \mathcal{W}_t(w) \wedge k.$$

Subtracting these, we get

$$\mathcal{W}_t(w \vee k) - \mathcal{W}_t(w \wedge k) \geq |\mathcal{W}_t(w) - k|,$$

which can be rewritten as

$$|\mathcal{W}_t(w) - k| \leq |w - k| + \mu(1-\theta)\mathcal{Q}(w). \tag{4.28}$$

Also the operator $\mathcal{T}_t$ is monotone with $\mathcal{T}_t(k) = k$, therefore for $z \in \ell^1$,

$$\mathcal{T}_t(z \vee k) - \mathcal{T}_t(z \wedge k) \geq |\mathcal{T}_t(w) - k|,$$

which can be rewritten as

$$|z - k| + \mu\theta\mathcal{Q}(z) \geq |\mathcal{T}_t(z) - k|. \tag{4.29}$$

Setting $w = u^0$ in (4.28) and $z = \mathcal{W}_t(u^0)$ in (4.29), we get

$$\left|v^2(t) - k\right| \leq \left|u^0 - k\right| + \mu(1-\theta)\mathcal{Q}(u^0) + \mu\theta\mathcal{Q}\left(\mathcal{W}_t\left(u^0\right)\right). \tag{4.30}$$

Using this, we rewrite A in (4.27) as

$$\begin{aligned}
A &= \int_0^s \int_{\mathbb{R}} \left(\eta(v^2) - \eta(u^0)\right)\phi_t - r\left(v^2(x,t)\right)\phi_{xx} \\
&\qquad\qquad - \left[\eta\big(v^2(x,s)\big) - \eta\big(u^0(x)\big)\right]\frac{1}{s}\phi(x,s)\,dx\,dt \\
&= \int_0^s \int_{\mathbb{R}} \left(\eta(v^2) - \eta(u^0)\right)\phi_t - r\big(v^2(x,t)\big)\phi_{xx} \\
&\qquad\qquad - \left[\eta\big(v^2(x,s)\big) - \eta\big(u^0(x)\big)\right]\frac{1}{s}\phi(x,t)\,dx\,dt \\
&\quad + \int_0^s \int_{\mathbb{R}} \left[\eta\big(v^2(x,s)\big) - \eta\big(u^0(x)\big)\right]\frac{1}{s}\left[\phi(x,s) - \phi(x,t)\right]dx\,dt \\
&\geq -\underbrace{\int_0^s \int_{\mathbb{R}} \left|v^2(t) - u^0\right|\phi_t\,dx\,dt}_{B} \\
&\quad - \underbrace{\int_0^s \int_{\mathbb{R}} r\big(v^2(x,t)\big)\phi_{xx}(x,t) - \frac{1}{\Delta x^2}\left[(1-\theta)Q(u^0) + \theta Q(z)\right]\phi(x,t)\,dx\,dt}_{C} \\
&\quad + \underbrace{\int_0^s \int_{\mathbb{R}} \left[\eta\left(v^2(x,s)\right) - \eta\left(u^0(x)\right)\right]\frac{1}{s}\int_t^s \phi_t(x,\tau)\,d\tau\,dx\,dt}_{D},
\end{aligned}$$

where we recall that $z = \mathcal{W}_s(u^0)$ and $\mu = s/\Delta x^2$, and we have set

$$Q(u)(x) = \sum_i (\mathcal{Q}(u))_i \, \chi_{I_i}(x).$$

Now we can bound B by

$$B \le \int_0^s \int_{\mathbb{R}} \left|v^2(x,t) - u^0(x)\right| |\phi_t| \, dx \, dt \le s\omega(s;v^2) \, \|\phi_t\|_{L^\infty(\mathbb{R}\times[0,s])} \tag{4.31}$$

by (4.21) and Remark 3.5. We also find that

$$\begin{aligned} D &\le \int_0^s \int_{\mathbb{R}} \left|v^2(x,s) - v^2(x,0)\right| \frac{1}{s} \int_t^s |\phi_t(x,\tau)| \, d\tau \, dx \\ &\le \frac{1}{2} s\,\omega(s;v^2) \, \|\phi_t\|_{L^\infty(\mathbb{R}\times[0,s])} \,. \end{aligned}$$

Therefore, keeping in mind that $B + D = \langle {\mathcal{E}_{\mathrm{I}}}^2, \phi_t \rangle$ in the notation of Chapter 3, we can choose the measure $d{\mu_{\mathrm{I}}}^2$ bounding ${\mathcal{E}_{\mathrm{I}}}^2$ as $d{\mu_{\mathrm{I}}}^2 = \left|v^2(\,\cdot\,,s) - u^0\right|$. Then

$$\int_{\mathbb{R}} d{\mu_{\mathrm{I}}}^2 \le \omega(s;v^2) \le \mathrm{Const}\,\Delta t^{1/2}. \tag{4.32}$$

To estimate C, for $w = z$ or $w = u^0$ we calculate

$$\begin{aligned} I(w;\phi) &= \int_{\mathbb{R}} r\left(v^2(x,t)\right) \phi_{xx}(x,t) - \frac{1}{\Delta x^2} Q(w)\phi(x,t)\, dx \\ &= \sum_i \Big[r\left(v_i^2(t)\right) \left[\phi_x\left(x_{i+1/2},t\right) - \phi_x\left(x_{i-1/2},t\right)\right] \\ &\qquad\qquad - \frac{1}{\Delta x} \left[\left(r\left(w_{i+1}\right) - r\left(w_i\right)\right) - \left(r\left(w_i\right) - r\left(w_{i-1}\right)\right)\right] \phi_i(t) \Big] \\ &= -\sum_i \Big[\left[r\left(v_{i+1}^2(t)\right) - r\left(v_i^2(t)\right)\right] \phi_x\left(x_{i+1/2},t\right) \\ &\qquad\qquad\qquad\qquad - \left[r\left(w_{i+1}\right) - r\left(w_i\right)\right] \frac{\phi_{i+1}(t) - \phi_i(t)}{\Delta x} \Big] \\ &= -\sum_i \Big[\left[r\left(v_{i+1}^2(t)\right) - r\left(w_{i+1}\right)\right] - \left[r\left(v_i^2(t)\right) - r\left(w_i\right)\right] \Big] \phi_x\left(x_{i+1/2},t\right) \\ &\quad - \sum_i \left[\phi_x\left(x_{i+1/2},t\right) - \frac{\phi_{i+1}(t) - \phi_i(t)}{\Delta x} \right] \left[r\left(w_{i+1}\right) - r\left(w_i\right)\right] \\ &= \int_{\mathbb{R}} \left[r\left(v^2(x,t)\right) - r(w(x))\right] \phi_{xx}(x,t)\, dx \\ &\quad - \sum_i \left(\phi_x\left(x_{i+1/2},t\right) - \frac{1}{\Delta x^2} \int_{I_i} \phi(y+\Delta x,t) - \phi(y,t)\, dy \right) \\ &\qquad\qquad\qquad\qquad\qquad\qquad \times \left[r\left(w_{i+1}\right) - r\left(w_i\right)\right] \end{aligned}$$

$$
\begin{aligned}
&= \int_{\mathbb{R}} \left[r\left(v^2(x,t)\right) - r(w(x))\right]\phi_{xx}(x,t)\,dx \\
&\quad - \frac{1}{\Delta x^2}\sum_i \left(\int_{I_i}\int_y^{y+\Delta x} \phi_x\left(x_{i+1/2},t\right) - \phi_x(\alpha,t)\,d\alpha dy\right) \\
&\qquad\qquad\qquad\qquad\qquad\qquad\qquad \times\left[r\left(w_{i+1}\right) - r\left(w_i\right)\right] \\
&= \int_{\mathbb{R}} \left[r\left(v^2(x,t)\right) - r(w(x))\right]\phi_{xx}(x,t)\,dx \\
&\quad - \frac{1}{\Delta x^2}\sum_i \left(\int_{I_i}\int_y^{y+\Delta x}\int_\alpha^{x_{i+1/2}} \phi_{xx}(\beta,t)\,d\beta d\alpha dy\right)\left[r\left(w_{i+1}\right) - r\left(w_i\right)\right].
\end{aligned}
$$

Hence

$$
|I(w;\phi)| \le \|A\|_{\mathrm{Lip}} \int_{\mathbb{R}} \left|v^2(x,t) - w(x)\right| + |w(x+\Delta x) - w(x)|\,dx\,\|\phi_{xx}\|_{L^\infty(\mathbb{R}\times[0,s])}\,.
$$

Therefore, with $C = \langle {\mathcal{E}_{\mathrm{III}}}^2, \phi_{xx}\rangle$ in the notation of Chapter 3, and in view of the relation $v^2 - \mu\theta\mathcal{Q}(v^2) = z$, we get

$$
\begin{aligned}
C &\le \theta\left|I\left(z;\phi\right)\right| + (1-\theta)\left|I\left(u^0;\phi\right)\right| \\
&\le \|A\|_{\mathrm{Lip}}\,\|\phi_{xx}\|_{L^\infty(\mathbb{R}\times[0,s])} \int_{\mathbb{R}} \theta\left|v^2(x,t) - z(x)\right| + (1-\theta)\left|v^2(x,t) - u^0(x)\right| \\
&\qquad\qquad\qquad\qquad\qquad + \theta\left|z(x+\Delta x) - z(x)\right| \\
&\qquad\qquad\qquad\qquad\qquad + (1-\theta)\left|u^0(x+\Delta x) - u^0(x)\right|\,dx \\
&\le \|A\|_{\mathrm{Lip}}\,\|\phi_{xx}\|_{L^\infty(\mathbb{R}\times[0,s])} \int_{\mathbb{R}} 2\theta(1-\theta)\mu\,\|A\|_{\mathrm{Lip}}\left|z(x+\Delta x) - z(x)\right| \\
&\qquad\qquad\qquad\qquad\qquad + (1-\theta)\left|v^2(x,t) - u^0(x)\right| \\
&\qquad\qquad\qquad\qquad\qquad + \left|u^0(x+\Delta x) - u^0(x)\right|\,dx \\
&\le \|A\|_{\mathrm{Lip}}\,\|\phi_{xx}\|_{L^\infty(\mathbb{R}\times[0,s])} \int_{\mathbb{R}} \left(2\theta(1-\theta)\mu\,\|A\|_{\mathrm{Lip}} + 1\right)\left|u^0(x+\Delta x) - u^0(x)\right| \\
&\qquad\qquad\qquad\qquad\qquad + (1-\theta)\left|v^2(x,t) - u^0(x)\right|\,dx \\
&\le \|A\|_{\mathrm{Lip}}\,\|\phi_{xx}\|_{L^\infty(\mathbb{R}\times[0,s])} \\
&\qquad\qquad \times\left[\left(2\theta(1-\theta)\mu\,\|A\|_{\mathrm{Lip}} + 1\right)\nu(\Delta x;u^0) + (1-\theta)\omega(v^2;t)\right].
\end{aligned}
$$

Hence, the Radon measure bounding ${\mathcal{E}_{\mathrm{III}}}^2$ can be chosen such that

$$
\int_{\mathbb{R}} d{\mu_{\mathrm{III}}}^2 \le \mathrm{Const}\left(\Delta x + \Delta t^{1/2}\right). \tag{4.33}
$$

We have now shown that all the entropy discrepancies are bounded by finite Radon measures that tend to zero with the discretization parameters δ, Δx, and Δt. To conclude that the limit u is a weak solution we must show that $\partial_x A(u) \in L^2(\Pi_T)$.

Verification of $\partial_x A(u) \in L^2(\Pi_T)$. We will show this by proving that

$$\int_0^T \int_{\mathbb{R}} \Big(A\big(u(x+y,t)\big) - A\big(u(x,t)\big) \Big)^2 \, dx \, dt \le \text{Const } y^2.$$

To this end, we introduce the auxiliary function

$$\tilde{u}_{\Delta t}(x,t) = \mathcal{H}^{\Delta x}_{t-t_n} \circ \mathcal{S}^{\delta}_{\Delta t} u_{\Delta t}(\,\cdot\,, t_n) \quad \text{for } t \in [t_n, t_{n+1}). \tag{4.34}$$

Following the argument on page 53, we can show that this auxiliary function is close to $u_{\Delta t}$ since if $t \in [t_n, t_{n+1})$,

$$\|\tilde{u}_{\Delta t}(\,\cdot\,,t) - u_{\Delta t}(\,\cdot\,,t)\|_{L^1(\mathbb{R})} \le \text{Const} \begin{cases} \sqrt{t-t_n}, & \text{if } t \in [t_{n,1}, t_{n+1}), \\ \Delta t + \sqrt{t-t_n}, & \text{if } t \in [t_n, t_{n,1}). \end{cases}$$

Therefore $\lim_{\Delta t \to 0} \tilde{u}_{\Delta t} = u$.

First, we fix a real positive number y and an integer j_y such that

$$y = j_y \Delta x + h, \quad \text{where } j_y \ge 0 \text{ and } 0 \le h < \Delta x.$$

Then

$$\begin{cases} x \in \big[x_{i-1/2}, x_{i-1/2} + \Delta x - h\big) & \Rightarrow x + y \in I_{i+j_y}, \\ x \in \big[x_{i+1/2} - h, x_{i+1/2}\big) & \Rightarrow x + y \in I_{i+j_y+1}. \end{cases}$$

Henceforth, we will use the following shorthand notation: $z = \mathcal{S}^{\delta}_{\Delta t} u^n$, $w = \mathcal{W}_{\Delta t}(z)$, and $u^{n+1} = v = \mathcal{T}_{\Delta t}(w)$. Observe now that if $x \in I_i$ and $y \in I_j$ then

$$A\big(v(x+y)\big) - A\big(v(x)\big) = A\big(v_{i+j}\big) - A\big(v_i\big) = \sum_{k=i}^{i+j-1} \big[A\big(v_{k+1}\big) - A\big(v_k\big)\big].$$

Squaring the above we find that

$$\big[A\big(v(x+y)\big) - A\big(v(x)\big)\big]^2 \le (j-1) \sum_{k=i}^{i+j-1} \big[A\big(v_{k+1}\big) - A\big(v_k\big)\big]^2. \tag{4.35}$$

Set

$$j(x) = \begin{cases} j_y, & \text{if } x \in \big[x_{i-1/2}, x_{i-1/2} + \Delta x - h\big), \\ j_y + 1, & \text{if } x \in \big[x_{i+1/2} - h, x_{i+1/2}\big). \end{cases}$$

Integrating (4.35) with respect to x gives

$$\begin{aligned} &\int_{\mathbb{R}} \big[A\big(v(x+y)\big) - A\big(v(x)\big)\big]^2 \, dx \\ &\quad \le \sum_i \int_{I_i} (j(x) - 1) \sum_{k=i}^{i+j(x)-1} \big[A\big(v_{k+1}\big) - A\big(v_k\big)\big]^2 \, dx \end{aligned}$$

$$
\begin{aligned}
&= \sum_i \Bigg(\int_{x_{i-1/2}}^{x_{i+1/2}-h} (j_y - 1) \sum_{k=i}^{i+j_y-1} \big[A(v_{k+1}) - A(v_k)\big]^2 \, dx \\
&\qquad + \int_{x_{i+1/2}-h}^{x_{i+1/2}} j_y \sum_{k=i}^{i+j_y} \big[A(v_{k+1}) - A(v_k)\big]^2 \, dx \Bigg) \\
&= \sum_i \Bigg((\Delta x - h)\,(j_y - 1) \sum_{m=0}^{j_y-1} \big[A(v_{m+i+1}) - A(v_{m+i})\big]^2 \\
&\qquad + h j_y \sum_{m=0}^{j_y} \big[A(v_{m+i+1}) - A(v_{m+i})\big]^2 \Bigg) \\
&= \overbrace{\frac{1}{\Delta x}\left[(j_y - 1)\, j_y\, (\Delta x - h) + j_y\,(j_y + 1)\, h\right]}^{J(y)} \Delta x \sum_i \big[A(v_{i+1}) - A(v_i)\big]^2 .
\end{aligned}
$$

Next we will estimate the sum on the right-hand side. By definition v satisfies $v - \mu\theta\mathcal{R}(v) = w$, multiplying this pointwise by v and summing over i gives

$$
\begin{aligned}
\sum_i v_i\,(w_i - v_i) &= -\mu\theta \sum_i \big[A(v_{i+1}) - 2A(v_i) + A(v_{i-1})\big] v_i \\
&= \mu\theta \sum_i \big(A(v_{i+1}) - A(v_i)\big)\,(v_{i+1} - v_i) .
\end{aligned}
$$

Moreover, we have that

$$
v_i(w_i - v_i) = \tfrac{1}{2}\big[w_i^2 + v_i^2 - (w_i - v_i)^2\big] - v_i^2 \le \tfrac{1}{2}\big(w_i^2 - v_i^2\big).
$$

Now since $A' \ge 0$,

$$
\big[A(v_{i+1}) - A(v_i)\big]^2 \le \|A\|_{\mathrm{Lip}}\,(v_{i+1} - v_i)\,\big(A(v_{i+1}) - A(v_i)\big),
$$

and therefore

$$
\begin{aligned}
\Delta x \sum_i \big[A\,(v_{i+1}) - A(v_i)\big]^2 &\le \|A\|_{\mathrm{Lip}}\, \Delta x \sum_i\, (v_{i+1} - v_i)\,\big[A(v_{i+1}) - A(v_i)\big] \\
&\le \frac{\|A\|_{\mathrm{Lip}}}{\mu\theta} \Delta x \sum_i v_i\,(w_i - v_i) \le \frac{\|A\|_{\mathrm{Lip}}}{\mu\theta} \frac{\Delta x}{2} \sum_i \big(w_i^2 - v_i^2\big) \\
&= \frac{\|A\|_{\mathrm{Lip}}}{\mu\theta} \frac{\Delta x}{2} \sum_i \Big(\big(z_i + \mu(1-\theta)\,(\mathcal{R}(z))_i\big)^2 - v_i^2 \Big) \\
&= \frac{\|A\|_{\mathrm{Lip}}}{\mu\theta} \frac{\Delta x}{2} \sum_i \Big(\underbrace{z_i^2 - v_i^2}_{a_i} + \underbrace{2\mu(1-\theta) z_i\,(\mathcal{R}(z))_i}_{b_i} + \underbrace{\mu^2(1-\theta)^2\,(\mathcal{R}(z))_i^2}_{c_i} \Big).
\end{aligned}
\tag{4.36}
$$

Then

$$
\begin{aligned}
\Delta x \sum_i a_i &= \Delta x \sum_i \Big(z_i^2 - (u_i^n)^2 - \big((u_i^{n+1})^2 - (u_i^n)^2\big)\Big) \\
&\leq \|\tilde{u}_{\Delta t}\|_{L^\infty(\Pi_T)} \Delta x \sum_i \left| \left(\mathcal{S}^\delta_{\Delta t} u^n\right)_i - u_i^n \right| + \Delta x \sum_i \left((u_i^n)^2 - \left(u_i^{n+1}\right)^2 \right) \\
&\leq \|\tilde{u}_{\Delta t}\|_{L^\infty(\Pi_T)} \left\|f^\delta\right\|_{\mathrm{Lip}} \mathrm{T.V.}\,(u_0)\,\Delta t + \Delta x \sum_i \left((u_i^n)^2 - \left(u_i^{n+1}\right)^2 \right),
\end{aligned}
$$

and

$$
\begin{aligned}
\frac{\|A\|_{\mathrm{Lip}}}{\mu\theta} \frac{\Delta x}{2} \sum_i b_i &= \frac{(1-\theta)}{\theta} \|A\|_{\mathrm{Lip}} \Delta x \sum_i z_i \left(\mathcal{R}(z)\right)_i \\
&= -\frac{(1-\theta)}{\theta} \|A\|_{\mathrm{Lip}} \Delta x \sum_i (z_{i+1} - z_i) \left[A(z_{i+1}) - A(z_i) \right] \\
&\leq -\frac{(1-\theta)}{\theta} \Delta x \sum_i \left[A(z_{i+1}) - A(z_i) \right]^2.
\end{aligned}
$$

Regarding the last term in (4.36), we estimate this by

$$
\frac{\|A\|_{\mathrm{Lip}}}{\mu\theta} \frac{\Delta x}{2} \sum_i c_i \leq 2 \frac{(1-\theta)^2 \mu}{\theta} \|A\|_{\mathrm{Lip}} \Delta x \sum_i \left(A(z_{i+1}) - A(z_i) \right)^2,
$$

since $(\mathcal{R}(z))_i^2 \leq 2[A(z_{i+1}) - A(z_i)]^2 + 2[A(z_i) - A(z_{i-1})]^2$.

Combining the above estimates gives

$$
\begin{aligned}
\Delta x &\sum_i \left[(A(v_{i+1}) - A(v_i) \right]^2 \qquad (4.37) \\
&\leq \frac{\|A\|_{\mathrm{Lip}}}{\mu\theta} \frac{\Delta x}{2} \sum_i \left((u_i^n)^2 - (u_i^{n+1})^2 \right) \\
&\quad + \frac{\|A\|_{\mathrm{Lip}}}{2\mu\theta} \|\tilde{u}_{\Delta t}\|_{L^\infty(\Pi_T)} \left\|f^\delta\right\|_{\mathrm{Lip}} \mathrm{T.V.}\,(u_0)\,\Delta t \\
&\quad + \frac{1-\theta}{\theta} \Big(2 \|A\|_{\mathrm{Lip}} \mu (1-\theta) - 1 \Big) \Delta x \sum_i \left[A(z_{i+1}) - A(z_i) \right]^2.
\end{aligned}
$$

Note that by the CFL-condition (4.20) the last term is nonpositive. Then, recalling that $\mu = \Delta t / \Delta x^2$, we see that

$$
\begin{aligned}
\int_{\mathbb{R}} \Big(A(u^{n+1}(x+y)) - A(u^{n+1}(x)) \Big)^2 dx &\leq J(y) \Delta x^2 \frac{\|A\|_{\mathrm{Lip}}}{2} \\
\times \Big[\frac{1}{\Delta t} \int_{\mathbb{R}} \left(u^n(x)\right)^2 - \left(u^{n+1}(x)\right)^2 dx &+ \|\tilde{u}_{\Delta t}\|_{L^\infty(\Pi_T)} \left\|f^\delta\right\|_{\mathrm{Lip}} \mathrm{T.V.}\,(u_0) \Big]. \qquad (4.38)
\end{aligned}
$$

If we let $\hat{u}_{\Delta t}$ denote the piecewise constant function

$$\hat{u}_{\Delta t}(\,\cdot\,,t) = \sum_n u^{n+1}(\,\cdot\,)\chi_{I^n}(t),$$

then (for $T = N\Delta t$)

$$\begin{aligned}
\|\hat{u}_{\Delta t} - \tilde{u}_{\Delta t}\|_{L^1(\Pi_T)} &= \sum_{n=0}^{N-1} \int_{t_n}^{t^{n+1}} \int_{\mathbb{R}} |\tilde{u}_{\Delta t}(x,t) - \tilde{u}_{\Delta t}(x,t_{n+1})|\,dx \\
&= \sum_{n=0}^{N-1} \int_{t^n}^{t_{n+1}} \text{Const}\sqrt{t_{n+1}-t}\,dt \\
&\le \text{Const} \sum_{n=0}^{N-1} (\Delta t)^{3/2} \le \text{Const}_T \sqrt{\Delta t}.
\end{aligned}$$

Therefore $A(\hat{u}_{\Delta t}) \to A(u)$ in $L^p(\Pi_T)$, $1 \le p < \infty$, as $\Delta t \to 0$. Furthermore,

$$\begin{aligned}
J(y)\Delta x^2 &= j_y \Delta x\big((\Delta x - h)(j_y - 1) + h\,(j_y + 1)\big) \\
&= j_y^2 \Delta x^2 + 2 j_y h \Delta x \\
&\le j_y^2 \Delta x^2 + 2 j_y h \Delta x + h^2 \\
&= (j_y \Delta x + h)^2 = y^2.
\end{aligned}$$

Multiplying (4.38) with Δt and summing over n, we get

$$\begin{aligned}
&\int_0^T \int_{\mathbb{R}} \big(A\big(\hat{u}_{\Delta t}(x+y,t)\big) - A\big(\hat{u}_{\Delta t}(x,t)\big)\big)^2\,dx\,dt \\
&\qquad \le y^2 \|A\|_{\text{Lip}} \left[\int_{\mathbb{R}} u_0^2(x)\,dx + \|\tilde{u}_{\Delta t}\|_{L^\infty(\Pi_T)} \left\|f^\delta\right\|_{\text{Lip}} \text{T.V.}\,(u_0)\,T\right],
\end{aligned}$$

assuming that (we have set things up so that) $\left\|f^\delta\right\|_{\text{Lip}} \le 2\,\|f\|_{\text{Lip}}$ and the initial data $\left\|u^0\right\|_{L^2(\mathbb{R})} \le 2\,\|u_0\|_{L^2(\mathbb{R})}$. Thus, letting $\Delta t \to 0$ above, we get

$$\int_0^T \int_{\mathbb{R}} \Big(A\big(u(x+y,t)\big) - A\big(u(x,t)\big)\Big)^2\,dx\,dt \le \text{Const}_{A,f,u_0,T}\, y^2. \tag{4.39}$$

Finally, we can let $y \to 0$ to conclude that $\partial_x A(u) \in L^2(\Pi_T)$.

Summing up, we have derived the following result:

Theorem 4.4. *Assume that $u_0 \in L^1(\mathbb{R}) \cap \mathrm{BV}(\mathbb{R})$, and that f and A are locally Lipschitz continuous with $A' \ge 0$ and $A(0) = 0$. Define a sequence of functions $\{u_{\Delta t}\}_{\Delta t \ge 0}$ by (4.18). There exists a subsequence of $\{\Delta t_j\}_{j\in\mathbb{Z}} \subset \{\Delta t\}$ such that*

$$u_{\Delta t_j} \to u \quad \textit{in } L^1_{\text{loc}}(\Pi_T) \textit{ as } \Delta t_j \to 0,$$

where u is an entropy weak solution of

$$u_t + f(u)_x = A(u)_{xx}, \qquad u(x,0) = u_0(x).$$

Remark 4.5. The above analysis can also be carried out if $\mathcal{H}^{\Delta}x_{\Delta t}$ is replaced by several smaller steps, i.e., instead of $\mathcal{H}^{\Delta x}_{\Delta t}$ we use

$$\left(\mathcal{H}^{\Delta x}_{\Delta t/m}\right)^m,$$

for some integer $m \geq 2$. This would relax the CFL-condition (4.20) to

$$(1-\theta)\mu \left\|A\right\|_{\text{Lip}} \leq \frac{m}{2}.$$

One can also replace the front-tracking method for the hyperbolic conservation law by other numerical methods and still be able to prove essentially the same results. Let us here mention only two alternatives. One alternative is to approximate the hyperbolic solution operator $\mathcal{S}_t$ by the *Euler characteristic Galerkin splitting method* [186]. For the parabolic operator, we can still use the implicit-explicit scheme (4.11) (the θ-scheme). Alternatively, one can employ a *second-order MUSCL method*, introduced by Bouchut, Bourdarias, and Perthame [31] for which the necessary entropy estimates have been proved, while using the θ-scheme analyzed in the previous section for the parabolic part.

4.3 Nonlinear error mechanisms

In the two examples in Section 4.2.1 and the rigorous mathematical analysis in the previous section we studied convergence in a global sense through the use of the L^1 norm. For many applications, convergence in the L^1 norm is not sufficient, and one is interested in both the qualitative behavior of the approximate solutions and the pointwise accuracy. This is dictated by the interaction of two types of errors: the discretization errors in each substep and the error resulting from the operator splitting. Here we will focus on errors caused by operator splitting, and start by an example which illustrates the major error mechanisms in the viscous splitting.

Example 4.6. *Let us simplify Example 4.2 and use Riemann initial data*

$$u_0(x) = \begin{cases} 1, & x < 0, \\ -1, & x > 0. \end{cases}$$

This corresponds to a single, stationary, viscous shock, where the width of the viscous shock is proportional to the scaling parameter ε. In the semi-discrete splitting the diffusive step amounts to solving the heat equation, or in other words,

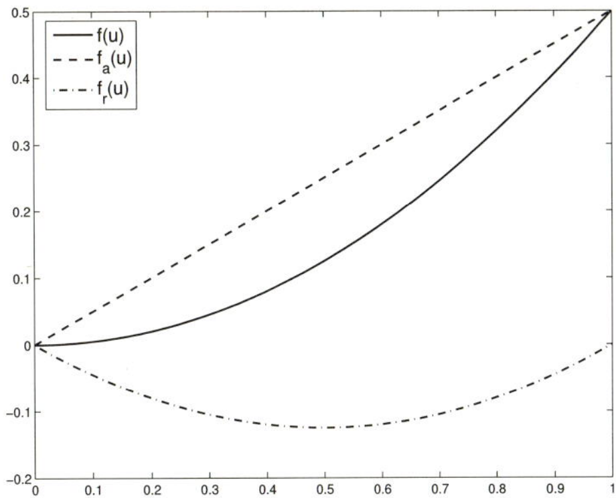

Figure 4.3. Convexification and residual flux for Burgers' equation for a shock between $u_L = 1$ and $u_R = 0$.

convolution with the heat kernel. It is easy to see from the form of the heat kernel that the viscous splitting will lead to a numerical front of width $\mathcal{O}(\sqrt{\varepsilon\Delta t})$. *To properly resolve the viscous front, the size of the splitting step should therefore be proportional to* ε.

To investigate the error mechanisms of the viscous splitting, let us examine the first step of the algorithm. In the hyperbolic step we solve the equation $u_t + f(u)_x = 0$ *for* $f(u) = u^2/2$. *The characteristics are negative for* $x > 0$ *and positive for* $x < 0$ *(satisfying* $x'(t) = -1$ *and* $x'(t) = 1$*, respectively). Hence, the solution consists of a stationary shock, as in the initial data; that is* $u(x,t) = u_0(x)$. *In other words, the effective hyperbolic step reads* $u_t = 0$. *Therefore, the viscous splitting means that we disregard the advective flux and solve only the diffusive part of the equation!*

Let us now consider a computation with multiple steps. In the first step, the heat operator (or its discretization) will give a smooth profile with $u > 0$ *for* $x < 0$ *and* $u < 0$ *for* $x > 0$. *The second hyperbolic step will have a sharpening effect on the smooth profile since* $f'(u) > 0$ *for* $x < 0$ *and* $f'(u) < 0$ *for* $x > 0$. *Thus, a new shock will form if the time-step is sufficiently large. The second diffusive step will smooth the approximate solution, and so on.*

Consider now instead the initial data

$$u_0(x) = \begin{cases} 1, & x < 0, \\ 0, & x > 0. \end{cases}$$

In the first hyperbolic step, the initial discontinuity will be translated as a shock with speed 0.5. Thus the effective hyperbolic step amounts to solving the linear equation $u_t + 0.5u_x = 0$. *In other words, the viscous splitting introduces a splitting of the flux into a linear advective flux and a nonlinear flux residual (see Figure 4.3)*

$$f(u) = \tfrac{1}{2}u^2 = \tfrac{1}{2}u + \tfrac{1}{2}u(u-1) := f_a(u) + f_r(u). \tag{4.40}$$

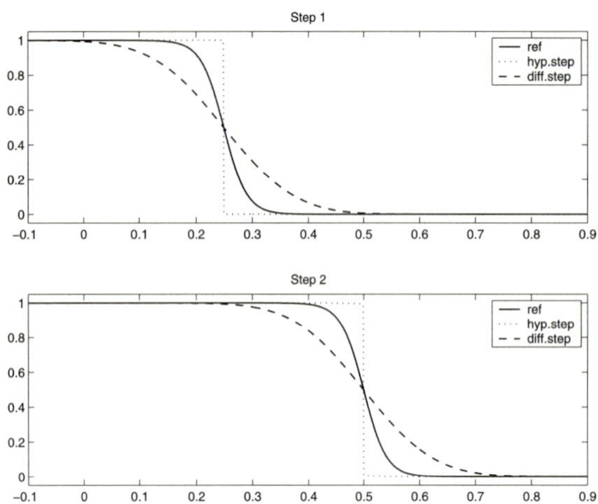

Figure 4.4. Two steps in the viscous operator-splitting method.

The residual flux $f_r(u)$ is a manifestation of the entropy condition imposed upon the hyperbolic equation through the local linearization of the flux (which is often referred to as Oleĭnik's convexification). Let us look more closely at the effect the residual flux has on the moving front for the hyperbolic subequation $u_t+f(u)_x = 0$. Since $f_r'(u)$ is positive for $u > 0.5$ and negative for $u < 0.5$, the residual flux will try to make u-values in the interval $[0.5, 1]$ move faster than the front velocity and values in the interval $[0, 0.5]$ move slower. In other words, the discontinuous front will overturn and become triple-valued. Therefore the effects of the residual flux cannot be included if the solution is to be single-valued.

If viscosity is added to the hyperbolic equation in the form of a second-order spatial derivative, the viscous forces will counteract and balance the self-sharpening effect in the nonlinear residual flux and ensure that the solution is single-valued. If the residual flux is disregarded in the operator splitting, the splitting scheme will therefore move the viscous front correctly, but overestimate its spatial width since the nonlinear self-sharpening mechanisms in the flux are neglected, see Figure 4.4.

The lesson learnt from the above example is the following one: once a discontinuity is formed in the hyperbolic step, the entropy condition will enforce a local linearization of the flux function in the form of a convexification. This linearization will keep the linear part of the flux that accounts for the transport and disregard the nonlinear part affecting the local shape of viscous fronts. Moreover, imposing the local convexification will generally lead to a loss of entropy, as we will see later when we give a rigorous mathematical analysis of the viscous operator splitting.

Motivated by similar observations as in Example 4.6, it is tempting to perform an *a priori* splitting of the flux function into an advective part and a nonlinear part, $f(u) = f_a(u) + f_r(u)$. Then, unsteady convection-diffusion equations of the form

$$u_t + f(u)_x = \varepsilon A(u)_{xx}$$

can be solved by a sequential splitting involving a hyperbolic transport step

$$u_t + f_a(u)_x = 0$$

and the parabolic, diffusive step

$$u_t + f_r(u)_x = \varepsilon A(u)_{xx}.$$

This suggestion may at a first glance seem pretty strange. However, by splitting off the transport effects in a separate step, the nonlinearity in the second-order equation can be reduced and this will speed up the convergence of the nonlinear solver used in any implicit formulation.

The idea of using in the diffusion step a "residual flux term" coming from a flux splitting was introduced in [92], and further developed in a series of papers [74–78] for the simulation of two-phase flow in porous media, where very efficient numerical schemes have been developed to resolve the varying balances of advective and diffusive forces. Some parts of this activity are summarized in [93].

Example 4.7. *Above we argued that the hyperbolic step of the operator-splitting method could be viewed as a splitting of the flux function; see* (4.40). *Let us therefore apply this flux splitting* a priori *to improve the viscous operator-splitting methods; that is, alternatively solve the two equations*

$$u_t + \tfrac{1}{2}u_x = 0, \qquad \text{and} \qquad u_t + \tfrac{1}{2}\big(u(u-1)\big)_x = \varepsilon u_{xx}.$$

In Figure 4.5 *we have recomputed the case in Figure* 4.4 *using the improved splitting algorithm. We see that the* a priori *flux splitting amends the previous deficiency in the viscous splitting and now produces correct solutions.*

A major restriction with the *a priori* splitting method is that it assumes that the solution has a certain structure. If, for instance, the constants 1 and 0 in the initial data of Example 4.7 were replaced by 1 and 1/2, the local speed of propagation would be 3/4 and correct splitting of the flux would change to

$$u_t + \tfrac{3}{4}u_x = 0, \qquad \text{and} \qquad u_t + \tfrac{1}{2}\big(u(u-\tfrac{3}{2})\big)_x = \varepsilon u_{xx}.$$

In the next section we will discuss how this potential problem can be circumvented by introducing an *a posteriori* flux splitting given by the local structure of the data *after* the hyperbolic substep.

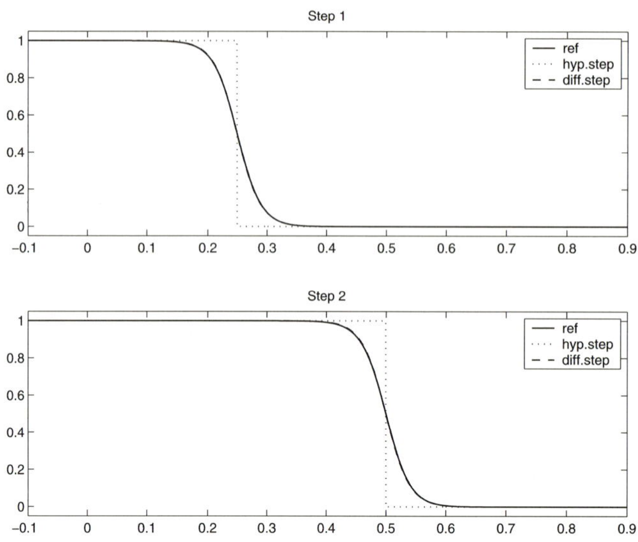

Figure 4.5. Two steps in the viscous operator-splitting method with *a priori* splitting of the flux.

4.4 Viscous splitting with *a posteriori* flux splitting

The idea behind *a priori* splitting of flux functions was taken one step further in [146]. The ideas introduced in [146] were developed further and applied in a series of papers [100, 140, 140–143].

In [146] the authors showed how to locally define *a posteriori* flux residuals to account for the entropy loss introduced in the hyperbolic steps. As above, the idea is that once a shock is formed, the evolution of the hyperbolic solution is governed locally by a *linear* transport equation. Let u^- and u^+ denote the constant states adjacent to the shock and σ be the local Rankine–Hugoniot shock speed $\sigma = [f(u^+) - f(u^-)]/[u^+ - u^-]$. Then, locally the solution is evolved according to (for simplicity we have translated discontinuity to $x = 0$)

$$v_t + \sigma v_x = 0, \qquad v(x,0) = \begin{cases} u^-, & x < 0, \\ u^+, & x > 0. \end{cases}$$

The local flux residual is defined as

$$f_r(u) = \begin{cases} f(u) - f(u^-) - \sigma(u - u^-), & u \in (u^-, u^+), \\ 0, & \text{otherwise}. \end{cases} \tag{4.41}$$

Since the discontinuity (u^-, u^+) satisfies the conservation law $u_t + f(u)_x = 0$ locally, it follows from standard theory that the function $f(u)$ must either lie

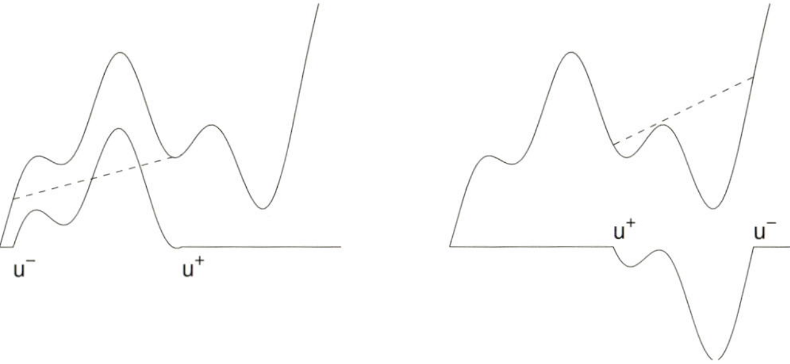

Figure 4.6. Flux residuals defined implicitly by the Oleĭnik convexification

entirely above or below the straight line connecting $f(u^-)$ and $f(u^+)$ for all values u between u^- and u^+. Moreover, it follows from Oleĭnik's convexification that if $u^- < u^+$ then $f_r'(u^-) > 0 > f_r'(u^+)$ and if $u^- > u^+$ then $f_r'(u^-) < 0 < f_r'(u^+)$; see Figure 4.6. This implies that $f_r(u)$ will have an overall self-sharpening effect on the monotone shock profile that counteracts smearing effects introduced by $A(u)$. Note, however, that $f_r(u)$ is not necessarily a strictly convex or concave function as seen in Figure 4.6.

In the particular case of front-tracking, identifying such flux residuals is straightforward, since the gist of the algorithm is to update a piecewise constant solution by solving local transport equations (tracking discontinuities). The flux residuals can therefore be obtained as a by-product of the solution algorithm in a straightforward manner. Although the flux residuals are defined for every discontinuity, they will only have a significant effect at "large" discontinuities. In a practical implementation one typically neglects flux residuals corresponding to discontinuities with strength below a certain user-defined threshold δ_u.

If we want to use the flux residuals in a local a posteriori splitting of the flux function, we must determine where the residuals should be introduced in physical space. This is typically achieved by using monotonicity intervals in the solution or a prescribed interval length. A detailed discussion of the construction of local flux residuals is given in [141, 142]. In this book we have chosen to use prescribed interval lengths given by a user-specified parameter δ_x that determines the width of the spatial interval in which each correction is applied.

Having defined the local residuals in the (x, u) space, they can be included in the diffusive step as outlined for the *a priori* splitting above; that is, we introduce a new parabolic substep given by

$$w_t + f_r(w, x)_x = \varepsilon A(w)_{xx}, \qquad w(x, 0) = w_0(x). \tag{4.42}$$

At first glance, this equation looks more complex than the equation we set out to solve since now the flux function also depends explicitly on spatial position. However, this spatial dependence is quite simple: $x \mapsto f_r(x, w)$ is a piecewise constant function for each fixed w and $w \mapsto f_r(x, w)$ is continuous for each fixed x.

Algorithm 4.4.1 The COS algorithm

Construct a piecewise constant initial function $u^0(x)$
Set $t = 0$ and $\Delta t = T/N$
For $n = 0 : N - 1$
  Use front tracking to compute solution $v(x, \Delta t)$ of
    $v_t + f(v)_x = 0, \quad v(x, 0) = u^n(x)$
  Extract from $v(x, \Delta t)$ all discontinuities $\{(v_i^-, v_i^+, x_i)\}$ that satisfy
    $|v_i^- - v_i^+| > \delta_u$
  Define *nonoverlapping* intervals $\{I_i\}$ such that $I_i \approx [x_i - \delta_x, x_i + \delta_x]$
  Let $u^{n+1/2}(x)$ be the projection of $v(x, \Delta t)$ onto a regular grid
  Use scheme (4.43) to compute solution $w(x, \Delta t)$ of
    $w_t + f_r(w, x)_x = \varepsilon A(w)_{xx}, \quad w(x, 0) = u^{n+1/2}(x)$
  with $f_r(w, x)$ given by (4.41) for $x \in I_i$ and zero otherwise
  Set $u^{n+1}(x)$ equal $w(x, \Delta t)$
end
Set $u(x, T) = u^N(x)$.

There are several ways to discretize this equation. For instance, one can use a standard central-difference approximation:

$$W_i^{n+1} = W_i^n - \tfrac{1}{2}\lambda\big[f_r(W_{i+1}^n) - f_r(W_{i-1}^n)\big] + \mu\big[A(W_{i+1}^n) - 2A(W_i^n) + A(W_{i-1}^n)\big].$$

This scheme is stable provided the local time-step k and the spatial discretization parameter Δx satisfy the conditions

$$k \le 0.5\Delta x^2/\varepsilon, \qquad \Delta x \max_w |f_r'(w)| \le 2\varepsilon.$$

The stability conditions may put severe restrictions on the discretization parameters, especially on Δx for small values of ε. The stability condition on Δx disappears if we use an upwind discretization of the flux. In the examples in this book we therefore use a Godunov upwind method

$$W_i^{n+1} = W_i^n - \lambda\big[F_{i+1/2}^n - F_{i-1/2}^n\big] + \mu\big[A(W_{i+1}^n) - 2A(W_i^n) + A(W_{i-1}^n)\big]. \tag{4.43}$$

To evaluate the Godunov flux, we approximate the Riemann problem at the cell interface by a single shock (and disregard the entropy fix, see Appendix A.6.2)

$$F_{i+1/2}^r = \begin{cases} f_r(W_i^n, x_i), & \text{if } \big[f_r(W_{i+1}^n, x_{i+1}) - f_r(W_i^n, x_i)\big]/(W_{i+1}^n - W_i^n) > 0, \\ f_r(W_{i+1}^n, x_{i+1}), & \text{otherwise.} \end{cases}$$

We have thus specified a new fully-discrete splitting method, which we will call *corrected operator splitting* (COS). The COS splitting method is summarized in Algorithm 4.4.1. Let us now revisit Examples 4.2 and 4.3 and apply the new corrected splitting approach.

Table 4.3. Estimated L^1-errors at time $t = 1.0$ computed by the original (OS) with $\nu = 1$ and $\nu = 16$ and the improved splitting method (COS) for $\nu = 16$.

	Example 4.2			Example 4.3		
N_x	OS(1)	OS(16)	COS	OS(1)	OS(16)	COS
75	4.29e-02	1.05e-01	4.59e-02	4.36e-02	1.18e-01	5.79e-02
150	2.08e-02	7.50e-02	2.07e-02	1.80e-02	7.30e-02	2.56e-02
300	1.05e-02	4.41e-02	1.03e-02	9.41e-03	3.73e-02	3.35e-02
600	5.19e-03	2.37e-02	7.55e-03	4.80e-03	1.88e-02	1.88e-02
1200	2.51e-03	1.15e-02	1.15e-02	2.29e-03	9.07e-03	9.07e-03
2400	1.19e-03	5.80e-03	5.80e-03	8.48e-04	4.54e-03	4.54e-03

Example 4.8. *In Table 4.3 we have recomputed parts of the grid-refinement study from Tables 4.1 and 4.2, now using the* a posteriori *flux residuals to improve the solution. To localize the corrections to only take effect at the origin, we set $\delta_u = 1.0$ and use a prescribed interval length equal $\sqrt{\varepsilon \Delta t}$. The local residuals have a pronounced effect on the coarse grids, where $\nu = 16$ corresponds to a few splitting steps. On the finest grids, however, the splitting steps are of the same size as ε and the residuals have little effect or are not included. Figure 4.7 shows plots of the OS and COS solutions for diffusion function $A(u)$.*

Correcting splitting errors in a single shock layer for a convex flux function, as in the example above, might not be too impressive. In fact, the correction algorithm works as a "black-box" for an (almost) arbitrary solution profile coming out from the hyperbolic substep. To demonstrate the capabilities of the algorithm we will now consider a more complex example, in which the flux function is nonconvex and the initial data is nonmonotone.

Example 4.9. *Consider the convection-diffusion equation*

$$u_t + f(u)_x = \varepsilon u_{xx}, \qquad u(x,0) = -\sin(\pi x)\chi_{[-1,2]}(x), \tag{4.44}$$

with $\varepsilon = 0.05$ and flux function $f(u) = u\sin(2\pi u) + u$. To emphasise the difference in performance between the original splitting (OS) and the a posteriori *splitting (COS) we use a single splitting step $\Delta t = 0.1$. The parameters δ_u and δ_x in the COS algorithm are both set to 0.1, quite arbitrarily.*

Figure 4.8 shows solutions computed by the OS and COS algorithms on a uniform grid with 100 cells together with a reference solution on a fine grid. Figure 4.9 shows that the COS algorithm correctly identifies residual fluxes for all eight shock layers. This COS algorithm gives improved resolution in all shock layers except for the eighth and rightmost layer, where the corresponding residual flux is small in magnitude. Moreover, when a residual flux is applied near a local critical point where the solution changes monotonicity, the correction may have a sharpening effect also for values outside the shock layer resulting in small "overshoots" in the solution, as can be observed in shock layers two, four and seven.

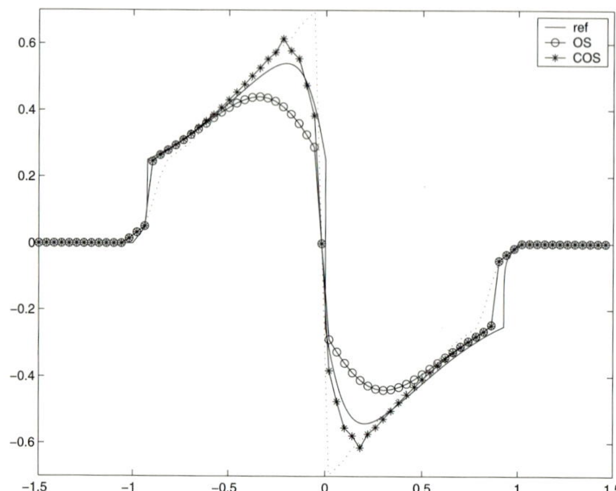

Figure 4.7. Viscous splitting solutions computed with two splitting steps on a grid with 75 grid cells for diffusion function $A(u)$ using the original (OS) and the improved splitting method (COS). The dotted line indicates the result of the last hyperbolic step in the COS computation.

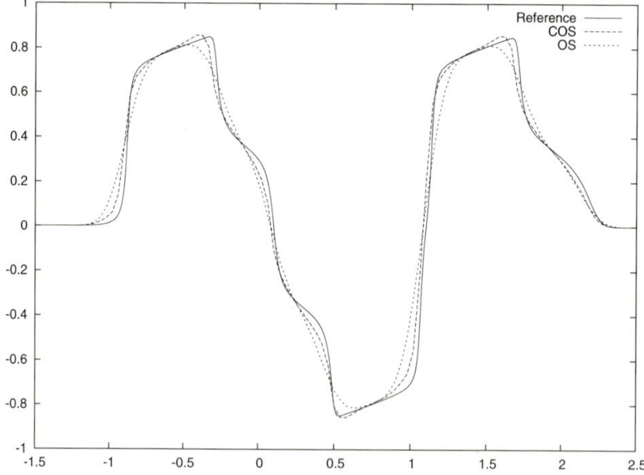

Figure 4.8. Solution at time $t = 0.1$ for equation (4.44) computed by the OS and the COS splittings. The original flux $f(u)$ is given by a solid line, the local linear flux by short dashes, and the residual by long dashes.

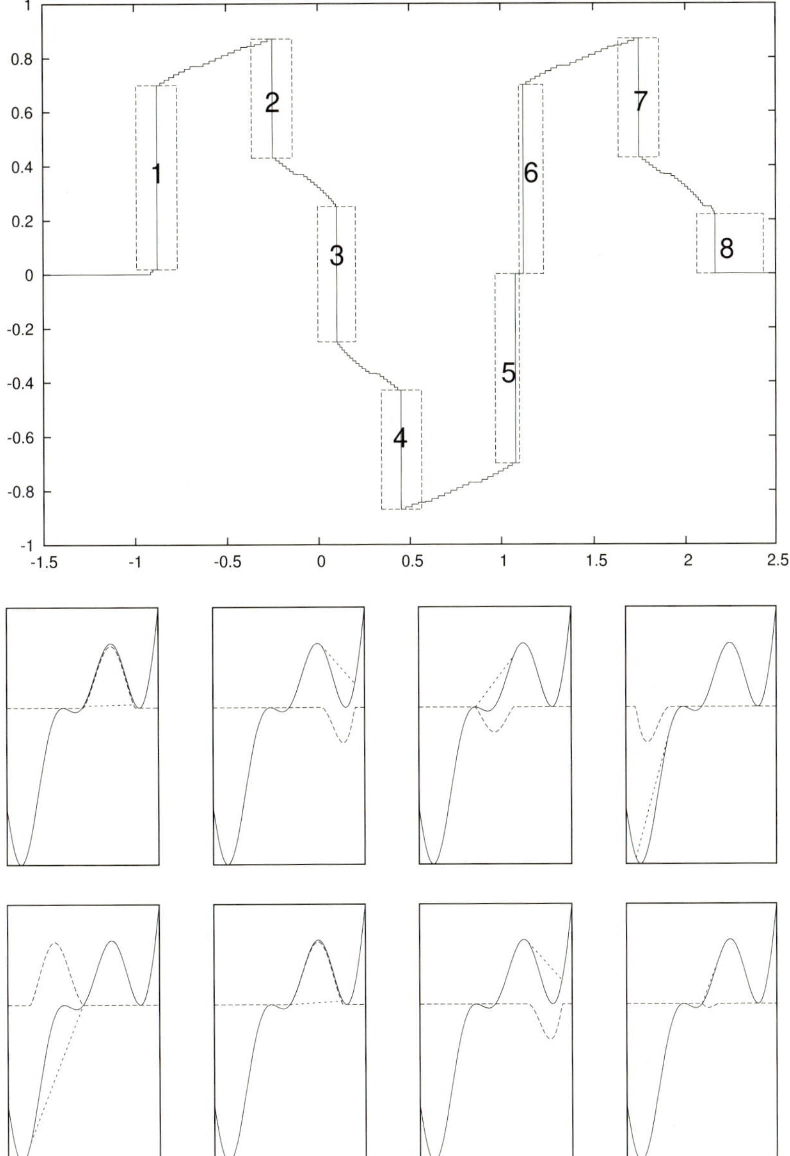

Figure 4.9. (Top) Identification of eight residual fluxes in the (x, u) plane. (Bottom) Plots of the eight residual fluxes (from top left to bottom right).

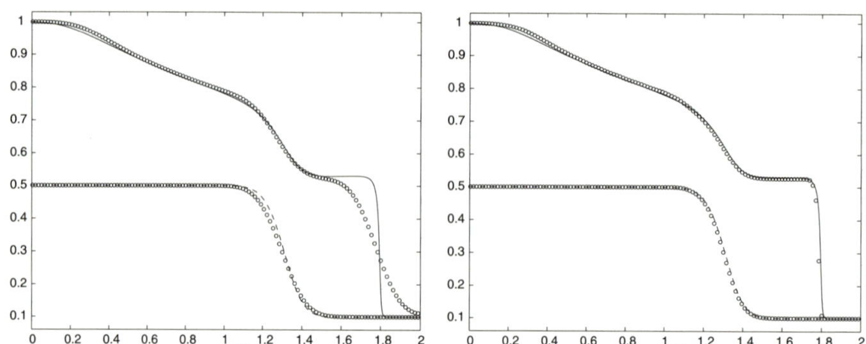

Figure 4.10. Buckley–Leverett profile for the polymer system computed by the OS (left) and COS (right) operator splittings. The plot shows s (sold line) and c (dotted line) plotted for every second grid point.

The same techniques can also be applied to systems of viscous conservation laws. As an example of such a system we consider the 2×2 polymer system modeling the flow of water and polymer in a porous medium,

$$\begin{aligned} \partial_t s + \partial_x f(s,c) &= \varepsilon \partial_x^2 s, \\ \partial_t [sc + a(c)] + \partial_x (c f(s,c)) &= \varepsilon \partial_x^2 [sc + a(c)]. \end{aligned} \tag{4.45}$$

Here, the unknowns are the water saturation s and the polymer concentration c. The fractional flow function f and the adsorption function a are given by

$$f(s,c) = \frac{s^2}{s^2 + 0.2(1+2c)(1-s)^2}, \qquad a(c) = \frac{0.2c}{1+c}. \tag{4.46}$$

The corresponding inviscid system is nonstrictly hyperbolic since the eigenvalues $\lambda_s = f_s$ and $\lambda_c = f/(s + a_c)$ coincide along a curve in state space. This means that there will be special cases where both the characteristics point into a shock.

Below we will present two examples of viscous operator splitting applied to this system. For a more thorough discussion, we refer the reader to [142].

Example 4.10. *In the first example we study the propagation of a so-called Buckley–Leverett profile that arises from the initial data*

$$(s_0, c_0)(x) = \begin{cases} (1.0, 0.5), & x \le 0.1, \\ (0.1, 0.1), & x > 0.1. \end{cases}$$

This corresponds to a single Riemann problem in the inviscid case, which is solved by a fast s-shock, followed by a c-shock and a rarefaction wave in s. Figure 4.10 shows approximations to the solution at time $t = 1.0$ for $\varepsilon = 0.005$ computed by a single step of the OS and COS algorithms on a uniform grid with 256 cells. The

Table 4.4. Estimated error in the relative L^1 norm and convergence rates at time $t = 1.0$ for the polymer system (4.45) with a Buckley–Leverett profile.

	$\varepsilon = 0.01$				$\varepsilon = 0.001$			
N_t	OS		COS		OS		COS	
1	4.42e-02	—	1.96e-02	—	1.88e-02	—	3.47e-03	—
2	2.90e-02	0.61	1.61e-02	0.29	1.26e-02	0.57	5.18e-03	-0.58
4	1.97e-02	0.56	1.22e-02	0.40	8.16e-03	0.63	4.30e-03	0.27
8	1.27e-02	0.63	8.53e-03	0.51	5.69e-03	0.52	3.27e-03	0.40
16	7.57e-03	0.75	5.65e-03	0.59	4.18e-03	0.44	2.52e-03	0.38
32	4.16e-03	0.86	3.58e-03	0.66	3.27e-03	0.35	2.22e-03	0.18
64	2.30e-03	0.86	2.45e-03	0.55	2.78e-03	0.23	2.17e-03	0.03
128	1.48e-03	0.63	1.53e-03	0.68	2.85e-03	-0.03	2.58e-03	-0.25
256	1.30e-03	0.19	1.31e-03	0.23	3.61e-03	-0.34	3.52e-03	-0.45

shock layer in c contains almost no self-sharpening effects and is resolved almost perfectly by the OS algorithm. The shock layer corresponding to the s-shock, on the other hand, contains strong self-sharpening effects and is only resolved accurately if the residual flux is included. Let us now study the effects of the temporal operator-splitting error. To eliminate the spatial error we fix the spatial discretization to 2^{10} cells and increase the number of splitting steps N_t in powers of 2. Table 4.4 reports errors for this convergence study measured in a relative L^1 norm measuring the deviance from a fine-grid solution computed by a central-difference scheme.

For $\varepsilon = 0.01$ the accuracy of both algorithms increases with increasing number of time-steps. As expected, the effect of the flux corrections in COS decreases as the number of time-steps increases. The case with $\varepsilon = 0.001$ behaves a bit differently. First of all, we observe that the error for COS increases when going from one to two splitting steps. This is because with two splitting steps the c-shock is not fully formed in the second hyperbolic step and hence the correction effect is reduced. Secondly, the errors for both algorithms increase as N_t is increased beyond 64. This is due to the numerical diffusion introduced when projecting the front-tracking solution onto a regular grid. This error increases with N_t and eventually overshadows the splitting error.

Example 4.11. *In the next example we consider a nonmonotone profile given by the initial data*

$$(s_0, c_0)(x) = \begin{cases} (0.45, 0.0), & x \le 0.1, \\ (0.2, 1.0), & x > 0.1. \end{cases}$$

The corresponding inviscid Riemann problem has a nonmonotone solution that consists of five different constant states separated by simple waves: $u^L \overset{c}{\rightarrow} u^1 \overset{s}{\rightarrow} u^2 \overset{c}{\rightarrow} u^3 \overset{s}{\rightarrow} u^R$. Figure 4.11 shows the saturation profile $s(x)$ and the solution in phase space (s, c). Figure 4.12 shows approximate solutions computed by the two

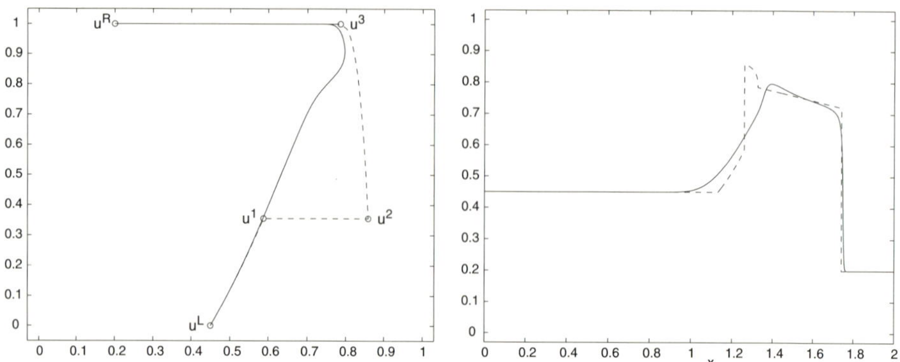

Figure 4.11. (Left) Solution in (s, c)-space: the solid line gives the solution for $\varepsilon = 0.0025$ at time $t = 1.0$ and the dashed line the inviscid solution. (Right) The s-component as a function of spatial coordinate x.

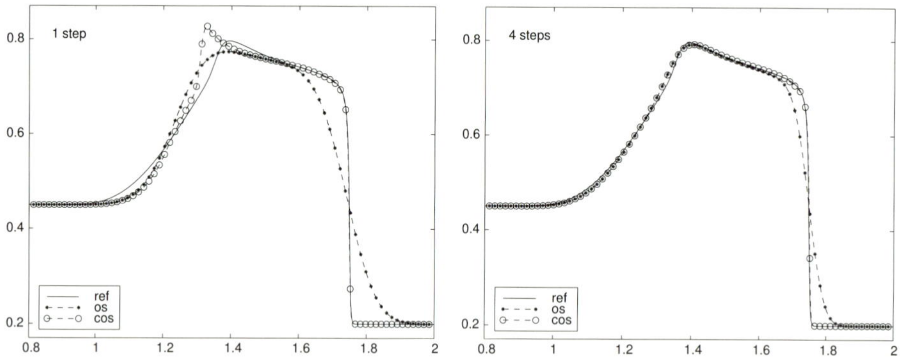

Figure 4.12. Comparison of OS (small dots) and COS (large dots) on the nonmonotone Riemann profile.

splitting algorithms on a fine grid with 2^{10} uniform cells. In the leading shock layer we observe the expected behavior: whereas OS smears the front, COS computes it accurately. In the first splitting step, the COS algorithm applies a residual flux also to the leftmost s-shock, giving an extra peak in the solution at the local maximum as we also observed for the nonmonotone profile in Figure 4.8. This effect disappears if we use two splitting steps, and with four splitting steps COS computes the whole solution with good resolution. (Notice that four splitting steps corresponds to a CFL number of around 300 for this particular case!)

In this section we have discussed how a standard viscous operator splitting generally will produce solutions that are much too diffusive in viscous shock layers. This is due to a temporal error that results from enforcing a local linearization in the hyperbolic substeps. To compensate for this error, we introduced an improved operator-splitting method that identifies the nonlinear parts of the flux that are ignored in the hyperbolic substep. These residual fluxes can then be applied as part of the parabolic substep to provide the necessary self-sharpening effects needed to produce the correct behavior of the solution in viscous shock layers.

We end the chapter by pointing out that there are a number of numerical approaches for degenerate convection-diffusion equations that are not based on operator splitting, see [6, 32, 44, 46, 51, 60, 96, 98, 99, 102, 111, 147, 164, 207, 216, 217].

5

Error Estimates for Hyperbolic Problems

In this chapter we develop an abstract error estimation theory for dimensional or source splitting methods in the context of weakly coupled systems of hyperbolic equations. To verify convergence of a specific splitting method one only has to check whether each split solver satisfies certain assumptions, whereupon convergence follows. More precisely, we will demonstrate how the approach introduced in Chapter 3 can be extended to yield not only convergence of splitting methods, but also precise error estimates, for weakly coupled systems of the form

$$u_t^\kappa + \sum_i f_i^\kappa (u^\kappa)_{x_i} = g^\kappa (U), \quad u^\kappa|_{t=0} = u_0^\kappa, \quad \kappa = 1, \dots, K. \tag{5.1}$$

Our theory of error estimates has its origin in the approximation theory of Kuznetsov [166, 167], which is founded on the work by Kružkov [161]. The influential work of Kuznetsov has been used and extended in a number of works, see for example [33, 64, 65, 139, 166, 167, 170, 177, 182, 194–196, 234, 253, 254, 257]. Relevant to the presentation that follows, in [33] the authors have formalized the method of Kružkov and Kuznetsov, so that the celebrated doubling of variables (see, e.g., [126, 161]) can be avoided when checking the convergence rate of a specific numerical scheme. We will present a version of the result from [33] adapted to dimensional splitting methods; when applied in a specific situation, this result avoids Kružkov's ingenious but cumbersome doubling of variables.

5.1 Multi-dimensional scalar conservation laws

We first turn our attention to error estimates for dimensional splitting methods in the case of a scalar conservation law

$$u_t + \sum_i f_i(u)_{x_i} = 0, \qquad u|_{t=0} = u_0. \tag{5.2}$$

We consider initial data u_0 in the class $L^1(\mathbb{R}^d) \cap L^\infty(\mathbb{R}^d)$. We shall assume that the exact entropy solution $u(x,t)$ is a measurable function in the same space as the initial data for each $t \leq T$ and that it satisfies

$$
\begin{aligned}
&\int_{\mathcal{B}_R(x_0)} |u(x+h,t) - u(x,t)| \, dx \leq \nu_{R,T}(|h|\,;u), \\
&\int_{\mathcal{B}_R(x_0)} |u(x,t+\tau) - u(x,t)| \, dx \leq \omega_{R,T}(|\tau|\,;u),
\end{aligned}
\tag{5.3}
$$

for some moduli of continuity $\nu_{R,T}(\,\cdot\,;u)$ and $\omega_{R,T}(\,\cdot\,;u)$, and

$$
\mathcal{B}_R(x) = \{\xi \in \mathbb{R}^d \mid |\xi - x| \leq R\}.
$$

In this setting $\ell = d$, $K = 1$, i.e., $U = u$, and $g = 0$. Let $v(t) = \mathcal{S}_t^i v_0$ be an approximate or exact weak solution of the one-dimensional problem

$$
v_t + f_i(v)_{x_i} = 0, \qquad v|_{t=0} = v_0.
$$

Consider the following (dimensional) splitting procedure

$$
\begin{aligned}
u^n &= \left[\mathcal{S}_{\Delta t}^d \circ \cdots \circ \mathcal{S}_{\Delta t}^1\right]^n u_0, \\
u^{n,i} &= \left[\mathcal{S}_{\Delta t}^i \circ \cdots \circ \mathcal{S}_{\Delta t}^1\right] u^n, \quad i = 1, \dots, d, \qquad u^{n,0} = u^n, \\
u_{\Delta t}(t) &= \mathcal{S}_{d(t - t_{n,i-1})}^i u^{n,i-1}, \qquad t \in \left[t_{n,i-1}, t_{n,i}\right),
\end{aligned}
\tag{5.4}
$$

for $n \in \mathbb{N}_0$ and $i = 1, \dots, d$. Assume that our solution operator satisfies (3.54); that is, for $i = 1, \dots, d$,

$$
\eta\left(v^i, k\right)_t + q_i\left(v^i, k\right)_{x_i} \leq \left(\mathcal{E}_{\mathrm{I}}{}^i\right)_t + \left(\mathcal{E}_{\mathrm{II}}{}^i\right)_{x_i} + \left(\mathcal{E}_{\mathrm{III}}{}^i\right)_{x_i x_i} + \mathcal{E}_{\mathrm{IV}}{}^i \text{ in } \mathcal{D}'(\Pi_T), \tag{5.5}
$$

where $\mathcal{E}_{\mathrm{I}}{}^i = \mathcal{E}_{\mathrm{I}}{}^i(x,t)$, $\mathcal{E}_{\mathrm{II}}{}^i = \mathcal{E}_{\mathrm{II}}{}^i(x,t)$, $\mathcal{E}_{\mathrm{III}}{}^i = \mathcal{E}_{\mathrm{III}}{}^i(x,t)$, $\mathcal{E}_{\mathrm{IV}}{}^i = \mathcal{E}_{\mathrm{IV}}{}^i(x,t)$ are distributions (cf. the notation of Chapter 3), and (η, q_i) is the Kružkov entropy pair

$$
\eta(u,k) = |u-k|\,, \quad q_i(u,k) = \operatorname{sign}(u-k)\,(f_i(u) - f_i(k))\,, \quad q = (q_1, \dots, q_d),
$$

where k is a constant. The distributions $\mathcal{E}_{\mathrm{I}i}$, $\mathcal{E}_{\mathrm{II}i}$, $\mathcal{E}_{\mathrm{III}i}$ and $\mathcal{E}_{\mathrm{IV}i}$ may depend on k, but we assume that they are dominated by nonnegative Radon measures that do not, i.e.,

$$
\left|\mathcal{E}_{\mathrm{I}}{}^i\right| \leq d\mu_{\mathrm{I}}{}^i, \quad \left|\mathcal{E}_{\mathrm{II}}{}^i\right| \leq d\mu_{\mathrm{II}}{}^i, \quad \left|\mathcal{E}_{\mathrm{III}}{}^i\right| \leq d\mu_{\mathrm{III}}{}^i, \qquad \left|\mathcal{E}_{\mathrm{IV}}{}^i\right| \leq d\mu_{\mathrm{IV}}{}^i. \tag{5.6}
$$

For ease of notation, we set

$$
\begin{aligned}
d\mu_{\mathrm{I}} &= \sum_i \chi_i^d d\mu_{\mathrm{I}}{}^i, & d\mu_{\mathrm{II}} &= \sum_i \chi_i^d d\mu_{\mathrm{II}}{}^i, \\
d\mu_{\mathrm{III}} &= \sum_i \chi_i^d d\mu_{\mathrm{III}}{}^i, & d\mu_{\mathrm{IV}} &= \sum_i \chi_i^d d\mu_{\mathrm{IV}}{}^i,
\end{aligned}
$$

where the characteristic functions χ_i^d are defined in (3.57). Furthermore, we assume that the approximate solution operators $\mathcal{S}_t^i$ are such that the splitting solution $u_{\Delta t}$ satisfies:

$$\begin{aligned} \|u_{\Delta t}(\,\cdot\,,t)\|_{L^\infty(\mathbb{R}^d)} &\le C,\\ \int_{\mathcal{B}_R(x_0)} |u_{\Delta t}(x+h,t)-u_{\Delta t}(x,t)|\,dx &\le \nu_{R,T}\left(|h|\,;u_{\Delta t}\right),\\ \int_{\mathcal{B}_R(x_0)} |u_{\Delta t}(x,t+\tau)-u_{\Delta t}(x,t)|\,dx &\le \omega_{R,T}\left(|\tau|\,;u_{\Delta t}\right), \end{aligned} \tag{5.7}$$

for some constant C not depending on Δt and for some moduli of continuity $\nu_{R,T}\left(\,\cdot\,;u_{\Delta t}\right)$ and $\omega_{R,T}\left(\,\cdot\,;u_{\Delta t}\right)$ not depending on Δt or t.

Theorem 5.1. *Let u be an exact weak entropy solution of* (5.2), *and let $u_{\Delta t}$ be the approximate solution defined by* (5.4). *Let $M \ge \max_i \|f_i\|_{\mathrm{Lip}}$ and let R be a positive constant, then for all sufficiently small positive constants ρ, h, and τ, we have*

$$\begin{aligned} \|u_{\Delta t}(\,\cdot\,,T)-u(\,\cdot\,,T)\|_{L^1(\mathcal{B}_R(x_0))} &\le \|u_{\Delta t}(\,\cdot\,,0)-u(\,\cdot\,,0)\|_{L^1(\mathcal{B}_{R+MT}(x_0))}\\ &\quad + \mathrm{Const}\,\left(\,E^{\mathrm{split}}(\rho,h,\tau)+E^{\mathrm{method}}(\rho,h,\tau)\,\right), \end{aligned} \tag{5.8}$$

where

$$\begin{aligned} E^{\mathrm{split}}(\rho,h,\tau) &= \omega_{R+MT,T}(\tau;u)+\nu_{R+MT,T}(\sqrt{d}\,h;u)\\ &\quad + \Delta t\Big(\omega_{R+MT,T}(\Delta t+\tau;u)+\omega_{R+MT,T}(\Delta t+\tau;u_{\Delta t})\Big)\Big(\frac{1}{h}+\frac{1}{\rho}\Big)\\ &\quad + \Delta t\,\nu_{R+MT,T}\big(\sqrt{d}\,h;u_{\Delta t}\big)\Big(\frac{1}{\rho}+\frac{1}{h}+\frac{1}{\rho^2}+\frac{1}{\rho\tau}+\frac{1}{\rho h}+\frac{1}{h\tau}\Big), \end{aligned} \tag{5.9}$$

$$\begin{aligned} E^{\mathrm{method}}(\rho,h,\tau) &= \Big(1+\frac{MT}{\rho}\Big)\sup_{t\in[0,T]}\int_{\mathcal{B}_{R+MT}(x_0)} d\mu_{I}(\,\cdot\,,t)\\ &\quad + \iint_{\tilde{\mathcal{B}}(x,t)} \Big(\frac{1}{\tau}d\mu_{I}+\Big(\frac{1}{h}+\frac{1}{\rho}\Big)d\mu_{II}+\Big(\frac{1}{h^2}+\frac{1}{h\rho}+\frac{1}{\rho^2}\Big)d\mu_{III}+d\mu_{IV}\Big), \end{aligned} \tag{5.10}$$

and $\tilde{\mathcal{B}}(x,t)$ denotes the set $\{(x,t) \mid |x-x_0| \le R+M(T-t),\, 0\le t\le T\}$. Furthermore, we also have

$$\begin{aligned} \|u_{\Delta t}(\,\cdot\,,T)-u(\,\cdot\,,T)\|_{L^1(\mathbb{R}^d)} &\le \|u_{\Delta t}(\,\cdot\,,0)-u(\,\cdot\,,0)\|_{L^1(\mathbb{R}^d)}\\ &\quad + \mathrm{Const}\,\left(\,E_{\mathrm{g}}^{\mathrm{split}}(h,\tau)+E_{\mathrm{g}}^{\mathrm{method}}(h,\tau)\,\right), \end{aligned} \tag{5.11}$$

where

$$E_{\mathrm{g}}^{\mathrm{method}}(h,\tau) = \sup_{t\in[0,T]} \int_{\mathbb{R}^d} d\mu_I(\,\cdot\,,t) + \iint\limits_{\Pi_T} \Big(\frac{1}{\tau} d\mu_I + \frac{1}{h} d\mu_{II} + \frac{1}{h^2} d\mu_{III} + d\mu_{IV}\Big), \tag{5.12}$$

and

$$\begin{aligned} E_{\mathrm{g}}^{\mathrm{split}}(h,\tau) = \omega_T(\tau;u) + \nu_T(\sqrt{d}\,h;u) + \omega_T(\tau;u_{\Delta t}) + \nu_T(\sqrt{d}\,h;u_{\Delta t}) \\ + \Delta t \left[\omega_T(\Delta t+\tau;u) + \omega_T\left(\Delta t+\tau;u_{\Delta t}\right)\right] \frac{1}{h} \\ + \Delta t\, \nu_T\left(\sqrt{d}\,h;u_{\Delta t}\right)\left(\frac{1}{h}+\frac{1}{h\tau}\right). \end{aligned} \tag{5.13}$$

Proof. We prove only (5.8) here. The proof of (5.11) is easier, and follows by choosing a simpler test function. The essential ingredients in this proof are found in Section 5.2 where we use this type of test function to show error estimates for balance laws. Now for the proof of (5.8), let

$$\Pi_{T,\varepsilon} = \mathbb{R}^d \times [-\varepsilon, T+\varepsilon]$$

and extend $u_{\Delta t}$ to $\Pi_{T,\varepsilon}$ by setting it equal to $u_{\Delta t}(\,\cdot\,,0)$ for $t<0$ and $u_{\Delta t}(T)$ for $t>T$; and similarly for u. Lemma 3.15 says that

$$-\iint\limits_{\Pi_{T,\varepsilon}} \Big(\eta\left(u_{\Delta t},k\right)\phi_t + q\left(u_{\Delta t},k\right)\cdot\nabla_x\phi\Big)\,dt\,dx \le \mathcal{E}\left(\Delta t;\phi\right) + I_1, \tag{5.14}$$

with

$$\begin{aligned} \mathcal{E}\left(\Delta t;\phi\right) = \sum_{n,i} \int\limits_{\mathbb{R}^d} \left|\phi\left(\,\cdot\,,t_{n,i}\right)\right| d\mu_{\mathrm{I}i}(\,\cdot\,,t_{n,i}) \\ + d\iint\limits_{\Pi_{T,\varepsilon}} \sum_i \chi_i^d\Big(\frac{1}{d}\left|\phi_t\right| d\mu_{\mathrm{I}i} + \left|\phi_{x_i}\right| d\mu_{\mathrm{II}i} + \left|\phi_{x_ix_i}\right| d\mu_{\mathrm{III}i} + \left|\phi\right| d\mu_{\mathrm{IV}i}\Big) \end{aligned} \tag{5.15}$$

$$I_1 = d\sum_i \iint\limits_{\Pi_{T,\varepsilon}} \Big(\chi_i^d - \frac{1}{d}\Big) q_i\left(u_{\Delta t},k\right)\phi_{x_i}\,dt\,dx. \tag{5.16}$$

Let the nonnegative test function ϕ be a function of two additional variables (y,s), that is, $\phi=\phi(x,t,y,s)$ with $\phi\in C_0^\infty((\mathbb{R}^d\times[-\varepsilon,\infty))^2)$. Fixing $(y,s)\in\mathbb{R}^d\times(-\varepsilon,\infty)$, we take $k=u(y,s)$. Integrating the entropy estimate (5.14) with respect to y,s over $\Pi_{T,\varepsilon}$ gives

$$\begin{aligned} -\iiiint\limits_{\Pi_{T,\varepsilon}\times\Pi_{T,\varepsilon}} \left[\eta\big(u_{\Delta t}(x,t),u(y,s)\big)\phi_t + q\big(u_{\Delta t}(x,t),u(y,s)\big)\cdot\nabla_x\phi\right] dt\,dx\,ds\,dy \\ \le R^s + R^\alpha, \end{aligned} \tag{5.17}$$

where the two error terms R^s and R^α are given as

$$
\begin{aligned}
R^s &= \iint\limits_{\Pi_{T,\varepsilon}} I_1 \, ds \, dy \\
&= d \sum_i \iiiint\limits_{\Pi_{T,\varepsilon} \times \Pi_{T,\varepsilon}} \Big(\chi_i^d(x,t) - \frac{1}{d} \Big) q_i \left(u_{\Delta t}(x,t), u(y,s) \right) \phi_{x_i} \, dt \, dx \, ds \, dy, \qquad (5.18) \\
R^\alpha &= \iint\limits_{\Pi_{T,\varepsilon}} \mathcal{E} \, ds \, dy \\
&= d \sum_{n,i} \iiiint\limits_{\Pi_{T,\varepsilon} \times \Pi_{T,\varepsilon}} \chi_{n,i}(x,t) \Big(\frac{1}{d} \left| \phi_t \right| d\mu_{\mathrm{I}i}(x,t) \\
&\qquad + \left| \phi_{x_i} \right| d\mu_{\mathrm{II}i}(x,t) + \left| \phi_{x_i x_i} \right| d\mu_{\mathrm{III}i}(x,t) + \left| \phi \right| d\mu_{\mathrm{IV}i}(x,t) \Big) \, ds \, dy \\
&\qquad + \sum_{n,i} \iiint\limits_{\Pi_{T,\varepsilon} \times \mathbb{R}^d} \left| \phi \left(x, t_{n,i}, y, s \right) \right| d\mu_{\mathrm{I}i}(x, t_{n,i}) \, ds \, dy. \qquad (5.19)
\end{aligned}
$$

Similarly, for each fixed $(x,t) \in \mathbb{R}^d \times (-\varepsilon, \infty)$, we take $k = u_{\Delta t}(x,y)$ in the entropy inequality for u. Integrating this with respect to x, t gives the inequality:

$$
- \iiiint\limits_{\Pi_{T,\varepsilon} \times \Pi_{T,\varepsilon}} \Big(\eta \left(u(y,s), u_{\Delta t}(x,t) \right) \phi_s + q \left(u(y,s), u_{\Delta t}(x,t) \right) \cdot \nabla_y \phi \Big) \, dt \, dx \, ds \, dy \leq 0. \qquad (5.20)
$$

By adding (5.17) and (5.20), we get

$$
\begin{aligned}
- \iiiint\limits_{\Pi_{T,\varepsilon} \times \Pi_{T,\varepsilon}} \Big(&\eta \left(u_{\Delta t}(x,t), u(y,s) \right) \left(\phi_t + \phi_s \right) \\
&+ q \left(u_{\Delta t}(x,t), u(y,s) \right) \cdot \left(\nabla_x \phi + \nabla_y \phi \right) \Big) \, dt \, dx \, ds \, dy \leq R^s + R^\alpha.
\end{aligned}
$$

Next specify the test function ϕ. Namely, for any $\varepsilon, \rho, h, \tau > 0$, we take ϕ to be of the form

$$
\begin{aligned}
\phi(x,t,y,s) &= \varphi_{\varepsilon,\rho}(x,t) \, \Omega_{h,\tau}(x-y, t-s), \\
\Omega_{h,\tau}(x-y, t-s) &= \delta_h(x-y) \, \tilde{\delta}_\tau(t-s),
\end{aligned} \qquad (5.21)
$$

where δ_h and $\tilde{\delta}_\tau$ are regularizing sequences (also denoted mollifiers) given by $\delta_h(x) = h^{-d} \delta(x_1/h) \cdots \delta(x_d/h)$ and $\tilde{\delta}_\tau(t) = \tau^{-1} \tilde{\delta}(t/\tau)$ for even and nonnegative C_0^∞ functions δ and $\tilde{\delta}$ with support in $(-1,1)$ and unit integral on the real line. Next, we choose $\varphi_{\varepsilon,\rho}$ to be a sequence of the form

$$
\begin{aligned}
\varphi_{\varepsilon,\rho}(x,t) &= \Psi_\rho(x,t) \, \psi_\varepsilon(t), \\
\Psi_\rho(x,t) &= 1 - \beta_\rho \left(|x - x_0| - (R + M(T-t)) \right), \\
\psi_\varepsilon(t) &= \beta_\varepsilon(t) - \beta_\varepsilon(t-T),
\end{aligned}
$$

where $\beta_\varepsilon(\lambda) = \int_{-\infty}^{\lambda} \delta_\varepsilon(s)\,ds$ for ε positive. Note that β_ε tends to the characteristic function of $[0,\infty)$ and $\psi_\varepsilon \in C^\infty((-\varepsilon, T+\varepsilon))$ with $0 \le \psi_\varepsilon \le 1$. We also have that Ψ_ρ is a smooth approximation to the characteristic function of the set

$$\big\{(x,t) \mid |x-x_0| \le R + M(T-t)\big\}.$$

It is easy to see that $\varphi_{\varepsilon,\rho} \in C_0^\infty(\mathbb{R}^d \times (-\varepsilon,\infty))$; $0 \le \varphi_{\varepsilon,\rho} \le 1$; $\varphi_{\varepsilon,\rho} = 1$ for $|x-x_0| \le R+M(T-t)-\rho$, $0 \le t \le T$; and $\varphi_{\varepsilon,\rho} = 0$ for $|x-x_0| \ge R+M(T-t)+\rho$ or $t \notin [-\varepsilon, T+\varepsilon]$. In other words, $\varphi_{\varepsilon,\rho}$ is an approximation to the characteristic function of the set $\tilde{\mathcal{B}}(x,t)$ in (5.10). Also

$$\chi_{D_-}(x,t) \le \Psi_\rho(x,t) \le \chi_{D_+}(x,t),$$

where

$$D_\pm = \{(x,t) \mid |x-x_0| \le R + M(T-t) \pm \rho\}.$$

Clearly

$$\phi_t + \phi_s = (\varphi_{\varepsilon,\rho})_t, \quad \nabla_x\phi + \nabla_y\phi = \nabla_x\varphi_{\varepsilon,\rho}.$$

Regarding derivatives, we have

$$\begin{aligned} \varphi_{\varepsilon,\rho}(x,t)_t &= \psi_\varepsilon'(t)\Psi_\rho(x,t) - M\psi_\varepsilon(t)\delta_\rho\big(|x-x_0| - R - M(T-t)\big), \\ \nabla_x\varphi_{\varepsilon,\rho}(x,t) &= -\frac{(x-x_0)}{|x-x_0|}\psi_\varepsilon(t)\delta_\rho\big(|x-x_0| - R - M(T-t)\big). \end{aligned}$$

Therefore, if $M \ge \max_i \|f_i\|_{\mathrm{Lip}}$,

$$\begin{aligned} &\eta\big(u_{\Delta t}(x,t), u(y,s)\big)\varphi_{\varepsilon,\rho}(x,t)_t + q\left(u_{\Delta t}(x,t), u(y,s)\right)\cdot\nabla_x\varphi_{\varepsilon,\rho}(x,t) \\ &\quad = \eta\big(u_{\Delta t}(x,t), u(y,s)\big)\psi_\varepsilon'(t)\Psi_\rho(x,t) - \psi_\varepsilon(t)\delta_\rho\big(|x-x_0| - R - M(T-t)\big) \\ &\qquad \times \Big(M\eta\left(u_{\Delta t}(x,t), u(y,s)\right) + \frac{1}{|x-x_0|}q\left(u_{\Delta t}(x,t), u(y,s)\right)\cdot(x-x_0)\Big) \\ &\quad \le \eta\big(u_{\Delta t}(x,t), u(y,s)\big)\psi_\varepsilon'(t)\Psi_\rho(x,t). \end{aligned}$$

It now follows that

$$-\iiiint\limits_{\Pi_{T,\varepsilon}\times\Pi_{T,\varepsilon}} \eta\left(u_{\Delta t}(x,t), u(y,s)\right)\psi_\varepsilon'(t)\Psi_\rho(x,t)\Omega_{h,\tau}(x-y,t-s)\,dt\,dx\,ds\,dy \le R^s + R^\alpha.$$

Writing

$$\begin{aligned} |u_{\Delta t}(x,t) - u(y,s)|\,\psi_\varepsilon'(t) &= |u_{\Delta t}(x,t) - u(x,t)|\,\psi_\varepsilon'(t) \\ &\quad + \big(\,|u_{\Delta t}(x,t) - u(y,s)| - |u_{\Delta t}(x,t) - u(x,t)|\,\big)\psi_\varepsilon'(t), \end{aligned}$$

and estimating the latter term using the triangle inequality, we get the key inequality

$$0 \le I + R^t + R^x + R^s + R^\alpha, \tag{5.22}$$

where

$$I = \iiiint\limits_{\Pi_{T,\varepsilon}\times\Pi_{T,\varepsilon}} \eta\left(u_{\Delta t}(x,t), u(x,t)\right) \psi_\varepsilon'(t)\Psi_\rho(x,t)\Omega_{h,\tau}(x-y,t-s)\,dt\,dx\,ds\,dy,$$

$$R^t = \iiiint\limits_{\Pi_{T,\varepsilon}\times\Pi_{T,\varepsilon}} \eta\left(u(y,t), u(y,s)\right) \left|\psi_\varepsilon'(t)\right| \Psi_\rho(x,t)\Omega_{h,\tau}(x-y,t-s)\,dt\,dx\,ds\,dy,$$

$$R^x = \iiiint\limits_{\Pi_{T,\varepsilon}\times\Pi_{T,\varepsilon}} \eta\left(u(x,t), u(y,t)\right) \left|\psi_\varepsilon'(t)\right| \Psi_\rho(x,t)\Omega_{h,\tau}(x-y,t-s)\,dt\,dx\,ds\,dy.$$

We need to estimate each term in (5.22). Consider first $|R^t|$. In this case we find that

$$\begin{aligned}
\left|R^t\right| \leq \iint\limits_{\Pi_{T,\varepsilon}} \int\limits_{-\varepsilon}^{T+\varepsilon} \int\limits_{\mathcal{B}_{R+M(T-t)+\rho}(x_0)} &|u(y,t)-u(y,s)| \\
&\times (\delta_\varepsilon(t)+\delta_\varepsilon(t-T))\delta_h(x-y)\tilde{\delta}_\tau(t-s)\,dt\,dx\,ds\,dy \\
\leq \int\limits_{-\varepsilon}^{\varepsilon} \iint\limits_{\Pi_{T,\varepsilon}} \int\limits_{\mathcal{B}_{R+M(T-t)+\rho}(x_0)} &|u(y,t)-u(y,s)|\,\delta_\varepsilon(t)\delta_h(x-y)\tilde{\delta}_\tau(t-s)\,dx\,ds\,dy\,dt \\
+ \int\limits_{T-\varepsilon}^{T+\varepsilon} \iint\limits_{\Pi_{T,\varepsilon}} \int\limits_{\mathcal{B}_{R+M(T-t)+\rho}(x_0)} &|u(y,t)-u(y,s)| \\
&\times \delta_\varepsilon(t-T)\delta_h(x-y)\tilde{\delta}_\tau(t-s)\,dx\,ds\,dy\,dt.
\end{aligned}$$

The first term can be estimated by

$$\begin{aligned}
\limsup_{\varepsilon\to 0} \int_{-\varepsilon}^{\varepsilon} \int_{-\tau-\varepsilon}^{\tau+\varepsilon} \Bigg[\delta_\varepsilon(t)\tilde{\delta}_\tau(t-s)\Big(\int_{\mathcal{B}_{R+M(T-t)+\sqrt{d}\,h+\rho}(x_0)} |u(y,t)-u(y,s)|\Big) \\
\times \Big(\int_{\mathcal{B}_{R+M(T-t)+\rho}(x_0)} \delta_h(x-y)\,dx\Big)\,ds\Bigg]\,dy\,dt \\
\leq \omega_{R+MT+\sqrt{d}\,h+\rho,T}(\tau;u).
\end{aligned}$$

A similar calculation for the second term yields for the sum

$$\limsup_{\varepsilon\to 0}\left|R^t\right| \leq 2\,\omega_{R+MT+\sqrt{d}\,h+\rho,T}(\tau;u). \tag{5.23}$$

From an analogous estimate we obtain

$$\limsup_{\varepsilon\to 0}|R^x| \leq 2\,\nu_{R+MT+\sqrt{d}\,h+\rho,T}\left(\sqrt{d}\,h;u\right). \tag{5.24}$$

The term I does not contain the absolute value of $\psi_\varepsilon'(t)$. Thus

$$\begin{aligned} I &= \iiiint\limits_{\Pi_{T,\varepsilon}\times\Pi_{T,\varepsilon}} \eta\left(u_{\Delta t}(x,t),u(x,t)\right)\left(\delta_\varepsilon(t)-\delta_\varepsilon(t-T)\right) \\ &\qquad\qquad \times \Psi_\rho(x,t)\Omega_{h,\tau}(x-y,t-s)\,dt\,dx\,ds\,dy \\ &\le \iint\limits_{\Pi_{T,\varepsilon}} \int_{-\varepsilon}^{\varepsilon}\int_{\mathcal{B}_{R+M(T-t)+\rho}(x_0)} \eta\left(u_{\Delta t}(x,t),u(x,t)\right)\delta_\varepsilon(t)\delta_h(x-y)\tilde{\delta}_\tau(t-s)\,dt\,dx\,ds\,dy \\ &\quad - \iint\limits_{\Pi_{T,\varepsilon}} \int_{T-\varepsilon}^{T+\varepsilon}\int_{\mathcal{B}_{R+M(T-t)-\rho}(x_0)} \eta\left(u_{\Delta t}(x,t),u(x,t)\right) \\ &\qquad\qquad \times \delta_\varepsilon(t-T)\delta_h(x-y)\tilde{\delta}_\tau(t-s)\,dt\,dx\,ds\,dy \\ &\le \int_{-\varepsilon}^{\varepsilon}\int_{|t-s|\le\tau} \delta_\varepsilon(t)\tilde{\delta}_\tau(t-s)\int_{\mathcal{B}_{R+M(T-t)+\rho}(x_0)} \eta\left(u_{\Delta t}(x,t),u(x,t)\right)\,dx\,ds\,dt \\ &\quad - \int_{T-\varepsilon}^{T+\varepsilon}\int_{|t-s|\le\tau} \delta_\varepsilon(t-T)\tilde{\delta}_\tau(t-s) \\ &\qquad\qquad \times \int_{\mathcal{B}_{R+M(T-t)-\rho}(x_0)} \eta\left(u_{\Delta t}(x,t),u(x,t)\right)\,dx\,ds\,dt. \end{aligned}$$

Thus we conclude

$$\begin{aligned} \limsup_{\varepsilon\to 0} I \le \int_{\mathcal{B}_{R+MT+\rho}(x_0)} &\left|u_{\Delta t}(x,0)-u(x,0)\right|\,dx \\ &- \int_{\mathcal{B}_{R-\rho}(x_0)} \left|u_{\Delta t}(x,T)-u(x,T)\right|\,dx. \end{aligned} \tag{5.25}$$

We then estimate the term R^α. In this case we have that (all test functions are nonnegative, their derivatives are not)

$$\begin{aligned} |R^\alpha| \le d\sum_{n,i} \iiiint\limits_{\Pi_{T,\varepsilon}\times\Pi_{n,i}} &\Big[\frac{1}{d}\big(\left|(\varphi_{\varepsilon,\rho})_t\right|\Omega_{h,\tau}+\varphi_{\varepsilon,\rho}\left|(\Omega_{h,\tau})_t\right|\big)d\mu_{\mathrm{I}}{}^i(x,t) \\ &+ \Big(\left|(\varphi_{\varepsilon,\rho})_{x_i}\right|\Omega_{h,\tau}+\varphi_{\varepsilon,\rho}\left|(\Omega_{h,\tau})_{x_i}\right|\Big)d\mu_{\mathrm{II}}{}^i(x,t) \\ + \Big(\left|(\varphi_{\varepsilon,\rho})_{x_ix_i}\right|\Omega_{h,\tau} &+ 2\left|(\varphi_{\varepsilon,\rho})_{x_i}\right|\left|(\Omega_{h,\tau})_{x_i}\right| + \varphi_{\varepsilon,\rho}\left|(\Omega_{h,\tau})_{x_ix_i}\right|\Big)d\mu_{\mathrm{III}}{}^i(x,t) \\ &+ \varphi_{\varepsilon,\rho}\Omega_{h,\tau}d\mu_{\mathrm{IV}}{}^i(x,t)\Big]\,ds\,dy. \end{aligned}$$

Regarding the individual terms in this integral we first find that

$$\iiiint\limits_{\Pi_{T,\varepsilon}\times\Pi_{n,i}} \varphi_{\varepsilon,\rho}\Omega_{h,\tau}d\mu_{\mathrm{IV}}{}^i(x,t)\,ds\,dy = \iint\limits_{\Pi_{n,i}} \varphi_{\varepsilon,\rho}(x,t)d\mu_{\mathrm{IV}}{}^i(x,t) \le \iint\limits_{\Pi_{n,i}\cap D_+} d\mu_{\mathrm{IV}}{}^i.$$

Furthermore,

$$\iiiint\limits_{\Pi_{T,\varepsilon}\times\Pi_{n,i}} \varphi_{\varepsilon,\rho}\big|(\Omega_{h,\tau})_{x_i}\big|\,d\mu_{\mathrm{I}}{}^i(x,t)\,ds\,dy \le \frac{1}{h^2}\iint\limits_{\Pi_{n,i}\cap D_+}\Big(\int\limits_{\mathcal{B}_{\sqrt{d}\,h}(x)} dy_i\Big)\,d\mu_{\mathrm{I}}{}^i(x,t) \le \frac{\mathrm{Const}}{h}\iint\limits_{\Pi_{n,i}\cap D_+} d\mu_{\mathrm{I}}{}^i.$$

Considering the term

$$\big|(\varphi_{\varepsilon,\rho})_t\big|\,\Omega_{h,\tau}d\mu_{\mathrm{I}}{}^i = \Omega_{h,\tau}\,|\psi_\varepsilon'(t)\Psi_\rho(x,t) - M\psi_\varepsilon(t)\delta_\rho(p(x,t))|\,d\mu_{\mathrm{I}}{}^i,$$

where $p(x,t) = |x-x_0| - (R+M(T-t))$, we find

$$\begin{aligned}\iiiint\limits_{\Pi_{T,\varepsilon}\times\Pi_{n,i}} &\big|(\varphi_{\varepsilon,\rho})_t\big|\Omega_{h,\tau}d\mu_{\mathrm{I}}{}^i\,ds\,dy \\ &\le \iint\limits_{\Pi_{n,i}\cap D_+}\big(\delta_\varepsilon(t)+\delta_\varepsilon(t-T)+M\delta_\rho(p(x,t))\big)d\mu_{\mathrm{I}}{}^i(x,t) \\ &\le \int_{t_{n,i-1}}^{t_{n,i}}\Big(\delta_\varepsilon(t)+\delta_\varepsilon(t-T)+\frac{M}{\rho}\Big)\,dt \sup_{t\in[t_{n,i-1},t_{n,i}]}\int_{\mathcal{B}_{R+MT+\rho}(x_0)} \chi_i^d\,d\mu_{\mathrm{I}}{}^i(\,\cdot\,,t).\end{aligned}$$

Summing over n and i we find

$$\sum_{n,i}\iiiint\limits_{\Pi_{T,\varepsilon}\times\Pi_{n,i}} \big|(\varphi_{\varepsilon,\rho})_t\big|\Omega_{h,\tau}d\mu_{\mathrm{I}}{}^i\,ds\,dy \le \Big(2+\frac{MT}{\rho}\Big)\sup_{t\in[0,T]}\int_{\mathcal{B}_{R+MT+\rho}(x_0)} d\mu_{\mathrm{I}}(\,\cdot\,,t).$$

By similar arguments we find that

$$\begin{aligned}\iiiint\limits_{\Pi_{T,\varepsilon}\times\Pi_{n,i}} \varphi_{\varepsilon,\rho}(x,t)\big|\Omega_{h,\tau}(x-y,t-s)_t\big|d\mu_{\mathrm{I}i}(x,t)\,ds\,dy &\le \frac{\mathrm{Const}}{\tau}\iint\limits_{\Pi_{n,i}\cap D_+} d\mu_{\mathrm{I}i},\\ \iiiint\limits_{\Pi_{T,\varepsilon}\times\Pi_{n,i}} \big|(\varphi_{\varepsilon,\rho})_{x_i}\big|\Omega_{h,\tau}(x-y,t-s)d\mu_{\mathrm{II}}{}^i(x,t)\,ds\,dy &\le \frac{\mathrm{Const}}{\rho}\iint\limits_{\Pi_{n,i}\cap D_+} d\mu_{\mathrm{II}}{}^i,\\ \iiiint\limits_{\Pi_{T,\varepsilon}\times\Pi_{n,i}} \varphi_{\varepsilon,\rho}(x,t)\big|\Omega_{h,\tau}(x-y,t-s)_{x_ix_i}\big|d\mu_{\mathrm{III}}{}^i(x,t)\,ds\,dy &\le \frac{\mathrm{Const}}{h^2}\iint\limits_{\Pi_{n,i}\cap D_+} d\mu_{\mathrm{III}}{}^i,\\ \iiiint\limits_{\Pi_{T,\varepsilon}\times\Pi_{n,i}} \big|\varphi_{\varepsilon,\rho}(x,t)_{x_i}\big|\,\big|\Omega_{h,\tau}(x-y,t-s)_{x_i}\big|d\mu_{\mathrm{III}}{}^i(x,t)\,ds\,dy &\le \frac{\mathrm{Const}}{h\rho}\iint\limits_{\Pi_{n,i}\cap D_+} d\mu_{\mathrm{III}}{}^i,\\ \iiiint\limits_{\Pi_{T,\varepsilon}\times\Pi_{n,i}} \big|\varphi_{\varepsilon,\rho}(x,t)_{x_ix_i}\big|\Omega_{h,\tau}(x-y,t-s)d\mu_{\mathrm{III}}{}^i(x,t)\,ds\,dy &\le \frac{\mathrm{Const}}{\rho^2}\iint\limits_{\Pi_{n,i}\cap D_+} d\mu_{\mathrm{III}}{}^i.\end{aligned}$$

Hence

$$\limsup_{\varepsilon\to 0} |R^\alpha| \le \text{Const}\left[\left(1+\frac{\max_i \|f_i\|_{\text{Lip}} T}{\rho}\right) \sup_t \int_{\mathcal{B}_{R+MT+\rho}(x_0)} d\mu_{\text{I}}(\,\cdot\,,t) \right.$$
$$\left. + \iint_{D_+} \Big(\frac{1}{\tau} d\mu_{\text{I}} + \big(\frac{1}{h}+\frac{1}{\rho}\big) d\mu_{\text{II}} + \Big(\frac{1}{h^2}+\frac{1}{h\rho}+\frac{1}{\rho^2}\Big) d\mu_{\text{III}} + d\mu_{\text{IV}}\Big)\right]. \quad (5.26)$$

It remains to estimate R^s. We use the same method that we used in the proof of Theorem 3.17. Let $\hat{p}_i(t)$ be given as

$$\hat{p}_i(t) = q_i\left(u_{\Delta t}(x,t), u(y,s)\right)\, \phi(x,y,t,s)_{x_i}.$$

Thus

$$\begin{aligned} R^s &\le \iint_{\Pi_{T,\varepsilon}} \int_{\mathcal{B}_{R+MT+\rho}(x_0)} \sum_{n,i} \Big(d \int_{t_{n,i-1}}^{t_{n,i}} \hat{p}_i(t)\, dt - \int_{t_n}^{t_{n+1}} \hat{p}_i(t) dt\Big)\, dx\, ds\, dy \\ &= \sum_{n,i} \iint_{\Pi_{T,\varepsilon}} \int_{\mathcal{B}_{R+MT+\rho}(x_0)} \Big(d \int_{t_{n,i-1}}^{t_{n,i}} \left(\hat{p}_i(t) - \hat{p}_i\left(t_n\right)\right) dt \\ &\qquad\qquad - \int_{t_n}^{t_{n+1}} \left(\hat{p}_i(t) - \hat{p}_i\left(t_n\right)\right) dt\Big)\, dx\, ds\, dy \\ &= \sum_{n,i} \big(d\, I_{n,i} - I_n\big), \end{aligned}$$

where

$$\begin{aligned} I_{n,i} &= \iint_{\Pi_{T,\varepsilon}} \int_{\mathcal{B}_{R+MT+\rho}(x_0)} \int_{t_{n,i-1}}^{t_{n,i}} \left(\hat{p}_i(t) - \hat{p}_i\left(t_n\right)\right) dt\, dx\, ds\, dy, \\ I_n &= \iint_{\Pi_{T,\varepsilon}} \int_{\mathcal{B}_{R+MT+\rho}(x_0)} \int_{t_n}^{t_{n+1}} \left(\hat{p}_i(t) - \hat{p}_i\left(t_n\right)\right) dt\, dx\, ds\, dy. \end{aligned}$$

We have that

$$\begin{aligned} I_{n,i} = &\iint_{\Pi_{T,\varepsilon}} \int_{\mathcal{B}_{R+MT+\rho}(x_0)} \int_{t_{n,i-1}}^{t_{n,i}} \Big(q_i\big(u_{\Delta t}(x,t), u(y,s)\big) - q_i\big(u_{\Delta t}\left(x,t_n\right), u(y,s)\big)\Big) \\ &\qquad\qquad \times \left(\varphi_{\varepsilon,\rho}\Omega_{h,\tau}\right)_{x_i}(t_n)\, dt\, dx\, ds\, dy \\ &+ \iint_{\Pi_{T,\varepsilon}} \int_{\mathcal{B}_{R+MT+\rho}(x_0)} \int_{t_{n,i-1}}^{t_{n,i}} q_i(u_{\Delta t}(x,t), u(y,s)) \end{aligned}$$

$$\times \Big(\big(\varphi_{\varepsilon,\rho}\Omega_{h,\tau}\big)_{x_i} - \big(\varphi_{\varepsilon,\rho}\Omega_{h,\tau}\big)_{x_i}(t_n) \Big)\, dt\, dx\, ds\, dy$$

$$=: I^1_{n,i} + I^2_{n,i}.$$

Analogously,

$$I_n = I^1_n + I^2_n.$$

Consider the first term, namely

$$\begin{aligned} \left| I^1_{n,i} \right| \le \| f_i \|_{\text{Lip}} \iint\limits_{\Pi_{T,\varepsilon}} \int\limits_{\mathcal{B}_{R+MT+\rho}(x_0)} \int_{t_{n,i-1}}^{t_{n,i}} & |u_{\Delta t}(x,t) - u_{\Delta t}(x,t_n)| \\ & \times \left(|(\varphi_{\varepsilon,\rho})_{x_i}(t_n)|\, \Omega_{h,\tau}(t_n) + \varphi_{\varepsilon,\rho}(t_n)\, |(\Omega_{h,\tau})_{x_i}(t_n)| \right) dt\, dx\, ds\, dy. \end{aligned}$$

Each term can be estimated as before, yielding

$$\begin{aligned} \limsup_{\varepsilon\to 0} \left| I^1_{n,i} \right| &\le \text{Const}\, \frac{\Delta t}{d}\, \omega_{R+MT+\rho,T}\left(\Delta t; u_{\Delta t}\right) \Big(\frac{1}{h} + \frac{1}{\rho} \Big), \\ \limsup_{\varepsilon\to 0} \left| I^1_n \right| &\le \text{Const}\, \Delta t\, \omega_{R+MT+\rho,T}\left(\Delta t; u_{\Delta t}\right) \Big(\frac{1}{h} + \frac{1}{\rho} \Big). \end{aligned} \tag{5.27}$$

To estimate $I^2_{n,i}$, we note that since ϕ is a test function, then

$$\int_{\mathbb{R}^d} q_i\big(u_{\Delta t}\left(y,t_n\right), u\left(y,t_n\right)\big) \phi(x,y,t,s)_{x_i}\, dx = 0.$$

Using this observation to estimate $I^2_{n,i}$, we find that

$$\begin{aligned} I^2_{n,i} &= \iint\limits_{\Pi_{T,\varepsilon}} \int\limits_{\mathcal{B}_{R+MT+\rho}(x_0)} \int_{t_{n,i-1}}^{t_{n,i}} \left[q_i\big(u_{\Delta t}(x,t), u(y,s)\big) - q_i\big(u_{\Delta t}\left(y,t_n\right), u\left(y,t_n\right)\big) \right] \\ &\qquad \times \left((\varphi_{\varepsilon,\rho}\Omega_{h,\tau})_{x_i} - (\varphi_{\varepsilon,\rho}\Omega_{h,\tau})_{x_i}(t_n) \right) dt\, dx\, ds\, dy \\ &= \iint\limits_{\Pi_{T,\varepsilon}} \int\limits_{\mathcal{B}_{R+MT+\rho}(x_0)} \int_{t_{n,i-1}}^{t_{n,i}} \left[q_i\big(u_{\Delta t}(x,t), u(y,s)\big) - q_i\big(u_{\Delta t}(x,t), u\left(y,t_n\right)\big) \right] \\ &\qquad \times \left((\varphi_{\varepsilon,\rho}\Omega_{h,\tau})_{x_i} - (\varphi_{\varepsilon,\rho}\Omega_{h,\tau})_{x_i}(t_n) \right) dt\, dx\, ds\, dy \\ &\quad + \iint\limits_{\Pi_{T,\varepsilon}} \int\limits_{\mathcal{B}_{R+MT+\rho}(x_0)} \int_{t_{n,i-1}}^{t_{n,i}} \left[q_i\big(u_{\Delta t}(x,t), u\left(y,t_n\right)\big) - q_i\big(u_{\Delta t}\left(y,t_n\right), u\left(y,t_n\right)\big) \right] \\ &\qquad \times \left((\varphi_{\varepsilon,\rho}\Omega_{h,\tau})_{x_i} - (\varphi_{\varepsilon,\rho}\Omega_{h,\tau})_{x_i}(t_n) \right) dt\, dx\, ds\, dy \\ &=: I^{2,1}_{n,i} + I^{2,2}_{n,i}, \end{aligned}$$

and similarly

$$I^2_n = I^{2,1}_n + I^{2,2}_n.$$

It follows that

$$\begin{aligned} \limsup_{\varepsilon\to 0} \left|I_{n,i}^{2,1}\right| &\leq \text{Const}\,\frac{\Delta t}{d}\,\omega_{R+MT+\rho+\sqrt{d}\,h,T}\left(\Delta t+\tau;u\right)\left(\frac{1}{h}+\frac{1}{\rho}\right),\\ \limsup_{\varepsilon\to 0} \left|I_{n}^{2,1}\right| &\leq \text{Const}\,\Delta t\,\omega_{R+MT+\rho+\sqrt{d}\,h,T}\left(\Delta t+\tau;u\right)\left(\frac{1}{h}+\frac{1}{\rho}\right). \end{aligned} \tag{5.28}$$

We write

$$\begin{aligned} I_{n,i}^{2,2} &= \iint\limits_{\Pi_{T,\varepsilon}} \int\limits_{\mathcal{B}_{R+MT+\rho}(x_0)} \int_{t_{n,i-1}}^{t_{n,i}} \left[q_i\big(u_{\Delta t}(x,t),u\left(y,t_n\right)\big)-q_i\big(u_{\Delta t}\left(x,t_n\right),u\left(y,t_n\right)\big)\right]\\ &\qquad\qquad \times \left(\left(\varphi_{\varepsilon,\rho}\Omega_{h,\tau}\right)_{x_i}-\left(\varphi_{\varepsilon,\rho}\Omega_{h,\tau}\right)_{x_i}(t_n)\right)\,dt\,dx\,ds\,dy\\ &\quad+\iint\limits_{\Pi_{T,\varepsilon}} \int\limits_{\mathcal{B}_{R+MT+\rho}(x_0)} \int_{t_{n,i-1}}^{t_{n,i}} \Big[q_i\big(u_{\Delta t}\left(x,t_n\right),u\left(y,t_n\right)\big)\\ &\qquad\qquad\qquad\qquad\qquad - q_i\big(u_{\Delta t}\left(y,t_n\right),u\left(y,t_n\right)\big)\Big]\\ &\qquad\qquad \times \left(\left(\varphi_{\varepsilon,\rho}\Omega_{h,\tau}\right)_{x_i}-\left(\varphi_{\varepsilon,\rho}\Omega_{h,\tau}\right)_{x_i}(t_n)\right)\,dt\,dx\,ds\,dy\\ &=: I_{n,i}^{2,2,1}+I_{n,i}^{2,2,2}. \end{aligned}$$

We also split $I_n^{2,2}$ in a similar way,

$$I_n^{2,2}=I_n^{2,2,1}+I_n^{2,2,2}.$$

We have already established the bounds

$$\begin{aligned} \limsup_{\varepsilon\to 0} \left|I_{n,i}^{2,2,1}\right| &\leq \text{Const}\,\frac{\Delta t}{d}\,\omega_{R+MT+\rho,T}\left(\Delta t;u_{\Delta t}\right)\left(\frac{1}{\rho}+\frac{1}{h}\right),\\ \limsup_{\varepsilon\to 0} \left|I_{n}^{2,2,1}\right| &\leq \text{Const}\,\Delta t\,\omega_{R+MT+\rho,T}\left(\Delta t;u_{\Delta t}\right)\left(\frac{1}{\rho}+\frac{1}{h}\right). \end{aligned} \tag{5.29}$$

What remains is to estimate

$$\begin{aligned} &I_{n,i}^{2,2,2}-\frac{1}{d}I_n^{2,2,2}\\ &\quad=\iint\limits_{\Pi_{T,\varepsilon}} \int\limits_{\mathcal{B}_{R+MT+\rho}(x_0)} \left[q_i\big(u_{\Delta t}\left(x,t_n\right),u\left(y,t_n\right)\big)-q_i\big(u_{\Delta t}\left(y,t_n\right),u\left(y,t_n\right)\big)\right]\\ &\qquad\quad\times\left[\int_{t_{n,i-1}}^{t_{n,i}} \left(\left(\varphi_{\varepsilon,\rho}\Omega_{h,\tau}\right)_{x_i}-\left(\varphi_{\varepsilon,\rho}\Omega_{h,\tau}\right)_{x_i}(t_n)\right)\,dt\right.\\ &\qquad\qquad\left.-\frac{1}{d}\int_{t_n}^{t_{n+1}} \left(\left(\varphi_{\varepsilon,\rho}\Omega_{h,\tau}\right)_{x_i}-\left(\varphi_{\varepsilon,\rho}\Omega_{h,\tau}\right)_{x_i}(t_n)\right)\,dt\right]\,dx\,ds\,dy\\ &\quad=\iint\limits_{\Pi_{T,\varepsilon}} \int\limits_{\mathcal{B}_{R+MT+\rho}(x_0)} \left[q_i\big(u_{\Delta t}\left(x,t_n\right),u\left(y,t_n\right)\big)-q_i\big(u_{\Delta t}\left(y,t_n\right),u\left(y,t_n\right)\big)\right] \end{aligned}$$

$$\times \left[\int_{t_{n,i-1}}^{t_{n,i}} (\varphi_{\varepsilon,\rho}\Omega_{h,\tau})_{x_i}\, dt - \frac{1}{d}\int_{t_n}^{t_{n+1}} (\varphi_{\varepsilon,\rho}\Omega_{h,\tau})_{x_i}\, dt \right] dx\, ds\, dy.$$

For any differentiable function $h(t)$

$$\begin{aligned} \int_{t_{n,i-1}}^{t_{n,i}} h(t)\, dt - \frac{1}{d}\int_{t_n}^{t_{n+1}} h(t)\, dt &= \frac{1}{d}\int_{t_n}^{t_{n+1}} \left(h((t-t_n)/d + t_{n,i-1}) - h(t) \right) dt \\ &= \frac{1}{d}\int_{t_n}^{t_{n+1}} \int_t^{(t-t_n)/d+t_{n,i-1}} h'(\xi)\, d\xi\, dt. \end{aligned}$$

Thus

$$\begin{aligned} \left| \int_{t_{n,i-1}}^{t_{n,i}} h(t)\, dt - \frac{1}{d}\int_{t_n}^{t_{n+1}} h(t)\, dt \right| &\le \frac{1}{d}\int_{t_n}^{t_{n+1}} \int_{t_n}^{t_{n+1}} |h'(\xi)|\, d\xi\, dt \\ &= \frac{\Delta t}{d}\int_{t_n}^{t_{n+1}} |h'(t)|\, dt. \end{aligned}$$

Using $h(t) = \left(\varphi_{\varepsilon,\rho}\Omega_{h,\tau}\right)_{x_i}$, we find

$$h'(t) = \left(\varphi_{\varepsilon,\rho}\right)_{x_i t}\Omega_{h,\tau} + \left(\varphi_{\varepsilon,\rho}\right)_{x_i}\left(\Omega_{h,\tau}\right)_t + \left(\varphi_{\varepsilon,\rho}\right)_t\left(\Omega_{h,\tau}\right)_{x_i} + \varphi_{\varepsilon,\rho}\left(\Omega_{h,\tau}\right)_{x_i t}.$$

Then also $I_{n,i}^{2,2,2} - \frac{1}{d}I_n^{2,2,2}$ will consist of the corresponding four terms. We now estimate the first of these. Recall that

$$\left(\varphi_{\varepsilon,\rho}\right)_{x_i t}(t) = \psi_\varepsilon(t)\delta_\rho'(p(x,t))M + \left(\delta_\varepsilon(t) - \delta_\varepsilon(t-T)\right)\delta_\rho(p(x,t)).$$

Furthermore

$$0 \le \sum_{n,i}\int_{t_n}^{t_{n+1}} |\delta_\varepsilon(t) - \delta_\varepsilon(t-T)|\, dt \le 2d.$$

Using the bounds

$$\left|\delta_\rho'(p(x,t))\right| \le \frac{\text{Const}}{\rho^2}, \quad \text{and} \quad |\delta_\rho(p(x,t))| \le \frac{\text{Const}}{\rho},$$

we find that

$$\begin{aligned} \limsup_{\varepsilon\to 0} \sum_{n,i} \Bigg| \iint\limits_{\Pi_{T,\varepsilon}} \int\limits_{\mathcal{B}_{R+MT+\rho}(x_0)} &\left[q_i\left(u_{\Delta t}(x,t_n), u(y,t_n)\right) - q_i\left(u_{\Delta t}(y,t_n), u(y,t_n)\right) \right] \\ &\times \frac{1}{d}\int_{t_n}^{t_{n+1}} \int_t^{(t-t_n)/d+t_{n,i-1}} \left(\varphi_{\varepsilon,\rho}\right)_{x_i t}(x,\xi)\Omega_{h,\tau}(x,\xi,y,s)\, d\xi\, dt\, dx\, ds\, dy \Bigg| \\ \le \frac{\Delta t}{d}\sum_{n,i} \iint\limits_{\Pi_{T,\varepsilon}} \int\limits_{\mathcal{B}_{R+MT+\rho}(x_0)} &|u_{\Delta t}(x,t_n) - u_{\Delta t}(y,t_n)| \end{aligned}$$

$$\times \int_{t_n}^{t_{n+1}} \left|(\varphi_{\varepsilon,\rho})_{x_i t}(x,t)\right| \Omega_{h,\tau}(x,t,y,s)\,dt\,dx\,ds\,dy$$
$$\leq \text{Const}\ \Delta t\, \nu_{R+MT+\rho,T}(\sqrt{d}\,h; u_{\Delta t}) \left(\frac{1}{\rho} + \frac{1}{\rho^2}\right).$$

Using analogous arguments for the remaining three terms, we find that

$$\limsup_{\varepsilon\to 0} \sum_{n,i} \left| I_{n,i}^{2,2,2} - \frac{1}{d} I_n^{2,2,2} \right|$$
$$\leq \text{Const}\ \Delta t\, \nu_{R+MT+\rho,T}(\sqrt{d}\,h; u_{\Delta t}) \left(\frac{1}{\rho} + \frac{1}{\rho^2} + \frac{1}{\rho\tau} + \frac{1}{h} + \frac{1}{\rho h} + \frac{1}{h\tau}\right). \quad (5.30)$$

Summing up, we have that

$$\limsup_{\varepsilon\to 0} |R^s| = \limsup_{\varepsilon\to 0} \sum_{n,i} |d\, I_{n,i} - I_n| \quad (5.31)$$
$$\leq \limsup_{\varepsilon\to 0} \sum_{n,i} \Big(d\big(|I_{n,i}^1| + \left|I_{n,i}^{2,1}\right| + \left|I_{n,i}^{2,2,1}\right| \big)$$
$$+ \big(|I_n^1| + |I_n^{2,1}| + |I_n^{2,2,1}| \big) + \left| d\, I_{n,i}^{2,2,2} - I_n^{2,2,2} \right| \Big)$$
$$\leq \text{Const} \Bigg[\Big(\omega_{R+MT+\rho+\sqrt{d}\,h,T}(\Delta t + \tau; u) + \omega_{R+MT+\rho+\sqrt{d}\,h,T}(\Delta t + \tau; u_{\Delta t}) \Big)$$
$$\times \left(\frac{1}{h} + \frac{1}{\rho}\right)$$
$$+ \Delta t\, \nu_{R+MT+\rho,T}\left(\sqrt{d}\,h; u_{\Delta t}\right) \left(\frac{1}{\rho} + \frac{1}{\rho^2} + \frac{1}{\rho\tau} + \frac{1}{h} + \frac{1}{\rho h} + \frac{1}{h\tau}\right)\Bigg].$$

Combining (5.24), (5.25), and (5.26), and using the continuity of the norms to absorb small constants into one R, finishes the proof of the theorem. □

Remark 5.2. If we set $\mathcal{S}^i$ to be the exact solution operators, the error term E^{method} vanishes. Furthermore, if u_0 is of bounded variation, then

$$\nu_{R,T}(h; u_{\Delta t}) \leq \|u_0\|_{\mathrm{BV}} h, \quad \nu_{R,T}(h; u) \leq \|u_0\|_{\mathrm{BV}} h,$$
$$\omega_{R,T}(h; u) \leq \text{Const}\ \|u_0\|_{\mathrm{BV}} h, \quad \text{and} \quad \omega_{R,T}(\Delta t; u_{\Delta t}) \leq \text{Const}\ \|u_0\|_{\mathrm{BV}} \Delta t.$$

Then choosing $\tau = \rho = h$ we find that

$$\|u(\,\cdot\,,T) - u_{\Delta t}(\,\cdot\,,T)\|_{L^1(\mathbb{R}^d)} \leq \text{Const} \left(\frac{\Delta t}{h} + \Delta t + h\right).$$

Minimizing this with respect to h, we find that

$$\|u(\,\cdot\,,T) - u_{\Delta t}(\,\cdot\,,T)\|_{L^1(\mathbb{R}^d)} \leq \text{Const}\ \sqrt{\Delta t},$$

as $\Delta t \to 0$. Hence, dimensional splitting using exact solution operators has a convergence rate of 1/2, as shown in [139, 257].

We shall here present two realizations of $\mathcal{S}_t^i$ that fit into the convergence framework developed in Chapter 3; namely, the front-tracking method and the characteristic Galerkin method. Let us first mention that the monotone difference schemes also fit into our convergence framework. We omit, however, the details since dimensional splitting combined with monotone schemes is analyzed in [71].

The front-tracking method. Our first example is provided by the front-tracking method [117, 118, 126]. As explained in Chapter 4 the front-tracking method for a scalar conservation law is based on replacing the flux function with a continuous and piecewise linear approximation and the initial data by a piecewise constant approximation, and then solving the resulting perturbed problem exactly.

Next, let us briefly describe dimensional splitting combined with front tracking, see, e.g., [125] for details. Consider a uniform Cartesian (rectangular) grid defined by the grid size $\Delta x > 0$ and let π be the usual first-order projection (grid block averaging) operator defined on this grid. Furthermore, let f_i^δ be the piecewise linear approximations to f_i, and $\tilde{\mathcal{S}}_t^{i,\delta}$ the solution operator associated with the corresponding one-dimensional equation. We shall assume that δ and Δx are related to Δt so that all three tend to zero together. The fully discrete dimensional splitting formula is defined by letting $\mathcal{S}_t^{i,\delta} = \pi \circ \tilde{\mathcal{S}}_t^{i,\delta}$ and $u^0 = \pi u_0$.

We now use the convergence framework developed in Chapter 3 and in this section to investigate the convergence properties of the product formula (5.4). First, it is well-known that the solution operators $\mathcal{S}_t^{i,\delta}$ do not introduce new extrema so that $\|u_{\Delta t}\|_{L^\infty}$ is uniformly bounded by $\|u_0\|_\infty$. Consequently, the local integrability condition (3.45) holds. In fact, the following stronger (non-local) integrability condition holds: $\|u_{\Delta t}(\,\cdot\,,t)\|_1 \le \|u_0\|_1$.

After [161], it is well-known that the operators $\mathcal{S}_t^i$ are L^1 contractions. This is also true for the operator π. Thanks to translation invariance, it follows for any ρ that

$$\int_{\mathbb{R}^d} |u_{\Delta t}\,(x+\rho e_i,t) - u_{\Delta t}(x,t)|\;dx \le \int_{\mathbb{R}^d} |u_0\,(x+\rho e_i) - u_0(x)|\;dx.$$

Since u_0 is integrable, it follows that it has a modulus of continuity. Hence

$$\sum_i \int_{\mathbb{R}^d} |u_{\Delta t}\,(x+\rho e_i,t) - u_{\Delta t}(x,t)|\;dx \\ \le \sum_i \int_{\mathbb{R}^d} |u_0\,(x+\rho e_i) - u_0(x)|\;dx =: \nu\left(|\rho|\,;u_0\right).$$

Thus the space estimate (3.46) holds.

Next, let us show that the weak time estimate (3.47) holds. To this end, let v be a weak solution of the one-dimensional problem

$$v_t + g(v)_x = 0, \quad x \in \mathbb{R},\ t > 0, \qquad v|_{t=0} = v_0. \tag{5.32}$$

Let α_h be a smooth approximation to the characteristic function of the interval $[t, s]$ and set $\varphi_h(x,t) = \alpha_h(t)\varphi(x)$ for some test function φ. Inserting this into the weak formulation of (5.32), and letting $h \to 0$, we find that

$$\int_{\mathbb{R}} \varphi(x)\,(v(x,s) - v(x,t))\,dx + \int_{\mathbb{R}}\int_t^s \varphi_x g(v)\,dt\,dx = 0.$$

Thus

$$\left|\int_{\mathbb{R}} \varphi(x)\,(v(x,s) - v(x,t))\,dx\right| \le \|g\|_{\mathrm{Lip}}\,\|v\|_{L^1(\mathbb{R})}\,\|\varphi_x\|_\infty\,|s-t|\,. \tag{5.33}$$

In view of this linear estimate, we can easily deduce

$$\left|\int_{\mathcal{B}_r} (u_{\Delta t}(x,t+\tau) - u_{\Delta t}(x,t))\,\phi(x)\,dx\right| \le \mathrm{Const}_r\,\|\nabla\phi\|_{L^\infty(\mathbb{R}^d)}\,|\tau|\,,$$

and thus (3.47) holds. In passing, we note that if v is of uniformly bounded variation, then the above reasoning yields

$$\begin{aligned}\|v(\,\cdot\,,s) - v(\,\cdot\,,t)\|_{L^1(\mathbb{R})} &\le \sup_{|\varphi|\le 1}\int_{\mathbb{R}} \varphi(x)(v(x,s) - v(x,t))\,dx\\ &= \sup_{|\varphi|\le 1}\int_{\mathbb{R}}\int_t^s \varphi_x(x)g(v)\,dt\,dx \le \|g\|_{\mathrm{Lip}}\,\|v\|_{\mathrm{BV}}\,|t-s|\,.\end{aligned}$$

By Remark 3.5, $u_{\Delta t}$ also possesses a temporal modulus of continuity,

$$\omega\,(h; u_{\Delta t}) = \mathrm{Const}\,\left(\sqrt{h} + \nu\left(\sqrt{h}; u_{\Delta t}\right)\right).$$

Using Theorem 3.13, there exists a subsequence of $\{u_{\Delta t}\}$ that converges in $L^1_{\mathrm{loc}}(\Pi_T)$ to some limit u.

To ensure that the limit function is a solution of (5.2), we need to derive an entropy estimate for the front-tracking method and verify that the error terms in this estimate tend to zero in a suitable manner as $\Delta t \to 0$. This is identical to the estimates (4.23)–(4.25) and (4.26), which in this case read

$$\begin{aligned}&\int_{\mathbb{R}^d} d{\mu_{\mathrm{I}}}^i(\,\cdot\,,t) \le \nu\,(\Delta x; u_0)\,,\\ &\int_{\mathbb{R}^d} d{\mu_{\mathrm{II}}}^i \le \left\|f_i^\delta\right\|_{\mathrm{Lip}}\nu\,(\Delta x; u_0) + 3\,\|u_0\|_{L^1(\mathbb{R}^d)}\,\|f_i\|_{\mathrm{Lip}}\,\delta,\\ &\int_{\mathbb{R}^d} d{\mu_{\mathrm{IV}}}^i(\,\cdot\,,t) \le \nu\,(\Delta x; u_0)\,.\end{aligned} \tag{5.34}$$

For convergence rates, the 'method' term (5.12) is

$$\begin{aligned}
E_g^{\text{method}}(h,\tau) &= \int_{\mathbb{R}^d} d\mu_{\mathrm{I}} + \iint_{\Pi_T} \frac{1}{\tau} d\mu_{\mathrm{I}} + \frac{1}{h} d\mu_{\mathrm{II}} + d\mu_{\mathrm{IV}} \\
&\le (T+1)\nu(\Delta x; u_0) \\
&\quad + T\left(\frac{\nu(\Delta x; u_0)}{\tau} + \frac{3\max_i \|f_i\|_{\mathrm{Lip}} \left(\nu\left(\Delta x; u_0\right) + \|u_0\|_{L^1(\mathbb{R}^d)}\, \delta\right)}{h}\right) \qquad (5.35) \\
&\le \mathrm{Const}_T \left(\nu(\Delta x; u_0)\left(1 + \frac{1}{\tau} + \frac{1}{h}\right) + \frac{\delta}{h}\right),
\end{aligned}$$

where we have assumed that $\left\|f_i^\delta\right\|_{\mathrm{Lip}} \le 3\|f_i\|_{\mathrm{Lip}}$. In our case, since $\nu_T(h;u) \le \nu(h;u_0)$ and $\nu_T(h;u_{\Delta t}) \le \nu(h;u_0)$, the 'split' term (5.13) is

$$\begin{aligned}
E_g^{\text{split}}(h,\tau) &= \omega_T(\tau; u) + 2\nu\left(\sqrt{d}h; u_0\right) + \omega_T(\tau; u_{\Delta t}) \\
&\quad + \Delta t\left[\omega_T(\Delta t + \tau; u) + \omega_T(\Delta t + \tau; u_{\Delta t})\right]\frac{1}{h} + \Delta t \nu\left(\sqrt{d}h; u_0\right)\left(\frac{1}{h} + \frac{1}{h\tau}\right).
\end{aligned}$$

The moduli of continuity ν and ω can be chosen to satisfy $\nu(\sqrt{d}h;u) \le \mathrm{Const}_d \nu(h,u)$ and $\omega(\Delta t+\tau;u) \le \mathrm{Const}(\omega(\Delta t;u)+\omega(\tau;u))$. Furthermore, we have that the initial error satisfies

$$\left\|u_0 - u^0\right\|_{L^1(\mathbb{R}^d)} \le \nu(\Delta x; u_0).$$

If we use this and $h=\tau$, we find that

$$\begin{aligned}
\|u_{\Delta t}(\,\cdot\,,T) - u(\,\cdot\,,T)\|_{L^1(\mathbb{R}^d)} &\le \mathrm{Const}\left(E_g^{\text{method}}(h) + E_g^{\text{split}}(h)\right) \\
&\le \mathrm{Const}\Bigg[\left(\omega_T(h;u) + \omega_T(h; u_{\Delta t})\right)\left(1 + \frac{\Delta t}{h}\right) \\
&\qquad + \left(\omega_T(\Delta t; u) + \omega_T(\Delta t; u_{\Delta t})\right)\frac{\Delta t}{h} \\
&\qquad + \left(\nu(h; u_0)\frac{\Delta t}{h} + +\nu(\Delta x; u_0)\right)\left(1 + \frac{1}{h}\right) + \frac{\delta}{h}\Bigg].
\end{aligned}$$

Now, if the initial data are of bounded variation, all the moduli of continuity can be chosen to be linear, and we end up with

$$\|u_{\Delta t}(\,\cdot\,,T) - u(\,\cdot\,,T)\|_{L^1(\mathbb{R}^d)} \le \mathrm{Const}\left(\Delta x + \Delta t + h + \frac{\Delta x + \Delta t + \Delta t^2 + \delta}{h}\right),$$

and minimizing with respect to h gives

$$\|u_{\Delta t}(\,\cdot\,,T) - u(\,\cdot\,,T)\|_{L^1(\mathbb{R}^d)} \le \mathrm{Const}\sqrt{\Delta t + \Delta x + \delta}. \qquad (5.36)$$

If u_0 is less regular than $B.V.$, we can then obtain other convergence rates for the front-tracking and dimensional-splitting method. Summing up, we have proved the following theorem:

Theorem 5.3. *Suppose* $u_0 \in L^1(\mathbb{R}^d) \cap L^\infty(\mathbb{R}^d)$ *and* $f_1, \dots, f_d \in \mathrm{Lip}_{\mathrm{loc}}(\mathbb{R})$. *Define a sequence of functions* $\{u_{\Delta t}\}$ *by* (5.4), *where the front-tracking method is used to solve one-dimensional conservation laws. Then there exists a subsequence of* $\{u_{\Delta t}\}$ *that converges to an entropy weak solution of* (5.2). *If* $u_0 \in L^\infty(\mathbb{R}^d) \cap L^1(\mathbb{R}^d) \cap \mathrm{BV}(\mathbb{R}^d)$, *then the convergence rate is* $1/2$.

Remark 5.4. Convergence of dimensional splitting with front tracking was first established in [125] in the case $u_0 \in L^1(\mathbb{R}^d) \cap L^\infty(\mathbb{R}^d) \cap \mathrm{BV}(\mathbb{R}^d)$. Convergence rate estimates for dimensional splitting methods were first proved independently in [257] and [139]. In Theorem 5.3, we obtain convergence in the case $u_0 \in L^1(\mathbb{R}^d) \cap L^\infty(\mathbb{R}^d)$. However, we can relax this requirement to $u_0 \in L^\infty(\mathbb{R}^d)$. If we (for simplicity) consider the semi-discrete splitting, then this can be done as follows:

In [161] it is proved that the solution operators $\mathcal{S}_t^i$ are L^1 contractions on any ball $\mathcal{B}_r$. Let $L > 0$ denote the common Lipschitz constant of the f_i's. We can then use the localized L^1 contraction result [161] to obtain

$$\sum_i \int_{\mathcal{B}_r} |u_{\Delta t}(x + \rho e_i, t) - u_{\Delta t}(x,t)|\, dx \\ \leq \sup_{|\varepsilon| \leq |\rho|} \sum_i \int_{\mathcal{B}_r + Lt} |u_0(x + \varepsilon e_i) - u_0(x)|\, dx =: \nu_r(|\rho|; u_0).$$

Since $u_0 \in L^1(\mathcal{B}_r + Lt)$, it follows that $\nu_r \colon [0,\infty) \to \mathbb{R}$ is a continuous function with $\nu_r(0) = 0$. Hence, the space estimate (3.46) holds. Similarly to the proof of Theorem 5.3, we can use this space estimate to conclude the desired convergence.

Dimensional splitting in practice. Dimensional splitting is a simple and inexpensive way to extend the high-resolution methods introduced for one-dimensional conservation laws in Appendix A to multi-dimensions. In two dimensions we temporarily use the notation

$$u_t + f(u)_x + g(u)_y = 0 \quad u(x,y,0) = u_0(x,y).$$

In Algorithms 5.1.1 and 5.1.2 we detail the dimensional splitting algorithms for front tracking and finite volume, respectively. In Algorithm 5.1.2 the numerical flux functions are denoted by F and G.

A particularly efficient splitting method is obtained when front tracking is used to approximate the one-dimensional subequation (5.2), see [116, 124, 125, 139, 180, 182]. In its simplest form, the splitting method is defined over a uniform rectangular grid onto which the one-dimensional, piecewise constant front-tracking solutions are projected after each directional sweep. If we let π denote the projection operator and $\mathcal{S}_t^{i,\delta}$ the one-dimensional front-tracking operator, the corresponding fully discrete operator splitting reads

$$u(x_1, \dots, x_m, n\Delta t) \approx \left[\pi \mathcal{S}_{\Delta t}^{m,\delta} \cdots \pi \mathcal{S}_{\Delta t}^{1,\delta}\right]^n \pi u_0(x_1, \dots, x_m). \tag{5.37}$$

Algorithm 5.1.1 Dimensional splitting with front tracking in 2-D

Define a uniform grid $[x_{i-1/2}, x_{i+1/2}] \times [y_{j-1/2}, y_{j+1/2}]$.
Construct a piecewise constant initial function $u^0(x, y) = \pi u_0(x, y)$
Set $t = 0$ and $\Delta t = T/N$
For $n = 0 : N - 1$
    For each row $j = 1, \dots, N_y$
        Use front tracking to compute solution $v(x, \Delta t)$ of
            $v_t + f(v)_x = 0, \quad v(x, 0) = u^n(x, y_j)$
        Project solution back onto grid:
            $u^{n+1/2}(x, y) = \pi_x^j v(x, \Delta t), \quad y_{j-1/2} < y < y_{j+1/2}$
    end
    For each column $i = 1, \dots, N_x$
        Use front tracking to compute solution $v(y, \Delta t)$ of
            $v_t + g(v)_y = 0, \quad v(y, 0) = u^n(x_i, y)$
        Project solution back onto grid:
            $u^{n+1}(x, y) = \pi_y^i v(y, \Delta t), \quad x_{i-1/2} < x < x_{i+1/2}$
    end
end
Set $u(x, y, T) = u^N(x, y)$.

Algorithm 5.1.2 Dimensional splitting with a conservative scheme in 2-D

Define a uniform grid $[x_{i-1/2}, x_{i+1/2}] \times [y_{j-1/2}, y_{j+1/2}]$.
Construct an initial function $u^0_{i,j}$
Set $t = 0$ and $\Delta t = T/N$
For $n = 0 : N - 1$
    For each row $j = 1, \dots, N_y$
        For $i = 1, \dots, N_x$
            $u^{n+1/2}_{i,j} = u^n_{i,j} - \lambda\big[F^n_{i+1/2} - F^n_{i-1/2}\big]$
    end
    For each column $i = 1, \dots, N_x$
        For $j = 1, \dots, N_y$
            $u^{n+1}_{i,j} = u^{n+1/2}_{i,j} - \lambda\big[G^{n+1/2}_{j+1/2} - G^{n+1/2}_{j-1/2}\big]$
    end
end
Set $u(x, y, T) = u^N(x, y)$.

Recall that there are three discretization parameters: the time-step Δt, the grid size Δx_i, and the parameter δ determining the approximation of Riemann fans. For scalar problems δ is the interval length in the piecewise linear approximation of the flux function and for systems, δ determines the size of the jumps used to approximate continuous rarefaction waves, see Appendix A.7. The front-tracking algorithm is unconditionally stable and this gives a natural freedom when choosing the size of the splitting step in (5.37).

Now we shall exhibit several concrete examples, attempting to verify the convergence of dimensional spitting combined with front tracking, as described in Theorem 5.3. In Chapter 6 we will use dimensional splitting for systems of equations, in which case there is no theory available.

Example 5.5. *In the first example, we consider a linear advection equation describing rotation around the origin*

$$u_t - yu_x + xu_y = 0, \qquad u(x,y,0) = u_0(x,y). \tag{5.38}$$

Strictly speaking, this equation is not covered by our theory, since the flux functions depend on the location, $f(u) = -yu$ and $g(u) = xu$. However, when we apply the splitting we solve $u_t - yu_x = 0$ for y constant, and $u_t + xu_y$ for x constant. Hence for this example the arguments proving Theorem 5.3 remain valid. Since the equation is linear, better convergence rates can be obtained depending on the regularity of the initial data. Also, since the flux functions are linear (and especially piecewise linear) the parameter δ is superfluous.

We will use two sets of initial data: a smooth Gaussian bell function

$$u_0^1(x,y) = \exp\bigl[-20\bigl((x-0.6)^2 + y^2\bigr)\bigr], \tag{5.39}$$

and a discontinuous profile

$$u_0^2(x,y) = \begin{cases} 1, & \text{if } (x-0.6)^2 + y^2 \le 0.16, \\ 0, & \text{otherwise.} \end{cases} \tag{5.40}$$

To compute approximations we use the front-tracking method (5.37). *Table* 5.1 *gives the result of a grid-refinement study of the solution after one revolution for the two initial functions u_0^1 and u_0^2. The errors are measured in the discrete L^1 norm using a numerical quadrature rule. For smooth initial data, the operator splitting converges with a rate equal to its formal accuracy of order 1 in L^1. For discontinuous data, the convergence rate drops to $\frac{1}{2}$. This indicates that for nonlinear equations, where discontinuities develop independently of the regularity of the initial data, this rate is optimal. However, for many non-linear equations discontinuities can be "self-sharpening", and this can increase the observed rate.*

Example 5.5 and numerous other experiments run by the authors verify that the estimate of order $1/2$ is sharp for *linear equations*. In the current algorithm the

Table 5.1. Estimated L^1 errors and convergence rates after one revolution on an $n \times n$ grid for initial data (5.39) (top) and (5.40) (bottom).

n	$\nu = 2.0$		$\nu = 4.0$		$\nu = 8.0$		$\nu = 16.0$	
25	1.420e-01	—	1.023e-01	—	8.624e-02	—	2.715e-01	—
50	9.362e-02	0.60	5.976e-02	0.78	3.870e-02	1.16	5.913e-02	2.20
100	5.621e-02	0.74	3.376e-02	0.82	1.936e-02	1.00	1.703e-02	1.80
200	3.158c-02	0.83	1.797e-02	0.91	9.960e-03	0.96	6.727e-03	1.34
400	1.690e-02	0.90	9.323e-03	0.95	5.024e-03	0.99	3.132e-03	1.10
800	8.798e-03	0.94	4.764e-03	0.97	2.535e-03	0.99	1.536e-03	1.03
25	5.042e-01	—	3.829e-01	—	3.203e-01	—	8.718e-01	—
50	3.578e-01	0.49	2.558e-01	0.58	1.879e-01	0.77	1.736e-01	2.33
100	2.531e-01	0.50	1.815e-01	0.49	1.293e-01	0.54	9.713e-02	0.84
200	1.797e-01	0.49	1.278e-01	0.51	9.132e-02	0.50	6.497e-02	0.58
400	1.274e-01	0.50	9.036e-02	0.50	6.424e-02	0.51	4.577e-02	0.51
800	9.030e-02	0.50	6.390e-02	0.50	4.535e-02	0.50	3.216e-02	0.51

low order comes from the smearing introduced by applying the projection operator to discontinuous data, which is of order $\mathcal{O}(\Delta x^{1/2})$. Since the equation is linear, there will be no self-sharpening mechanisms inherent in the equation to counteract this smearing (see the discussion in Section 4.3). For nonlinear equations, a self-sharpening mechanism will usually counteract the smearing induced by the projections. Convergence rates closer to 1 will therefore be observed in practice, as will be seen in the next two examples. Further numerical experiments for dimensional splitting with front tracking for scalar conservation laws have been reported in [125, 180, 182]. Similar grid-refinement studies for systems of conservation laws can be found in [116, 124].

Example 5.6. *Consider the inviscid Burgers' equation,*

$$u_t + \left(\tfrac{1}{2}u^2\right)_x + \left(\tfrac{1}{2}u^2\right)_y = 0, \qquad u(x, y, 0) = u_0(x, y).$$

As initial data we choose two Riemann problems,

$$u_0^1(x) = \begin{cases} 0.5, & x > 0, y > 0, \\ 0.25, & x < 0, y > 0, \\ -0.5, & x < 0, y < 0, \\ 0.25, & x > 0, y < 0, \end{cases} \quad \text{and} \quad u_0^2(x) = \begin{cases} -0.5, & x > 0, y > 0, \\ 0.25, & x < 0, y > 0, \\ 0.5, & x < 0, y < 0, \\ 0.25, & x > 0, y < 0. \end{cases}$$

Here $u_0^1(x)$ gives a pattern where the four original constant states are separated by two pairs of rarefaction waves. For each pair, the two rarefaction waves meet in a sharp kink along the line $y = x$, see Figure 5.1. For $u_0^2(x)$, the wave pattern consists of the four constant states separated by six shocks forming two triple points at $(3t/8, -t/8)$ and $(-t/8, 3t/8)$.

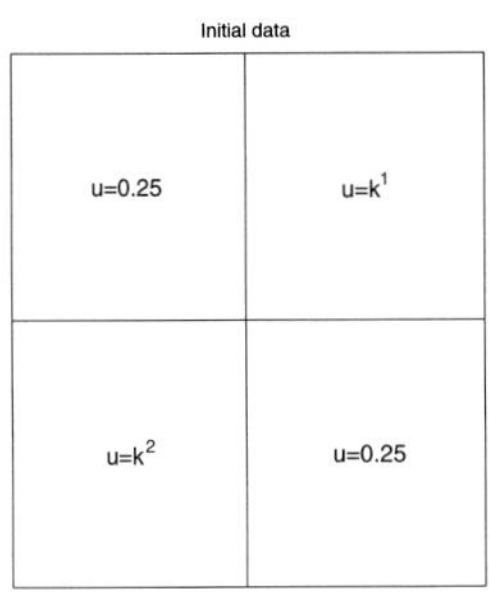

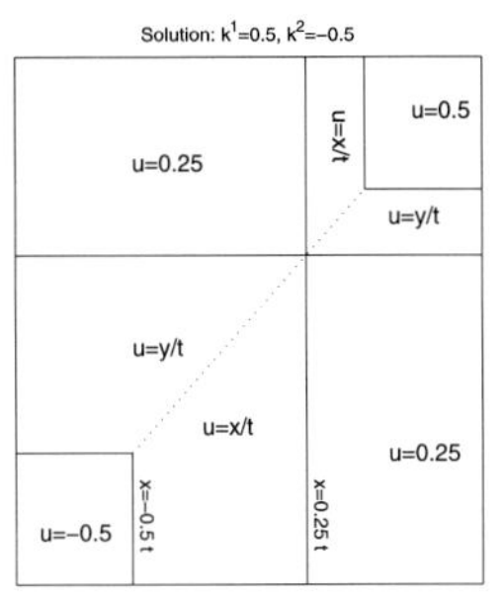

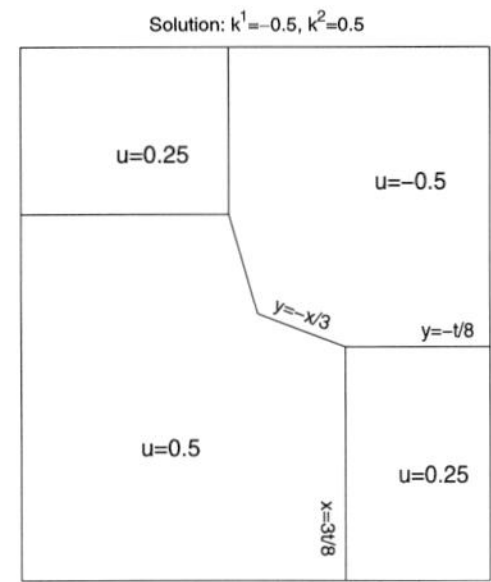

Figure 5.1. Initial data and exact solution for the two two-dimensional Riemann problems.

Table 5.2. Estimated L^1-errors and convergence rates at time $t = 0.5$ for Burgers' equation with initial functions u_0^1 (top) and u_0^2 (bottom).

N_x	$\nu = 1$		$\nu = 2$		$\nu = 4$		$\nu = 8$		$\nu = 16$	
32	1.48e-02	—	1.03e-02	—	7.34e-03	—	7.27e-03	—	—	—
64	9.52e-03	0.63	6.63e-03	0.64	5.18e-03	0.50	3.67e-03	0.99	3.65e-03	—
128	5.99e-03	0.67	4.14e-03	0.68	3.32e-03	0.64	2.58e-03	0.51	1.83e-03	0.99
256	3.67e-03	0.71	2.53e-03	0.71	2.07e-03	0.68	1.63e-03	0.66	1.25e-03	0.55
512	2.19e-03	0.74	1.50e-03	0.75	1.27e-03	0.71	1.00e-03	0.70	7.70e-04	0.70
1024	1.28e-03	0.78	8.70e-04	0.79	7.53e-04	0.75	6.07e-04	0.73	4.91e-04	0.65
32	8.36e-03	—	6.22e-03	—	5.86e-03	—	1.17e-02	—	—	—
64	4.30e-03	0.96	3.12e-03	1.00	2.93e-03	1.00	5.86e-03	1.00	1.17e-02	—
128	2.17e-03	0.99	1.56e-03	1.00	1.46e-03	1.00	2.93e-03	1.00	5.86e-03	1.00
256	1.09e-03	1.00	7.79e-04	1.00	7.32e-04	1.00	1.46e-03	1.00	2.93e-03	1.00
512	5.43e-04	1.00	3.90e-04	1.00	3.66e-04	1.00	7.32e-04	1.00	1.46e-03	1.00
1024	2.72e-04	1.00	1.95e-04	1.00	1.83e-04	1.00	3.66e-04	1.00	7.32e-04	1.00

Table 5.2 shows estimated L^1 errors and convergence rates for a grid-refinement study using the dimensional-splitting method with front tracking. For the discretization of the flux function we have used parameter $\delta_u = 0.01$. Here we observe slightly higher rates compared with Example 5.5. For u_0^2, the strong self-sharpening effects present near the shocks will counteract the smearing introduced by the projection operator leading to a convergence rate of approximately 1 for all ν under consideration. For the smooth solution resulting from u_0^1, the self-sharpening effects are much weaker, thereby giving a lower convergence rate.

Example 5.7. *In the next example, we consider initial data which equal -1 and 1 inside two circles of radius 0.4 centred at $(0.5, 0.5)$ and $(-0.5, -0.5)$, respectively, and zero elsewhere inside the square $[-1, 1] \times [-1, 1]$. Each circle will give rise to a leading circular shock wave and a lagging rarefaction wave propagating towards the origin. As the two leading shocks interact, they form a stationary shock aligned with the line $y = -x$. To form a more complex wave pattern, we rescale the y-flux of Burgers' equation slightly,*

$$u_t + \left(\tfrac{1}{2}u^2\right)_x + \left(\tfrac{2}{5}u^2\right)_y = 0.$$

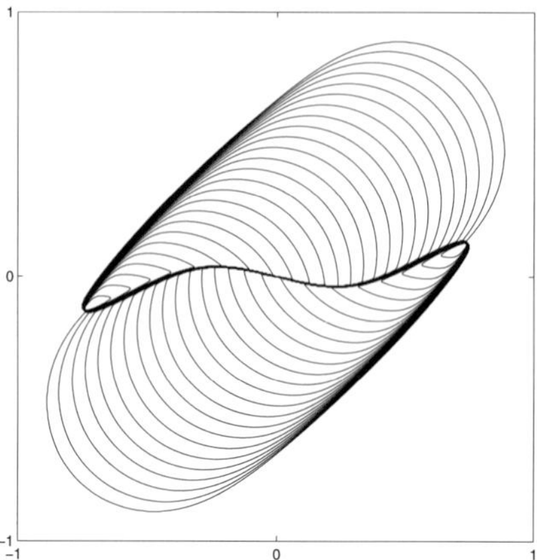

Figure 5.2. Reference solution at time $t = 2.0$ to the "almost Burgers" equation computed on a $2^{11} \times 2^{11}$ grid.

Table 5.3. Estimated L^1-errors and convergence rates at time $t = 2.0$ for the "almost Burgers" equation. The errors are computed relative to a reference solution computed on a $2^{11} \times 2^{11}$ grid.

N_x	$\nu = 2$		$\nu = 4$		$\nu = 8$		$\nu = 16$		$\nu = 32$	
32	1.71e-01	—	1.58e-01	—	2.23e-01	—	3.07e-01	—	4.99e-01	—
64	1.09e-01	0.65	9.75e-02	0.69	1.30e-01	0.78	2.15e-01	0.52	3.12e-01	0.68
128	6.64e-02	0.71	5.67e-02	0.78	7.28e-02	0.83	1.19e-01	0.85	2.14e-01	0.54
256	3.70e-02	0.84	2.97e-02	0.93	3.76e-02	0.95	6.18e-02	0.95	1.14e-01	0.90
512	2.04e-02	0.86	1.48e-02	1.01	1.84e-02	1.03	3.12e-02	0.99	5.87e-02	0.96

As a result, the leading shock waves will 'finger through' the outskirts of the opposite rarefaction region, leaving behind a secondary curved shock, as seen in Figure 5.2. Table 5.3 shows the result of a grid-refinement study. As above, the observed order of convergence is above the theoretical value of 1/2, which can be explained by the presence of strong shelf-sharpening effects in the shock regions.

Error mechanisms and computational efficiency. Two sources of errors contribute to the total error in dimensional splitting. Temporal splitting errors arise since we split the multi-dimensional equation into a sequence of one-dimensional problems. Loosely speaking, we can say that the temporal error determines how well we are able to resolve the dynamics of the problem. Obviously, the temporal error decreases with decreasing splitting steps Δt. The other source of error comes from the discretization of each individual subequation. We will henceforth refer to this error as the spatial error. The name comes from the observation that if the discretization method is sufficiently sophisticated to resolve the dynamics of the one-dimensional subequation, the associated discretization error will be reflected

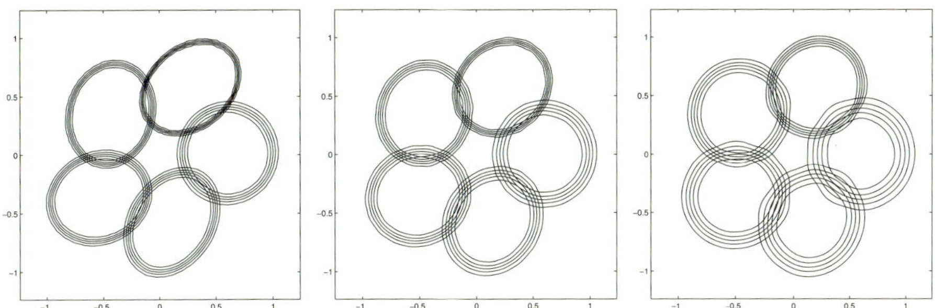

Figure 5.3. Contour plots of the rotating cylinder at time $t_n = 2n\pi/5$ for $n = 1, \dots, 5$. The number of time-steps is 20, 40 and 80 from left to right corresponding to CFL numbers 15.7, 7.9, and 3.9, respectively.

in the spatial resolution of the waves present in the approximate solution.

For the splitting method (5.37) using front tracking, the main spatial error contribution comes from the projection onto the regular grid. This error *increases* with decreasing Δt since the number of times the projection is applied is inversely proportional to Δt. The accuracy of the one-dimensional front-tracking algorithm, on the other hand, grows linearly with $\Delta t\delta$, cf., (5.35). Altogether this means that there are two error mechanisms that work in opposite directions: the temporal error decreases with decreasing Δt and the spatial error increases with decreasing Δt. To minimise the overall error we must therefore find the minimum where the temporal error balances the splitting error.

To illustrate the two error mechanisms we revisit Example 5.5.

Example 5.8. *Consider the rotation of a cylinder around the origin as described by* (5.38) *with initial data given by* (5.39)*. Since the flux function is linear, the front-tracking algorithm will be exact, ($f = f^\delta$ and $g = g^\delta$) and the only sources for error are the projection and the directional splitting of the equation. Figure* 5.3 *shows the temporal evolution of the cylinder computed with twenty, forty and eighty uniform time-steps on a uniform* 100×100 *grid. With only twenty time-steps, the cylinder is quite deformed during the rotation but retains a sharp profile. Eighty time-steps, on the other hand, gives an adequate representation of the rotating cylinder but also introduces considerable smearing due to the increased number of projections. Let us also revisit the grid-refinement study reported in Table* 5.1*. For fixed grid size with* $n > 25$*, the error after one rotation* decreases *with increasing CFL number. This gives a clear indication that the projection error gives the major contribution to the total error at time* $t = 2\pi$*. However, judging from Figure* 5.3 *it is likely that the splitting error has reached a local temporal minimum because of cancellations due to the symmetry in the problem. If we now plot the error versus runtime as in Figure* 5.4*, we observe that the highest computational efficiency is observed for* $\nu = 16$*. The only exception is on*

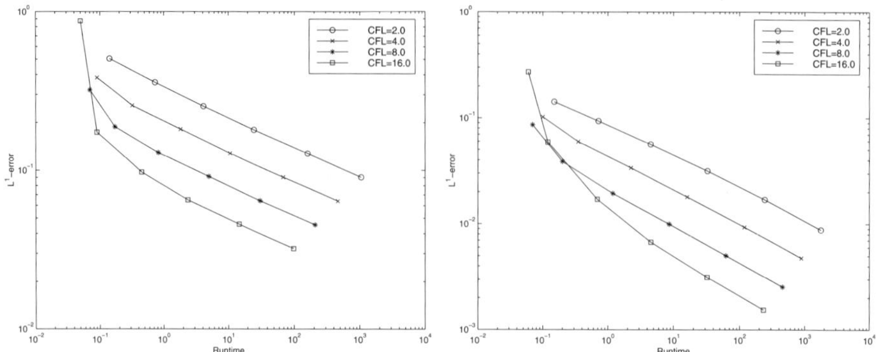

Figure 5.4. Errors in L^1 norm versus runtime for the grid refinement study in Table 5.1.

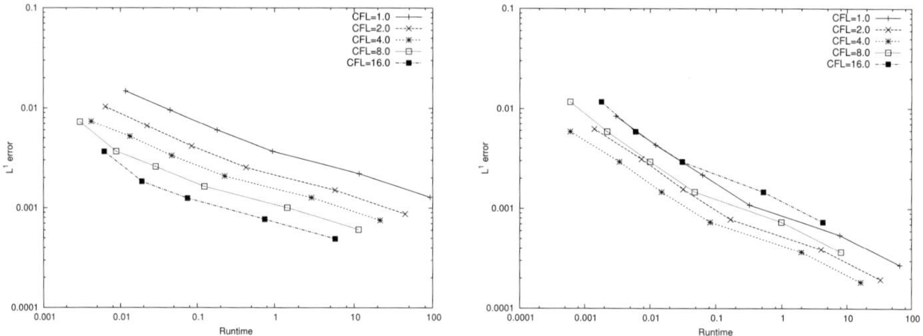

Figure 5.5. Errors in L^1 norm versus runtime for the grid refinement study in Table 5.2 for initial data u_0^1 (left) and u_0^2 (right).

the 25×25 grid, where the five splitting steps corresponding to $\nu = 16$ are not sufficient to resolve the dynamics of the problem properly.

Example 5.8 demonstrated that the dimensional-splitting method based upon front tracking has the ability to take large time-steps for the linear advection equation. Let us now revisit Example 5.6 to see if the same is true for the nonlinear Burgers' equation.

Example 5.9. *Figure 5.5 shows plots of the L^1 error versus runtime for the grid-refinement studies reported in Table 5.2 for initial functions u_0^1 and u_0^2. Initial function u_0^1 gives a smooth Riemann solution where the four constant states are separated by two pairs of rarefaction waves. Since the solution is piecewise smooth, we might have expected to obtain first-order convergence. This is clearly not the case in the upper half of Table 5.2. The reason is the influence from the projection operator. Due to the special symmetry of the problem, the solution can in fact be resolved by a single Godunov splitting step. With a single step, we also obtain*

Table 5.4. Convergence of the operator-splitting solution for u_0^1 with a single time-step on a sequence of $N_x \times N_x$ grids.

N_x	32	64	128	256	512	1024
error	7.27e-03	3.65e-03	1.83e-03	9.15e-04	4.58e-04	2.29e-04
rate	—	0.994	0.997	0.999	0.999	1.000

Table 5.5. Convergence of the operator-splitting solution with N_t splitting steps on a 1024×1024 grid. Here we have used $\delta_u = 10^{-3}$.

N_t	64	32	16	8	4	2
error	7.63e-04	6.38e-04	5.20e-04	4.16e-04	3.24e-04	2.29e-04
rate	—	0.258	0.293	0.324	0.361	0.498

the expected first-order convergence, see Table 5.4. When the rarefaction wave from the first hyperbolic step in the x-direction is projected onto the grid, the linear wave is replaced by a staircase function. After the next hyperbolic step in the x-direction, the solution will consist of a set of constant states separated by rarefaction waves unless the step is sufficiently large. This means that after a few projections, the linear rarefaction profile will be represented by an irregular step function that appears to have small kinks. For a fixed grid, the error will therefore increase with the number of splitting steps, as can be seen from Table 5.5. In the grid-refinement study for u_0^1 in Table 5.2, the highest efficiency is thus observed for $\nu = 16.0$, since the runtime increases with increasing number of time-steps.

Initial function u_0^2 gives a discontinuous solution where the four constant states are separated by six shocks forming two triple points. Resolution of the triple points and the shocks that are not aligned with the grid directions improves with increasing number of splitting steps. This is reflected in the error in the lower half of Table 5.2, which has a convex behavior as a function of ν for fixed grid size. With many splitting steps, the error is dominated by the smearing of the discontinuities caused by the projections, and with few splitting steps the error is dominated by the splitting error; The highest efficiency is observed for $\nu = 4.0$.

A key point when using an operator-splitting algorithm is how to choose the splitting step. For splitting methods based upon explicit schemes such as Algorithm 5.1.2, it is natural to let the size of the splitting step equal the time-step used in the one-dimensional schemes. This (local) time-step is restricted by a CFL stability condition and the size of the splitting step will therefore be related to and restricted by the spatial discretization. The front-tracking method in Algorithm 5.1.1 is unconditionally stable and does not offer a "natural" choice for the splitting steps. Hence, we can let the size of the splitting step be determined by the underlying dynamics of the equation and not by the spatial discretization. Often this allows for splitting steps corresponding to CFL numbers well above

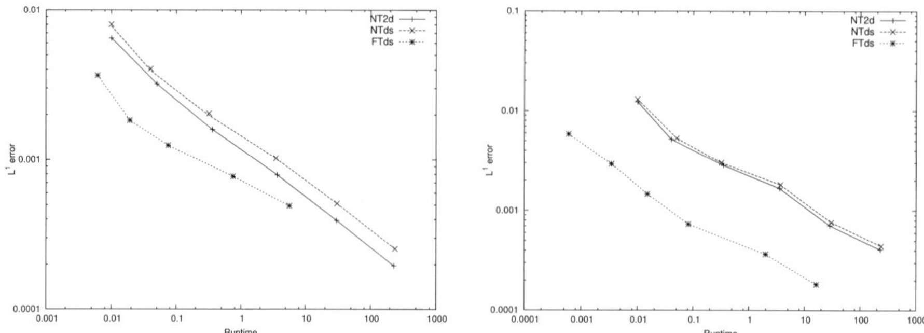

Figure 5.6. Comparison of errors in L^1 norm versus runtime for the front-tracking method (FTds), dimensional splitting with the Nessyahu–Tadmor scheme (NTds) and the multi-dimensional Nessyahu–Tadmor scheme (NT2d). The initial functions are Riemann problems given by u_0^1 (left) and u_0^2 (right).

unity, which in turn means that the front-tracking method will be significantly faster than a conventional finite-volume method. This claim has been supported by numerous computer experiments, some of which are reported in for instance [120, 180, 182]. Here we illustrate the point by once again revisiting Example 5.6.

Example 5.10. *Figure* 5.6 *shows a plot of the error versus runtime for the most efficient front-tracking runs from Figure* 5.5 *compared with two central-difference methods. The method NTds is the second-order Nessyahu–Tadmor scheme (see* (A.26) *in Appendix* A.6.1*) extended to two spatial dimensions by use of dimensional splitting as in Algorithm* 5.1.2*. The method NT2d is the multi-dimensional version of the Nessyahu–Tadmor scheme, see* (A.26) *in Appendix* A.6.1*. Several points can be observed. First of all, we notice that the front-tracking method is superior for the case with six shocks. For the rarefaction case, we see that although the front-tracking method is more efficient on coarse grids, the central schemes converges faster and will eventually be more efficient. Comparing the two central schemes, we see that the splitting scheme gives almost the same resolution as the unsplit scheme. However, since the operational count of NTds is higher than for NT2d, the unsplit scheme is more efficient. Notice that this is atypical behavior for finite-volume schemes.*

As we have seen above, there are two sources of errors for dimensional splitting with front tracking: splitting errors that are also present when using a finite-volume method in each direction, and smearing errors introduced by the projections from the irregular front-tracking description and back onto the underlying grid. For CFL numbers around unity, the latter is dominant. Increasing the CFL number to well above unity reduces that diffusion, but does not increase the splitting errors significantly. Typically, we observe feasible CFL numbers in the range of 10–20. Dimensional splitting with front-tracking is therefore superior to the finite-volume method for cases that are dominated by strong shocks, but where the underlying dynamics is relatively simple. More complex examples will be given in the next section, where we study applications to porous media flow, which has been a driving force behind our development of computational methodologies and the subsequent mathematical analysis.

The Euler characteristic Galerkin method. In this method [186] one seeks an approximation to the weak entropy solution of the one-dimensional conservation law

$$u_t + f(u)_x = 0,$$

at time t_n of the form

$$u^n = \sum_j u_j^n \phi_j,$$

where $\{\phi_j\}$ is a family consisting of piecewise constant or piecewise linear functions. In the standard ECG method, properties of the characteristics of the conservation law are used to establish the relation between the values u_n and u_{n+1}. Then this relation and the Galerkin projection are employed to generate the ECG scheme of the form

$$\langle u^{n+1} - u^n, \phi_i \rangle = -\Delta t \, \langle f'(u^n), \Phi_i^n \rangle,$$

where

$$\Phi_i^n(\alpha) = \frac{1}{f'(u^n(\alpha))\,\Delta t} \int_{x(\alpha)}^{y_n(\alpha)} \phi_i(s)\,ds, \qquad y_n(\alpha) = x(\alpha) + f'(u^n(\alpha))\,\Delta t \tag{5.41}$$

where $(x(\alpha), u^n(\alpha))$ is a continuous parametrization of (x, u^n) in the (x,u)-plane. For the purpose of obtaining a higher order of accuracy, u^n on the right-hand side of (5.41) is replaced by a physically more acceptable function $\tilde{u}_n$. The authors [186] describe two procedures for constructing $\tilde{u}_n$: continuous linear recovery and discontinuous linear recovery. The convergence of the methods discussed is examined in $L^\infty([0,T], L^1(\mathbb{R}))$, $0 < T < \infty$, and it is proved that the limit function of the approximation constructed in the ECG scheme, with either of the recoveries, is an admissible solution of the conservation law. There are also some computational examples presenting the results of the ECG schemes with and without recovery for the linear advection equation and the inviscid Burgers equation. In [187] the authors extend the method to two and three space dimensions, and show the same convergence.

The details of the method are beyond the scope of this exposition, we only note that by a result in [39], it holds that

$$\eta(u)_t + \sum_i q_i(u)_{x_i} \le \text{Const} \sup|u|\, t =: C, \quad t \in [0,T], \tag{5.42}$$

where the constant depends only on f_i and $\|\eta\|_{\text{Lip}}$. Using this in Theorem 5.1, we see

$$E^{\text{method}} = \mathcal{O}(\Delta t). \tag{5.43}$$

For simplicity we have ignored the error produced by discretization of the initial data as this will not affect the overall result.

Remark 5.11. If we use Theorem 5.1 *without* dimensional splitting, that is, using the fully d-dimensional method as the only splitting step, we find that

$$E^{\mathrm{split}} = \omega_{R,T}(\tau; u) + \nu_{R,T}(\sqrt{d}h; u) \\ + \left[\omega_{R,T}(\Delta t + \tau; u) + \omega_{R,T}(\Delta t + \tau, u_{\Delta t})\right]\left(\frac{1}{h} + \frac{1}{\rho}\right).$$

Using this and (5.43) we can calculate convergence rates depending on the regularity of the initial data, in particular, if $u_0 \in \mathrm{BV}$, then the convergence rate of the (unsplit) ECG method is 1/2. This was not reported in [187] or [186].

Using the split scheme, we similarly find the following result.

Theorem 5.12. *Suppose $u_0 \in L^1(\mathbb{R}^d) \cap L^\infty(\mathbb{R}^d)$ and $f_1, \dots, f_d \in \mathrm{Lip}_{\mathrm{loc}}(\mathbb{R})$. Define a sequence of functions $\{u_{\Delta t}\}$ by (5.4), where the ECG scheme is used to solve one-dimensional conservation laws. Then there exists a subsequence of $\{u_{\Delta t}\}$ that converges to an entropy weak solution of (5.2). If $u_0 \in L^\infty(\mathbb{R}^d) \cap L^1(\mathbb{R}^d) \cap \mathrm{BV}(\mathbb{R}^d)$, then the convergence rate is 1/2.*

5.2 Weakly coupled systems of conservation laws

In this section we study the splitting framework applied to weakly coupled systems of conservation laws

$$u_t^\kappa + \sum_i f_i^\kappa(u^\kappa)_{x_i} = g^\kappa(U), \quad u^\kappa|_{t=0} = u_0^\kappa, \quad \kappa = 1, \dots, K, \tag{5.44}$$

or

$$U_t + \sum_i F_i(U)_{x_i} = G(U), \quad U|_{t=0} = U_0,$$

where

$$F_i(U) = \left(f_i^1(u^1), \dots, f_i^K(u^K)\right), \quad G(U) = \left(g^1(U), \dots, g^K(U)\right).$$

As was the case for scalar conservation laws, we consider initial data U_0 in $L^1(\mathbb{R}^d) \cap L^\infty(\mathbb{R}^d)$, and assume that the solution remains in the same space for each $t > 0$. Furthermore, we assume that the unique weak solution to (5.44) has nondecreasing moduli of continuity

$$\begin{aligned} \int_{\mathbb{R}^d} |U(x+h,t) - U(x,t)|\,dx &\le \nu(|h|; U), \\ \int_{\mathbb{R}^d} |U(x,t+\tau) - U(x,t)|\,dx &\le \omega(|\tau|; U), \end{aligned} \tag{5.45}$$

uniformly for all t and $t+\tau$ in $[0,T]$. This assumption is satisfied if $U_0 \in \mathrm{BV}(\mathbb{R}^d)$.

We propose splitting this equation into K scalar conservation laws, and one ordinary differential equation (parametrized by x). To fix the notation, let $U = \mathcal{S}_t U_0 = (u^1, \dots, u^K)$ be the solution of the uncoupled system of conservation laws

$$u_t^\kappa + \sum_i f_i^\kappa (u^\kappa)_{x_i} = 0, \quad \kappa = 1, \dots, K. \tag{5.46}$$

Similarly let $\mathcal{O}_t$ be the solution operator for the system of ordinary differential equations

$$\frac{dU}{dt} = G(U). \tag{5.47}$$

For simplicity we use the term 'solution operator' to denote either an exact solution operator or a numerical method.

In this case $\ell = 2$ (cf. (3.57)). We define our approximation $U_{\Delta t}(x,t) = (u_{\Delta t}^1(x,t), \dots, u_{\Delta t}^K(x,t))$ as follows: (recall that $t_{n,1} = (n+1/2)\Delta t$)

$$\begin{aligned} U_{\Delta t}^n &= \left[\mathcal{O}_{\Delta t}\mathcal{S}_{\Delta t}\right]^n U_0^0, \\ U_{\Delta t}(\,\cdot\,,t) &= \mathcal{S}_{2(t-t_n)} U_{\Delta t}^n && \text{for } t_n \le t \le t_{n,1}, \\ U_{\Delta t}^{n,1} &= U_{\Delta t}\left(\,\cdot\,, t_{n,1}\right), \\ U_{\Delta t}(\,\cdot\,,t) &= \mathcal{O}_{2(t-t_{n,1})} U_{\Delta t}^{n,1} && \text{for } t_{n,1} \le t \le t_{n+1}. \end{aligned} \tag{5.48}$$

In order to avoid blow-up, we require that G and the numerical approximation to (5.47) are such that, for all positive r,

$$\mathcal{O}_t\left(\mathcal{B}_r\right) \subset \mathcal{B}_{e(t,r)} \tag{5.49}$$

for some continuous function $e\colon [0,\infty) \times [0,\infty) \to [0,\infty)$ with $e(0,r) = r$ for all r. Furthermore, we assume that $G(0) = 0$, so that if $\operatorname{supp}(U_0)$ is bounded, then by finite speed of propagation $\|G(U(\,\cdot\,,t))\|_{L^1} \le \text{Const}\, \|U(\,\cdot\,,t)\|_{L^1}$.

As before we require that the methods $\mathcal{S}$ and $\mathcal{O}$ have error terms that are partial derivatives:

$$\begin{aligned} &\eta\left(u^\kappa\right)_t + \sum_i q^\kappa\left(u^\kappa\right)_{x_i} \le \left(\mathcal{E}_{\mathrm{I}}{}^{1,\kappa}\right)_t + \sum_i \left(\mathcal{E}_{\mathrm{II}}{}_i^{1,\kappa}\right)_{x_i} + \sum_{i,j}\left(\mathcal{E}_{\mathrm{III}}{}_{ij}^{1,\kappa}\right)_{x_i x_j} + \mathcal{E}_{\mathrm{IV}}{}^{1,\kappa}, \\ &\eta\left(u^\kappa\right)_t - \eta'\left(u^\kappa\right) G\left(U\right) \le \left(\mathcal{E}_{\mathrm{I}}{}^{2,\kappa}\right)_t + \mathcal{E}_{\mathrm{IV}}{}^{2,\kappa}, \quad \kappa = 1,\dots,K, \end{aligned} \tag{5.50}$$

where the error terms $\mathcal{E}_{\mathrm{I}}{}^{l,\kappa}, \dots, \mathcal{E}_{\mathrm{IV}}{}^{l,\kappa}$ are bounded by positive Radon measures as in (5.6). We also assume that the approximation is essentially bounded and has uniform moduli of continuity, cf. (5.7).

When we establish error estimates for conservation laws without source, we essentially add entropy estimates for the exact solution using the approximate solution as a "constant" and for the approximate solution using the exact solution as a "constant", cf. Section 5.1. In the case with source, the entropy functional contains η' equal to the sign function. This is a discontinuous function, and it is difficult to obtain estimates in terms of $\omega(U_{\Delta t};\Delta t)$ from terms of the type

$$\operatorname{sign}(U_{\Delta t}(\,\cdot\,,t+\Delta t)-U)-\operatorname{sign}(U_{\Delta t}(\,\cdot\,,t)-U).$$

There are a couple of ways of circumventing this obstacle. One is to rewrite the conservation law with source as a conservation law with (x,t)–dependent flux function [170]. The main advantage of this is that a better convergence rate is obtained for exact solution operators. The disadvantage is that it does not exactly fit into the present framework. The other strategy is to use a symmetric test function $\phi(x,t,y,s)$, such that $\phi(x,t+\sigma,y,s+\sigma)=\phi(x,t,y,s)$ and $\phi(x,y,t,s)=\phi(y,x,s,t)$, see [254]. The advantage of this is that it fits very well with the present framework, but the disadvantage is that it gives a lower order for exact solution operators. Furthermore, this strategy does not give local (in the sense of $L^1\,(\mathcal{B}_R(x_0))$) estimates. Nevertheless, we opt to follow the second strategy here.

Before we state the main result, we present the important Kuznetsov's lemma. To this end, define the functional $\Lambda(u,v,\phi)$ as

$$\begin{aligned}\Lambda(u,v,\phi)=-\iiiint\limits_{\Pi_T\times\Pi_T}\left[\eta(u(x,t),v(y,s))\phi_t+q(u(x,t),v(y,s))\cdot\nabla_x\phi\right]dt\,dx\,ds\,dy\\+\iint\limits_{\Pi_T}\int\limits_{\mathbb{R}^d}\left(|u(x,T)-v(y,s)|-|u(x,0)-v(y,s)|\right)\phi\,dx\,ds\,dy.\end{aligned}\tag{5.51}$$

Here $\eta(u,v)=|u-v|$ and

$$\begin{aligned}q(u,v)&=(q_1(u,v),\dots,q_d(u,v))=\operatorname{sign}(u-v)\big(f(u)-f(v)\big),\\f(u)&=(f_1(u),\dots,f_d(u))\end{aligned}$$

for Lipschitz continuous functions f_i.

Lemma 5.13 (Kuznetsov's lemma [166, 167]). *Let the test function ϕ be given by (cf. (5.21))*

$$\phi(x,t,y,s)=\Omega_{h,\tau}(x,y,t,s)=\delta_h(x-y)\tilde{\delta}_\tau(t-s).\tag{5.52}$$

Assuming that $u,v\in L^1(\Pi_T)\cap L^\infty(\Pi_T)$ have moduli of continuity in space and time, then

$$\begin{aligned}\|u(\,\cdot\,,T)-v(\,\cdot\,,T)\|_{L^1(\mathbb{R}^d)}\le\|u(\,\cdot\,,0)-v(\,\cdot\,,0)\|_{L^1(\mathbb{R}^d)}+\Lambda(u,v,\phi)+\Lambda(v,u,\phi)\\+\nu(\sqrt{d}\,h;u)+\nu(\sqrt{d}\,h;v)+\omega(\tau;u)+\omega(\tau;v),\end{aligned}\tag{5.53}$$

where ω and ν denote temporal and spatial moduli of continuity, respectively.

For a proof of this lemma, see [166, 167] or [126, Theorem 4.5].

Lemma 5.14. *Let the test function ϕ be given by (5.52). Assume that $U = (u^1, \dots, u^K)$ is the entropy solution of (5.44), and $V = (v^1, \dots, v^K)$ is a function in $L^1(\Pi_T; \mathbb{R}^K) \cap L^\infty(\Pi_T; \mathbb{R}^K)$ that has a temporal modulus of continuity. Then for $\kappa = 1, \dots, K$,*

$$\Lambda(u^\kappa, v^\kappa, \phi) \leq 2 \iint\limits_{\Pi_T} \iint\limits_{\Pi_T} \chi_2^2(x,t)\, \mathrm{sign}(u^\kappa(x,t) - v^\kappa(y,s)) g^\kappa(U(x,t)) \phi \, dt\, dx\, ds\, dy$$
$$+ \mathrm{Const}_T \Big(\Delta t + \omega(\Delta t; v^\kappa) + \big(\Delta t + \omega(\Delta t; u^\kappa) + \omega(\Delta t; v^\kappa)\big) \Big(\frac{1}{\tau} + \frac{1}{h}\Big)\Big), \quad (5.54)$$

where we recall that χ_2^2 is the characteristic function on $\mathbb{R}^d \times \cup_n (t_{n,1}, t_{n+1})$, and where the constant Const_T depends on the $L^1(\mathbb{R}^d)$ norms of u^κ and v^κ and T.

Proof. For simplicity we write $u = u^\kappa$, $v = v^\kappa$ and $q = (q_1^\kappa, \dots, q_d^\kappa)$ for some $\kappa = 1, \dots, K$. Thus

$$\Lambda(u,v,\phi) = -2 \sum_n \int\limits_{t_{n,1}}^{t_{n+1}} \int\limits_{\mathbb{R}^d} \iint\limits_{\Pi_T} \big(\eta(u,v)\phi_t + q(u,v)\cdot\nabla_x \phi\big)\, dt\, dx\, dy\, ds$$
$$+ 2 \sum_n \int\limits_{t_{n,1}}^{t_{n+1}} \int\limits_{\mathbb{R}^d} \int\limits_{\mathbb{R}^d} \big(\eta(u(x,T),v)\phi - \eta(u(x,0),v)\phi\big)\, dx\, dy\, ds \quad (5.55a)$$
$$+ \sum_n \int\limits_{t_{n,1}}^{t_{n+1}} \int\limits_{\mathbb{R}^d} \iint\limits_{\Pi_T} \big(\eta(u,v)\phi_t + q(u,v)\cdot\nabla_x \phi\big)\, dt\, dx\, ds\, dy$$
$$- \sum_n \int\limits_{t_n}^{t_{n,1}} \int\limits_{\mathbb{R}^d} \iint\limits_{\Pi_T} \big(\eta(u,v)\phi_t + q(u,v)\cdot\nabla_x \phi\big)\, dt\, dx\, dy\, ds \quad (5.55b)$$
$$- \sum_n \int\limits_{t_{n,1}}^{t_{n+1}} \int\limits_{\mathbb{R}^d} \int\limits_{\mathbb{R}^d} \big(\eta(u(x,T),v)\phi - \eta(u(x,0),v)\phi\big)\, dx\, dy\, ds$$
$$+ \sum_n \int\limits_{t_n}^{t_{n,1}} \int\limits_{\mathbb{R}^d} \int\limits_{\mathbb{R}^d} \big(\eta(u(x,T),v)\phi - \eta(u(x,0),v)\phi\big)\, dx\, dy\, ds. \quad (5.55c)$$

Since u is an entropy solution, (5.55a) will give the first term on the right of (5.54), cf. (3.30). We now have to bound (5.55b) and (5.55c). Consider first a part of

(5.55b), more precisely, let

$$I := \sum_n \int_{t_{n,1}}^{t_{n+1}} \int_{\mathbb{R}^d} \iint_{\Pi_T} \eta(u,v)\phi_t \, dt\, dx\, dy\, ds - \sum_n \int_{t_n}^{t_{n,1}} \int_{\mathbb{R}^d} \iint_{\Pi_T} \eta(u,v)\phi_t \, dt\, dx\, dy\, ds. \tag{5.56}$$

To this end, change variables in the first integral, $t \mapsto t - \Delta t/2$, $s \mapsto s - \Delta t/2$, and note that under this change, $\phi_t(x,t,y,s) = \phi_t(x, t+\Delta t/2, y, s+\Delta t/2)$. Hence

$$\begin{aligned} I &= \sum_n \int_{t_n}^{t_{n,1}} \int_{\mathbb{R}^d} \int_{-\Delta t/2}^{0} \int_{\mathbb{R}^d} \eta(u(t+\Delta t/2), v(s+\Delta t/2))\phi_t \, dt\, dx\, dy\, ds \\ &\quad + \sum_n \int_{t_n}^{t_{n,1}} \int_{\mathbb{R}^d} \int_{0}^{T-\Delta t/2} \int_{\mathbb{R}^d} \Big(|u(x,t+\Delta t/2) - v(y,s+\Delta t/2)| \\ &\qquad\qquad\qquad - |u(x,t) - v(y,s)| \Big) \phi_t \, dx\, dt\, dy\, ds \\ &\quad - \sum_n \int_{t_n}^{t_{n,1}} \int_{\mathbb{R}^d} \int_{T-\Delta t/2}^{T} \int_{\mathbb{R}^d} \eta(u,v)\phi_t \, dx\, dt\, dy\, ds \\ &=: I_1 + I_2 + I_3. \end{aligned}$$

Now

$$|I_1| \le \iint_{\Pi_T} \int_0^{\Delta t/2} \int_{\mathbb{R}^d} (|u| + |v|) \left| \tilde{\delta}'_\tau(t-s) \right| \delta_h(x-y) \, dx\, dt\, dy\, ds.$$

Since both u and v are in $L^1(\Pi_T) \cap L^\infty(\Pi_T)$, we find that

$$|I_1| \le \mathrm{Const}_T \frac{\Delta t}{\tau},$$

and similarly,

$$|I_3| \le \mathrm{Const}_T \frac{\Delta t}{\tau}.$$

Also

$$\begin{aligned} |I_2| &\le \iint_{\Pi_T} \iint_{\Pi_T} \Big(|u(x,t+\Delta t/2) - u(x,t)| \\ &\qquad + |v(y,s+\Delta t/2) - v(y,s)| \Big) \left| \tilde{\delta}'_\tau(t-s) \right| \delta_h(x-y) \, dt\, dx\, ds\, dy \\ &\le \mathrm{Const}_T \left(\omega(\Delta t; u) + \omega(\Delta t; v) \right) \frac{1}{\tau}. \end{aligned}$$

The remaining part of (5.55b) (replacing $\eta\phi_t$ in (5.56) by $q \cdot \nabla_x \phi$) is estimated essentially the same way. Indeed we find that the absolute value of

$$\sum_n \int_{t_{n,1}}^{t_{n+1}} \int_{\mathbb{R}^d} \iint_{\Pi_T} q(u,v) \cdot \nabla_x \phi \, dt\, dx\, dy\, ds - \sum_n \int_{t_n}^{t_{n,1}} \int_{\mathbb{R}^d} \iint_{\Pi_T} q(u,v) \cdot \nabla_x \phi \, dt\, dx\, dy\, ds \tag{5.57}$$

is less than

$$M \iint_{\Pi_T} \int_0^{\Delta t/2} \int_{\mathbb{R}^d} (|u| + |v|) \left|\nabla_x \delta_h(x-y)\right| \tilde{\delta}_\tau(t-s)\, dx\, dt\, ds\, dy \tag{5.58a}$$

$$+ M \iint_{\Pi_T} \iint_{\Pi_T} \Big(|u(x,t+\Delta t/2) - u(x,t)| + |v(y,s+\Delta t/2) - v(y,s)| \Big) \times \left|\nabla_x \delta_h(x-y)\right| \tilde{\delta}_\tau(t-s)\, dt\, dx\, ds\, dy \tag{5.58b}$$

$$+ M \iint_{\Pi_T} \int_{T-\Delta t/2}^{T} \int_{\mathbb{R}^d} (|u| + |v|) \left|\nabla_x \delta_h(x-y)\right| \tilde{\delta}_\tau(t-s)\, dx\, dt\, ds\, dy, \tag{5.58c}$$

where $M = \|f\|_{\mathrm{Lip}}$. So (5.58a) and (5.58c) are bounded by

$$\mathrm{Const}_T \frac{\Delta t}{h},$$

and (5.58b) is bounded by

$$\mathrm{Const}_T \Big(\omega(\Delta t; u) + \omega(\Delta t; v) \Big) \frac{1}{h}.$$

Summing up, we have found the bound on (5.55b) to be

$$\mathrm{Const}_T \, \left(\Delta t + \omega(\Delta t; u) + \omega(\Delta t; v)\right) \left(\frac{1}{\tau} + \frac{1}{h} \right). \tag{5.59}$$

Now we estimate the "boundary terms" (5.55c) which consists of two identical parts evaluated for $t = T$ and $t = 0$, respectively. We consider the terms with $t = T$ only, namely

$$\sum_n \Bigg(\int_{t_{n,1}}^{t_n} \iint_{\mathbb{R}^d \times \mathbb{R}^d} \eta(u(x,T), v)\phi \, dx\, dy\, ds - \int_{t_{n,1}}^{t_{n+1}} \iint_{\mathbb{R}^d \times \mathbb{R}^d} \eta(u(x,T), v)\phi \, dx\, dy\, ds \Bigg). \tag{5.60}$$

By changing variables $s \mapsto s - \Delta t/2$ in the last integral, we find that the absolute value of (5.60) is bounded by

$$\Big| \sum_n \int_{t_n}^{t_{n,1}} \iint_{\mathbb{R}^d \times \mathbb{R}^d} \Big(|u(x,T) - v(y,s)| \, \phi(x,T,y,s)$$

$$- |u(x,T) - v(y,s+\Delta t/2)|\,\phi(x,T,y,s+\Delta t/2)\Big)\,dx\,dy\,ds\Big|$$

$$\le \sum_n \int_{t_n}^{t_{n,1}} \iint_{\mathbb{R}^d\times\mathbb{R}^d} \delta_h(x-y)\Big(|u(x,T)-v(y,s)|\,\Big|\tilde{\delta}_\tau(T-s)-\tilde{\delta}_\tau(T-s-\Delta t/2)\Big|$$
$$+ \Big|\,|u(x,T)-v(y,s)| - |u(x,T)-v(y,s+\Delta t/2)|\,\Big|$$
$$\times\, \tilde{\delta}_\tau(T-s-\Delta t/2)\Big)\,dx\,dy\,ds$$

$$\le \sum_n \int_{t_n}^{t_{n,1}} \iint_{\mathbb{R}^d\times\mathbb{R}^d} \delta_h(x-y)\,|u(x,T)-v(y,s)|$$
$$\times \Big|\tilde{\delta}_\tau(T-s)-\tilde{\delta}_\tau(T-s-\Delta t/2)\Big|\,dx\,dy\,ds$$
$$+\sum_n \int_{t_n}^{t_{n,1}} \iint_{\mathbb{R}^d\times\mathbb{R}^d} \delta_h(x-y)\,|v(y,s)-v(y,s+\Delta t/2)|\,\tilde{\delta}_\tau(T-s-\Delta t/2)\,dx\,dy\,ds$$
$$=: J_1 + J_2.$$

The term J_2 is estimated as usual; the integral over x equals unity using the approximate delta function $\delta_h(x-y)$. The y-integral of the difference in v is estimated using the time modulus ω, and the integral of $\tilde{\delta}_\tau$ with respect to s is estimated by unity. Thus

$$J_2 \le \omega(\Delta t; v).$$

To find a bound on J_1 we use that u and v are integrable. Hence

$$J_1 \le \sum_n \int_{t_n}^{t_{n,1}} \iint_{\mathbb{R}^d\times\mathbb{R}^d} \delta_h(x-y)\big(|u(x,T)|+|v(y,s)|\big)\left|\int_{T-s-\Delta t/2}^{T-s} \tilde{\delta}_\tau'(z)\,dz\right|\,dx\,dy\,ds$$
$$\le \sum_n \int_{t_n}^{t_{n,1}} \Big(\int_{\mathbb{R}^d} |u(x,T)|\,dx + \int_{\mathbb{R}^d} |v(y,s)|\,dy\Big)\int_0^{\Delta t/2} \Big|\tilde{\delta}_\tau'(z+T-s)\Big|\,dz\,ds$$
$$\le \Big(\int_{\mathbb{R}^d} |u(x,T)|\,dx + \sup_{0\le s\le T}\int_{\mathbb{R}^d} |v(y,s)|\,dy\Big)\int_0^T\int_0^{\Delta t/2} \Big|\tilde{\delta}_\tau'(z+T-s)\Big|\,dz\,ds$$
$$\le \Big(\int_{\mathbb{R}^d} |u(x,T)|\,dx + \sup_{0\le s\le T}\int_{\mathbb{R}^d} |v(y,s)|\,dy\Big)\int_0^\tau\int_0^{\Delta t/2} \frac{\text{Const}}{\tau^2}\,dz\,ds$$
$$\le \text{Const}_T\frac{\Delta t}{\tau}\big(\|u(T)\|_1 + \sup_{0\le t\le T}\|v(t)\|_1\big). \tag{5.61}$$

Collecting the bounds (5.59) and (5.61) finishes the proof. □

Then we have the following result:

Theorem 5.15. *Let U denote the weak entropy solution of (5.44), and $U_{\Delta t}$ an approximate solution given by (5.48). For all sufficiently small constants τ and h we have that*

$$\|U(\,\cdot\,,T) - U_{\Delta t}(\,\cdot\,,T)\|_1 \le \sqrt{K}\left(\|U(\,\cdot\,,0) - U_{\Delta t}(\,\cdot\,,0)\|_1 + E^{\text{method}} + E^{\text{split}}\right) \times \left(1 + 2\,\|G\|_{\text{Lip}}\, T \exp(2\,\|G\|_{\text{Lip}}\, T)\right), \quad (5.62)$$

where $\|G\|_{\text{Lip}}$ denotes the Lipschitz constant of G and

$$E^{\text{method}} = \text{Const}_T\left[\sup_{0\le t\le T}\int_{\mathbb{R}^d} d\mu_{I}(\,\cdot\,,t) + \iint_{\Pi_T}\left(\frac{1}{\tau}d\mu_{I} + \frac{1}{h}d\mu_{II} + \frac{1}{h^2}d\mu_{III} + d\mu_{IV}\right)\right] \quad (5.63)$$

and

$$E^{\text{split}} = \text{Const}_T\Big[\nu(\sqrt{d}\,h;U) + \nu(\sqrt{d}\,h;U_{\Delta t}) + \omega(\tau;U) + \omega(\tau;U_{\Delta t}) + \Delta t + \omega(\Delta t;U_{\Delta t}) + \left(\Delta t + \omega(\Delta t;U) + \omega(\Delta t;U_{\Delta t})\right)\left(\frac{1}{\tau} + \frac{1}{h}\right)\Big]. \quad (5.64)$$

Proof. Kuznetsov's lemma states that

$$\|u^\kappa(\,\cdot\,,T) - u^\kappa_{\Delta t}(\,\cdot\,,T)\|_1 \le \|u(\,\cdot\,,0) - u^\kappa_{\Delta t}(\,\cdot\,,0)\|_1 + \Lambda(u^\kappa, u^\kappa_{\Delta t},\phi) + \Lambda(u^\kappa_{\Delta t}, u, \phi) + \nu(\sqrt{d}\,h;u^\kappa) + \nu(\sqrt{d}\,h;u^\kappa_{\Delta t}) + \omega(\tau;u^\kappa) + \omega(\tau;u^\kappa_{\Delta t}). \quad (5.65)$$

Lemma 5.14 provides an estimate for $\Lambda(u^\kappa, u^\kappa_{\Delta t},\phi)$. More precisely, for $\kappa = 1,\dots,K$,

$$\begin{aligned}\Lambda(u^\kappa, u^\kappa_{\Delta t},\phi) \le{}& -2\iint_{\Pi_T}\iint_{\Pi_T}\chi_2^2(x,t)\operatorname{sign}(u^\kappa(x,t) - u^\kappa_{\Delta t}(y,s))g^\kappa(U(x,t))\phi\,dt\,dx\,ds\,dy\\ &+ \text{Const}_T\left(\Delta t + \omega(\Delta t;v^\kappa) + \left(\Delta t + \omega(\Delta t;u^\kappa) + \omega(\Delta t;u^\kappa_{\Delta t})\right)\left(\frac{1}{\tau} + \frac{1}{h}\right)\right). \end{aligned}\quad (5.66)$$

It remains to estimate $\Lambda(u^\kappa_{\Delta t}, u^\kappa, \phi)$. Lemma 3.15, applied to the present case, states that

$$-\iint_{\Pi_T}\left(\eta\left(u^\kappa_{\Delta t},k\right)\phi_t + q^\kappa\left(u^\kappa_{\Delta t},k\right)\cdot\nabla_x\phi\right)dt\,dx \le \iint_{\Pi_T}\eta'\left(u^\kappa_{\Delta t},k\right)g^\kappa\left(U_{\Delta t}\right)\phi\,dt\,dx + \mathcal{E}^\kappa\left(\Delta t;\phi\right) + \tilde{I}_1^\kappa + \tilde{I}_2^\kappa, \quad (5.67)$$

with

$$
\begin{aligned}
\mathcal{E}^\kappa(\Delta t;\phi) = \sum_{l=1}^{2}\Bigg[\iint_{\Pi_T} \chi_l^2 \Big(|\phi_t|\, d\mu_{\mathrm{I}}{}^{l\kappa} + 2\big(\sum_i |\phi_{x_i}|\, d\mu_{\mathrm{II}}{}_i^{l\kappa} \\
+ \sum_{i,j} \big|\phi_{x_i x_j}\big|\, d\mu_{\mathrm{III}}{}_{ij}^{l\kappa} + |\phi|\, d\mu_{\mathrm{IV}}{}^{l\kappa}\big)\Big) \\
+ \sum_{n=0}^{N-1} \int_{\mathbb{R}^d} |\phi(\,\cdot\,, t_{n,l})|\, d\mu_{\mathrm{I}}{}^{l\kappa}(\,\cdot\,, t_{n,l}) \Bigg],
\end{aligned}
\tag{5.68}
$$

$$
\tilde{I}_1^\kappa = 2\sum_{l=1}^{2} \iint_{\Pi_T} \Big(\chi_l^2 - \frac{1}{2}\Big) q^{l\kappa}\,(u_{\Delta t}^\kappa) \cdot \nabla_x \phi \, dt\, dx,
$$

$$
\tilde{I}_2^\kappa = 2\sum_{l=1}^{2} \iint_{\Pi_T} \Big(\chi_l^2 - \frac{1}{2}\Big) \eta'\,(u_{\Delta t}^\kappa)\, g^{l\kappa}\,(U_{\Delta t})\, \phi \, dt\, dx.
$$

Explicitly, we find

$$
\begin{aligned}
\mathcal{E}^\kappa(\Delta t;\phi) = \iint_{\Pi_T} \chi_1^2 \Big(|\phi_t|\, d\mu_{\mathrm{I}}{}^{1,\kappa} + 2\big(\sum_i |\phi_{x_i}|\, d\mu_{\mathrm{II}}{}_i^{1,\kappa} \\
+ \sum_{i,j} \big|\phi_{x_i x_j}\big|\, d\mu_{\mathrm{III}}{}_{ij}^{1,\kappa} + |\phi|\, d\mu_{\mathrm{IV}}{}^{1,\kappa}\big)\Big) \\
+ \sum_{n=0}^{N-1} \int_{\mathbb{R}^d} |\phi(\,\cdot\,, t_{n,1})|\, d\mu_{\mathrm{I}}{}^{1,\kappa}(\,\cdot\,, t_{n,1}) \\
+ \iint_{\Pi_T} \chi_2^2 \big(|\phi_t|\, d\mu_{\mathrm{I}}{}^{2,\kappa} + 2\,|\phi|\, d\mu_{\mathrm{IV}}{}^{2,\kappa}\big) \\
+ \sum_{n=0}^{N-1} \int_{\mathbb{R}^d} |\phi(\,\cdot\,, t_{n+1})|\, d\mu_{\mathrm{I}}{}^{2,\kappa}(\,\cdot\,, t_{n+1}) \\
= \iint_{\Pi_T} \Big(|\phi_t|\, d\mu_{\mathrm{I}}{}^{\kappa} + 2\Big(\sum_i |\phi_{x_i}|\, d\mu_{\mathrm{II}}{}_i^{\kappa} + \sum_{i,j} \big|\phi_{x_i x_j}\big|\, d\mu_{\mathrm{III}}{}_{ij}^{\kappa} + |\phi|\, d\mu_{\mathrm{IV}}{}^{\kappa}\Big)\Big) \\
+ \sum_{n=0}^{N-1} \sum_{l=1}^{2} \int_{\mathbb{R}^d} |\phi(\,\cdot\,, t_{n,l})|\, d\mu_{\mathrm{I}}{}^{l\kappa}(\,\cdot\,, t_{n,l}),
\end{aligned}
$$

where

$$
\begin{aligned}
d\mu_{\mathrm{I}}{}^{\kappa} &= \chi_1^2\, d\mu_{\mathrm{I}}{}^{1,\kappa} + \chi_2^2\, d\mu_{\mathrm{I}}{}^{2,\kappa}, & d\mu_{\mathrm{II}}{}^{\kappa} &= \chi_1^2\, d\mu_{\mathrm{II}}{}^{1,\kappa}, \\
d\mu_{\mathrm{III}}{}^{\kappa} &= \chi_1^2\, d\mu_{\mathrm{III}}{}^{1,\kappa}, & d\mu_{\mathrm{IV}}{}^{\kappa} &= \chi_1^2\, d\mu_{\mathrm{IV}}{}^{1,\kappa} + \chi_2^2\, d\mu_{\mathrm{IV}}{}^{2,\kappa}.
\end{aligned}
$$

Furthermore, using that $q^{2,\kappa} = 0$, the term $\tilde{I}_1^\kappa$ can be rewritten as

$$\tilde{I}_1^\kappa = \iint_{\Pi_T} \left(\chi_1^2 - \chi_2^2\right) q^\kappa \left(u_{\Delta t}^\kappa, k\right) \cdot \nabla_x \phi \, dt \, dx. \tag{5.69}$$

We collect the terms involving the source term g^κ, using $g^{1,\kappa} = 0$, and find

$$\iint_{\Pi_T} \eta' \left(u_{\Delta t}^\kappa, k\right) g^\kappa \left(U_{\Delta t}\right) \phi \, dt \, dx + \tilde{I}_2^\kappa = 2 \iint_{\Pi_T} \chi_2^2 \eta' \left(u_{\Delta t}^\kappa, k\right) g^\kappa \left(U_{\Delta t}\right) \phi \, dt \, dx.$$

Hence (5.67) reduces to (using that $\phi(x,t,y,s) = \phi(y,s,x,t)$)

$$\begin{aligned} -\iint_{\Pi_T} & \left(\eta \left(u_{\Delta t}^\kappa, k\right) \phi_t + q^\kappa \left(u_{\Delta t}^\kappa, k\right) \cdot \nabla_x \phi\right) dt \, dx \\ & \leq 2 \iint_{\Pi_T} \chi_2^2 \eta' \left(u_{\Delta t}^\kappa, k\right) g^\kappa \left(U_{\Delta t}\right) \phi \, dt \, dx + \mathcal{E}^\kappa \left(\Delta t; \phi\right) + \tilde{I}_1^\kappa, \end{aligned} \tag{5.70}$$

where $\mathcal{E}^\kappa$ and $\tilde{I}_1^\kappa$ are given by (5.68) and (5.69), respectively.

We now set $k = u^\kappa(y,s)$ and integrate over Π_T. Thus

$$\begin{aligned} \Lambda(u_{\Delta t}^\kappa, u^\kappa, \phi) \leq 2 & \iiiint_{\Pi_T \times \Pi_T} \chi_2^2(x,t) \, \eta' \left(u_{\Delta t}^\kappa, u^\kappa\right) g^\kappa \left(U_{\Delta t}\right) \phi \, dt \, dx \, ds \, dy \\ & + \iint_{\Pi_T} \mathcal{E}^\kappa \left(\Delta t; \phi\right) \, ds \, dy + \iint_{\Pi_T} \tilde{I}_1^\kappa \, ds \, dy. \end{aligned} \tag{5.71}$$

The term $\iint_{\Pi_T} \tilde{I}_1^\kappa \, ds \, dy$ is estimated as the corresponding expression (5.57), thus

$$\iint_{\Pi_T} \tilde{I}_1^\kappa \, ds \, dy \leq \mathrm{Const}_T \left(\Delta t + \omega(\Delta t; u_{\Delta t}^\kappa) + \omega(\Delta t; u^\kappa)\right) \frac{1}{h}. \tag{5.72}$$

Finally, the term $\iint_{\Pi_T} \mathcal{E}^\kappa \left(\Delta t; \phi\right) \, ds \, dy$ equals the term R^α given by (5.19), except that in the present case the test function ϕ is simpler, corresponding to $\varphi_{\varepsilon,\rho} = 1$. Hence we may use the estimate of R^α derived in the proof of Theorem 5.1, namely (5.26), with terms involving ρ absent. Thus

$$\begin{aligned} \sum_\kappa & \iint_{\Pi_T} \mathcal{E}^\kappa \left(\Delta t; \phi\right) \, ds \, dy \\ & \leq \mathrm{Const} \left[\sup_{0 \leq t \leq T} \int_{\mathbb{R}^d} d\mu_{\mathrm{I}}(\,\cdot\,, t) + \iint_{\Pi_T} \left(\frac{1}{\tau} d\mu_{\mathrm{I}} + \frac{1}{h} d\mu_{\mathrm{II}} + \frac{1}{h^2} d\mu_{\mathrm{III}} + d\mu_{\mathrm{IV}}\right) \right] \\ & = E^{\mathrm{method}}. \end{aligned} \tag{5.73}$$

Let now γ^κ be defined by

$$\gamma^\kappa(h,\tau,\Delta t) = \nu(\sqrt{d}\,h;u^\kappa) + \nu(\sqrt{d}\,h;u^\kappa_{\Delta t}) + \omega(\tau;u^\kappa) + \omega(\tau;u^\kappa_{\Delta t}) + \mathrm{Const}_T\Big(\Delta t + \omega(\Delta t;u^\kappa_{\Delta t}) + \big(\Delta t + \omega(\Delta t;u^\kappa) + \omega(\Delta t;u^\kappa_{\Delta t})\big)\big(\frac{1}{\tau} + \frac{1}{h}\big)\Big). \tag{5.74}$$

Combining (5.65), (5.66), and (5.71) we find

$$\|u^\kappa(\,\cdot\,,T) - u^\kappa_{\Delta t}(\,\cdot\,,T)\|_{L^1(\mathbb{R}^d)} \le \|u^\kappa(\,\cdot\,,0) - u^\kappa_{\Delta t}(\,\cdot\,,0)\|_{L^1(\mathbb{R}^d)} + \gamma^\kappa(h,\tau,\Delta t) + \iint\limits_{\Pi_T} \mathcal{E}^\kappa\,(\Delta t;\phi)\,ds\,dy$$
$$+ 2\iiiint\limits_{\Pi_T\times\Pi_T} \chi_2^2(x,t)\,\mathrm{sign}(u^\kappa(x,t) - u^\kappa_{\Delta t}(y,s))(g^\kappa(U) - g^\kappa(U_{\Delta t}))\phi\,dt\,dx\,ds\,dy.$$

The last term is estimated as

$$\begin{aligned}
&\Big|\iiiint\limits_{\Pi_T\times\Pi_T} \chi_2^2(x,t)\,\mathrm{sign}(u^\kappa(x,t) - u^\kappa_{\Delta t}(y,s))(g^\kappa(U) - g^\kappa(U_{\Delta t}))\phi\,dt\,dx\,ds\,dy\Big| \\
&\quad\le \|g^\kappa\|_{\mathrm{Lip}} \iiiint\limits_{\Pi_T\times\Pi_T} |U - U_{\Delta t}|\,\phi\,dt\,dx\,ds\,dy \\
&\quad\le \|g^\kappa\|_{\mathrm{Lip}} \sum_\ell \iiiint\limits_{\Pi_T\times\Pi_T} \big|u^\ell(x,t) - u^\ell_{\Delta t}(y,s)\big|\,\phi\,dt\,dx\,ds\,dy \\
&\quad\le \|g^\kappa\|_{\mathrm{Lip}} \sum_\ell \iiiint\limits_{\Pi_T\times\Pi_T} \big(\big|u^\ell(x,t) - u^\ell(y,t)\big| \\
&\qquad\qquad + \big|u^\ell(y,t) - u^\ell(y,s)\big| + \big|u^\ell(y,s) - u^\ell_{\Delta t}(y,s)\big|\,\big)\phi\,dt\,dx\,ds\,dy \\
&\quad\le \|g^\kappa\|_{\mathrm{Lip}} \sum_\ell \big(T\,\nu(\sqrt{d}\,h;u^\ell) + \omega(\tau;u^\ell) + \int_0^T \big\|u^\ell(\,\cdot\,,t) - u^\ell_{\Delta t}(\,\cdot\,,t)\big\|_{L^1(\mathbb{R}^d)}\,dt\big).
\end{aligned}$$

Now set

$$E^{\mathrm{split}}(h,\tau,\Delta t) = \sum_\kappa \Big(\gamma^\kappa(h,\tau,\Delta t) + 2\,\|g^\kappa\|_{\mathrm{Lip}}\Big(\omega(\tau;u^\kappa) + T\,\nu(\sqrt{d}\,h;u^\kappa)\Big)\Big)$$

and

$$N(t) = \sum_\kappa N^\kappa(t), \quad N^\kappa(t) = \|u^\kappa(\,\cdot\,,t) - u^\kappa_{\Delta t}(\,\cdot\,,t)\|_{L^1(\mathbb{R}^d)}\,.$$

Thus, we have established that

$$N(T) \le N(0) + 2\sup_\kappa \|g^\kappa\|_{\mathrm{Lip}} \int_0^T N(t)\,dt + E^{\mathrm{method}}(\Delta t) + E^{\mathrm{split}}(h,\tau,\Delta t).$$

Finally, we use Gronwall's inequality to conclude Theorem 5.15. □

Remark 5.16. If we use exact solution operators and have data with bounded variation, then the above theorem yields

$$\|U(\cdot,T) - U_{\Delta t}(\cdot,T)\|_1 \leq \mathrm{Const}_T\left(\Delta t + h + \tau + \frac{\Delta t}{h+\tau}\right),$$

which can be minimized to give a convergence rate of 1/2. This, as mentioned before, is not optimal.

Front tracking and Euler's method. We propose to use front tracking and dimensional splitting to solve the multi-dimensional conservation law which can be written as $u_t^\kappa + \sum_i f_i^\kappa(u^\kappa)_{x_i} = 0$ as in Section 5.1, and to use the simple Euler method to solve the ordinary differential equation $u_t^\kappa = g^\kappa(U)$. Showing that the resulting approximation satisfies the three basic requirements (3.45), (3.46), and (3.47) is now an easy exercise:

Since $\mathcal{S}_t$ is an L^1-contraction, we have that

$$\left\|U^{n+1}\right\|_{L^1(\mathbb{R}^d)} \leq \left(1 + \Delta t\,\|G\|_{\mathrm{Lip}}\right)\|U^n\|_{L^1(\mathbb{R}^d)}.$$

Hence (3.45) holds.

We then estimate the moduli of continuity of $U_{\Delta t}$ by Kružkov's interpolation lemma, Lemma 3.4. It is not difficult to see that

$$\left|\int_{\mathcal{K}_r}\phi(x)\left(U_{\Delta t}(x,t+\tau) - U_{\Delta t}(x,t)\right)dx\right| \\ \leq \mathrm{Const}_r\left(\|\nabla\phi\|_{L^\infty} + \|\phi\|_{L^\infty}\|G\|_{\mathrm{Lip}}\right)|\tau|,$$

which is (3.47). Furthermore,

$$\int_{\mathbb{R}^d}\left|U^{n+1}(x+h) - U^{n+1}(x)\right|dx \\ \leq \left(1 + \Delta t\,\|G\|_{\mathrm{Lip}}\right)\int_{\mathbb{R}^d}\left|U^{n,1}(x+h) - U^{n,1}(x)\right|dx.$$

Hence,

$$\nu\left(h;U^{n+1}\right) \leq \left(1 + \Delta t\,\|G\|_{\mathrm{Lip}}\right)\nu\left(h;U^n\right), \quad \text{and} \quad \nu\left(h;U_{\Delta t}\right) \leq \mathrm{Const}\,\nu\left(h;U_0\right),$$

which means that (3.46) holds. From this and the Kružkov lemma, it follows that $U_{\Delta t}$ has some temporal modulus of continuity. Furthermore, if U_0 is of bounded variation, this modulus is linear.

Thus we find that splitting using these methods produces a sequence that converges towards a weak entropy solution.

We have already shown that the operator $\mathcal{S}_t$ has error terms that are partial derivatives. The operator $\mathcal{O}_t$ is defined by

$$\mathcal{O}_\tau (U) = U + \tau G(U),$$

and consequently

$$\frac{d}{dt}\eta(\mathcal{O}_t U) = \eta'(\mathcal{O}_t U)\, G(U).$$

This means that the entropy error in Euler's method can be written

$$\begin{aligned}\eta_t(\mathcal{O}_t U) &= \eta'(\mathcal{O}_t U)\, G(\mathcal{O}_t U) + \eta'(\mathcal{O}_t U)\cdot(G(U) - G(\mathcal{O}_t U))\\ &\leq \eta'(\mathcal{O}_t U)\, G(\mathcal{O}_t U) + \|G\|_{\mathrm{Lip}}\,|\mathcal{O}_t U - U|\\ &=: \eta'(\mathcal{O}_t U)\, G(\mathcal{O}_t U) + {\mathcal{E}_{\mathrm{IV}}}^2(t).\end{aligned}$$

In the context of operator splitting, we see that ${\mathcal{E}_{\mathrm{IV}}}^2$ is bounded by the measure

$$d{\mu_{\mathrm{IV}}}^2 := \Delta t\, \|G\|_{\mathrm{Lip}}^2\, \|U_{\Delta t}\|_{L^\infty(\mathbb{R}^d)}\, dxdt,$$

since we are applying $\mathcal{O}_\tau$ only for $\tau \leq \Delta t$.

The remaining part of the E^{method} is the same as before, see (5.35). Hence in this case,

$$E^{\mathrm{method}}(h,\tau) \leq \mathrm{Const}_T\left(\nu(\Delta x; u_0)\left(1 + \frac{1}{\tau} + \frac{1}{h}\right) + \frac{\delta}{h}\right). \tag{5.75}$$

Setting $h = \tau$ and $\delta = \Delta x = \Delta t$, we find that

$$\begin{aligned}&\|U(\,\cdot\,,T) - U_{\Delta t}(\,\cdot\,,T)\|_{L^1(\mathbb{R}^d)} \leq\\ &\qquad \mathrm{Const}_T\Big(\Delta t + \omega(h;U) + \nu(h;U) + \omega(h;U_{\Delta t}) + \nu(h;U_{\Delta t})\\ &\qquad\qquad + \big(\Delta t + \omega(\Delta t;U) + \omega(\Delta t;U_{\Delta t}) + \nu(\Delta t;U_0)\big)\frac{1}{h}\Big).\end{aligned}$$

Summing up, we have shown:

Theorem 5.17. *Suppose* $U_0 \in L^1(\mathbb{R}^d) \cap L^\infty(\mathbb{R}^d)$ *and* $f_1, \dots, f_d \in \mathrm{Lip}_{\mathrm{loc}}(\mathbb{R})$, *and that* G *is Lipschitz continuous and satisfies the no-blow up condition* (5.49). *Define a sequence of functions* $\{U_{\Delta t}\}$ *by* (5.48), *where the front-tracking method and dimensional splitting is used to solve the* K *conservation laws and Euler's method is used to solve the ordinary differential equation. Then the sequence* $\{U_{\Delta t}\}$ *converges to the unique weak entropy solution of* (5.44). *If* $U_0 \in L^1(\mathbb{R}^d) \cap L^\infty(\mathbb{R}^d) \cap \mathrm{BV}(\mathbb{R}^d)$, *then the convergence rate in* Δt *is* $1/2$.

Remark 5.18. *The ECG scheme and Euler's method.* Recalling the error estimate (5.43), we easily show that the conclusions of the above Theorem 5.17 hold if the dimensional-splitting/front-tracking approximation is replaced by the characteristic Euler–Galerkin method.

Monotone methods and Euler's method. In this section we replace the front-tracking method by a monotone method for the conservation laws. For simplicity, we restrict ourselves to the case $K = d = 1$. A monotone, conservative, and consistent method for the conservation law

$$u_t + f(u)_x = 0$$

is a difference method that can be cast as

$$u_i^{n+1} = u_i^n - \lambda \left(F_i^n - F_{i-1}^n\right), \tag{5.76}$$

with $\lambda = \Delta x/\Delta t$, and where F_i^n is a numerical flux given by

$$F_i^n = F\left(u_{i-p}^n, \dots, u_{i+m}^n\right),$$

for some positive integers p and m. The method is consistent if

$$F(u, \dots, u) = f(u),$$

and monotone if

$$\frac{\partial u_i^{n+1}}{\partial u_j^n} \geq 0.$$

Monotone, conservative, and consistent methods converge to the entropy solution of conservation laws. However, they are at most (formally) first-order accurate, see, e.g., [175]. We include this example to illustrate in a simple case how the relevant measures are obtained, not to show how our results can be applied to state-of-the-art methods.

Let Q be given by

$$Q\left(u_1, \dots, u_{p+m+1}, k\right) = \\ F\left(u_1 \vee k, \dots, u_{p+m+1} \vee k\right) - F\left(u_1 \wedge k, \dots, u_{p+m+1} \wedge k\right) \tag{5.77}$$

where k is a constant and $a \vee b = \max(a, b)$, $a \wedge b = \min(a, b)$. Then Q is consistent with the entropy flux q in the sense that

$$Q(u, \dots, u, k) = q(u, k).$$

Furthermore, the result of a monotone scheme satisfies the discrete entropy inequality

$$\frac{1}{\Delta t}(\eta_i^{n+1} - \eta_i^n) + \frac{1}{\Delta x}(Q_i^n - Q_{i-1}^n) \leq 0, \tag{5.78}$$

for $n \geq 0$, where $\eta = \eta(u, k) = |u - k|$, and $\eta_i^n = |u_i^n - k|$ and similarly for Q_i^n. Let for the moment $u_{\Delta t}$ be defined as the approximate solution of the conservation law generated by the monotone method,

$$u_{\Delta t}(x, t) = u_i^n, \quad \text{for } x_{i-1/2} \leq x < x_{i+1/2} \text{ and } t_n \leq t < t_{n+1}. \tag{5.79}$$

We want to show (3.54) in this case that

$$
\begin{aligned}
-\iint\limits_{\Pi_T} &\left(\eta(u_{\Delta t},k)\phi_t + q(u_{\Delta t},k)\phi_x\right) dt\,dx \\
&= -\sum_{n=0}^{N-1}\sum_i \Bigg(\eta_i^n \int_{x_{i-1/2}}^{x_{i+1/2}} \left(\phi\left(x,t_{n+1}\right) - \phi\left(x,t_n\right)\right) dx \\
&\qquad + q_i^n \int_{t_n}^{t_{n+1}} \left(\phi\left(x_{i+1/2},t\right) - \phi(x_{i-1/2},t)\right) dt \Bigg) \\
&= \sum_{n=1}^{N}\sum_i \int_{x_{i-1/2}}^{x_{i+1/2}} \phi\left(x,t_n\right)\,dx\,\left(\eta_i^n - \eta_i^{n-1}\right) \\
&\qquad + \sum_i \left(\eta_i^0 \int_{x_{i-1/2}}^{x_{i+1/2}} \phi\left(x,0\right)\,dx - \eta_i^N \int_{x_{i-1/2}}^{x_{i+1/2}} \phi(x,T)\,dx \right) \\
&\qquad + \sum_{n=0}^{N-1}\sum_i \int_{t_n}^{t_{n+1}} \phi\left(x_{i-1/2},t\right)\,dt\,\left(q_i^n - q_{i-1}^n\right) \\
&= \sum_{n=0}^{N-1}\sum_i \Bigg[\left(\eta_i^{n+1} - \eta_i^n\right) \int_{x_{i-1/2}}^{x_{i+1/2}} \phi\left(x,t_{n+1}\right)\,dx \\
&\qquad\qquad + \left(q_i^n - q_{i-1}^n\right) \int_{t_n}^{t_{n+1}} \phi\left(x_{i-1/2},t\right)\,dt \Bigg] \\
&\qquad + \int\limits_{\mathbb{R}} \left(\eta(u_{\Delta t}(x,0),k)\phi(x,0) - \eta(u_{\Delta t}(x,T-),k)\phi(x,T)\right) dx.
\end{aligned}
\tag{5.80}
$$

Now let ϕ_i^n be the average of ϕ over the grid cell $[x_{i-1/2},x_{i+1/2}]\times[t_n,t_{n+1}]$. We multiply (5.78) by $\Delta x \Delta t \phi_i^n$ and sum over i and $n=0,\dots,N-2$, and subtract the resulting inequality from (5.80), yielding

$$
-\iint\limits_{\Pi_T} \left(\eta(u_{\Delta t},k)\phi_t + q(u_{\Delta t},k)\phi_x\right) dt\,dx
$$

$$
\leq \sum_{n=0}^{N-1}\sum_i \left(\eta_i^{n+1} - \eta_i^n\right)\left(\int_{x_{i-1/2}}^{x_{i+1/2}} \phi\left(x,t_{n+1}\right)\,dx - h\phi_i^n \right) \tag{5.81a}
$$

$$
+ \sum_{n=0}^{N-1}\sum_i \left(\left(q_i^n - q_{i-1}^n\right) \int_{t_n}^{t_{n+1}} \phi\left(x_{i-1/2},t\right)\,dt - \left(Q_i^n - Q_{i-1}^n\right)\Delta t \phi_i^n \right) \tag{5.81b}
$$

$$
+ \int_{\mathbb{R}} \left(\eta(u_{\Delta t}(x,0),k)\phi(x,0) - q(u_{\Delta t}(x,T-),k)\phi(x,T)\right) dx. \tag{5.81c}
$$

The first two terms, (5.81a) and (5.81b) will give the entropy error measures. First we estimate the right-hand side of (5.81a). This equals

$$\begin{aligned}\Bigg|\sum_{i,n}\frac{\eta_i^{n+1}-\eta_i^n}{\Delta t}\int_{t_n}^{t_{n+1}}\int_{x_{i-1/2}}^{x_{i+1/2}}\left(\phi\left(x,t_{n+1}\right)-\phi(x,t)\right)dt\,dx\Bigg| \\ \leq\sum_{i,n}\frac{\left|\eta_i^{n+1}-\eta_i^n\right|}{\Delta t}\int_{t_n}^{t_{n+1}}\int_{x_{i-1/2}}^{x_{i+1/2}}\int_{t}^{t_{n+1}}\left|\phi_t(x,s)\right|ds\,dx\,dt \\ \leq\sum_{i,n}\left|u_i^{n+1}-u_i^n\right|\int_{t_n}^{t_{n+1}}\int_{x_{i-1/2}}^{x_{i+1/2}}\left|\phi_t(x,s)\right|dx\,ds \\ =:\iint_{\Pi_T}\left|\phi_t(x,t)\right|d\mu_{\mathrm{I}}{}^1(x,t),\end{aligned}$$

which defines the error term $\mathcal{E}_{\mathrm{I}}{}^1$, i.e., $\mathcal{E}_{\mathrm{I}}{}^1$ is defined by its action on continuous functions e by

$$\langle\mathcal{E}_{\mathrm{I}}{}^1,e\rangle=\sum_{i,n}\left|u_i^{n+1}-u_i^n\right|\int_{t_n}^{t_{n+1}}\int_{x_{i-1/2}}^{x_{i+1/2}}e(x,t)\,dx\,dt.$$

It is easy to see that we can bound the corresponding measure $d\mu_{\mathrm{I}}{}^1$ by

$$\int_{\mathbb{R}}d\mu_{\mathrm{I}}{}^1(x,t)\leq\omega_T(\Delta t;u_{\Delta t}).$$

To estimate (5.81b), we write it as

$$\begin{aligned}\sum_{n,i}\Bigg[\left(q_i^n-q_{i-1}^n\right)\left(\int_{t_n}^{t_{n+1}}\phi\left(x_{i-1/2},t\right)\,dt-\Delta t\phi_i^n\right) \\ +\left((q-Q)_i^n-(q-Q)_{i-1}^n\right)\Delta t\phi_i^n\Bigg].\end{aligned}$$

The first term in this expression can be bounded by

$$\begin{aligned}\Bigg|\sum_{n,i}\frac{q_i^n-q_{i-1}^n}{\Delta x}\int_{t_n}^{t_{n+1}}\int_{x_{i-1/2}}^{x_{i+1/2}}\left(\phi\left(x_{i-1/2},t\right)-\phi(x,t)\right)dx\,dt\Bigg| \\ \leq\sum_{n,i}\frac{\left|q_i^n-q_{i-1}^n\right|}{\Delta x}\int_{t_n}^{t_{n+1}}\int_{x_{i-1/2}}^{x_{i+1/2}}\int_{x_{i-1/2}}^{x}\left|\phi_x(z,t)\right|dz\,dx\,dt\end{aligned}$$

$$
\begin{aligned}
&\leq \|f\|_{\mathrm{Lip}} \sum_{n,i} \left|u_i^n - u_{i-1}^n\right| \int\limits_{t_n}^{t_{n+1}} \int\limits_{x_{i-1/2}}^{x_{i+1/2}} |\phi_x(x,t)|\, dx\, dt \\
&:= \langle \mathcal{E}_{\mathrm{II}}{}^a, \phi_x \rangle,
\end{aligned}
$$

with the relevant measure bounded by

$$
\int_{\mathbb{R}} d\mu_{\mathrm{II}}{}^a(x,t) \leq \|f\|_{\mathrm{Lip}}\, \nu\left(\Delta x, u_0\right).
$$

To bound the next term, recall that $q(u) = Q(u, \ldots, u)$. Then the second part equals

$$
\begin{aligned}
-&\sum_{n,i} \frac{Q\left(u_i^n, \ldots, u_i^n\right) - Q\left(u_{i-p}^n, \ldots, u_{i+m}^n\right)}{\Delta x} \\
&\qquad \times \int_{t_n}^{t_{n+1}} \left[\int_{x_{i+1/2}}^{x_{i+3/2}} \phi(x,t)\, dx - \int_{x_{i-1/2}}^{x_{i+1/2}} \phi(x,t)\, dx\right] dt \\
&\leq \sum_{n,i} \frac{\left|Q\left(u_i^n, \ldots, u_i^n\right) - Q\left(u_{i-p}^n, \ldots, u_{i+m}^n\right)\right|}{\Delta x} \\
&\qquad \times \int_{t_n}^{t_{n+1}} \int_{x_{i-1/2}}^{x_{i+1/2}} \int_{x}^{x+\Delta x} |\phi_x(z,t)|\, dz\, dx\, dt \\
&\leq \sum_{n,i} \left|Q\left(u_i^n, \ldots, u_i^n\right) - Q\left(u_{i-p}^n, \ldots, u_{i+m}^n\right)\right| \int_{t_n}^{t_{n+1}} \int_{x_{i-1/2}}^{x_{i+3/2}} |\phi_x(x,t)|\, dx\, dt \\
&\leq \|F\|_{\mathrm{Lip}} \sum_{i,n} \sum_{j=i-p}^{i+m} \left|u_i^n - u_j^n\right| \int_{t_n}^{t_{n+1}} \int_{x_{i-1/2}}^{x_{i+3/2}} |\phi_x(x,t)|\, dx\, dt \\
&=: \langle \mathcal{E}_{\mathrm{II}}{}^b, \phi_x \rangle
\end{aligned}
$$

where F is the numerical flux function. The measure corresponding to $\mathcal{E}_{\mathrm{II}}{}^b$ can be bounded by

$$
\int_{\mathbb{R}} \mathcal{E}_{\mathrm{II}}{}^b(x,t) \leq \|F\|_{\mathrm{Lip}}\, \nu((m+p)\Delta x; u_0).
$$

Finally we set

$$
d\mu_{\mathrm{II}}{}^1 = d\mu_{\mathrm{II}}{}^a + d\mu_{\mathrm{II}}{}^b.
$$

This is all we need to establish the terms in E^{method} coming from the solution of the conservation law; the terms coming from the Euler method are as before. Hence the method dependent error reads

$$
E^{\mathrm{method}} = \sup_t \int d\mu_{\mathrm{I}}{}^1(\,\cdot\,, t) + \iint\limits_{\Pi_T} \left(\frac{1}{\tau} d\mu_{\mathrm{I}}{}^1 + \frac{1}{h} d\mu_{\mathrm{II}}{}^1 + d\mu_{\mathrm{IV}}{}^2\right). \tag{5.82}
$$

Furthermore, it is not difficult to show the bound

$$E^{\text{method}} \leq \text{Const}_T \left(\frac{1}{\tau}\omega\left(\Delta t; u_{\Delta t}\right) + \frac{1}{h}\left(\nu(\Delta x; u_{\Delta t}) + \Delta t\right) + \Delta t \right). \tag{5.83}$$

Setting $\tau = h$ and $\Delta x = \Delta t$, we find that

$$\begin{aligned} \|u(\,\cdot\,,T) - u_{\Delta t}(\,\cdot\,,T)\|_1 \leq \text{Const}_T \Big(&\nu\left(h;u\right) + \omega\left(h;u\right) + \nu\left(h;u_{\Delta t}\right) + \omega\left(h;u_{\Delta t}\right) \\ &+ \Delta t + \frac{1}{h}\left(\Delta t + \nu\left(\Delta t; u_{\Delta t}\right) + \omega\left(\Delta t; u_{\Delta t}\right)\right) \Big). \end{aligned} \tag{5.84}$$

Finally, we note that to show that these methods produce a convergent sequence, i.e., showing that (3.45), (3.46) and (3.47) hold, is (by now) routine. Hence, we have proved the following result. (Observe that in one dimension we have that $\text{BV}(\mathbb{R}) \subset L^\infty(\mathbb{R})$.)

Theorem 5.19. *Suppose $u_0 \in L^1(\mathbb{R}) \cap L^\infty(\mathbb{R})$ and $f \in \text{Lip}_{\text{loc}}(\mathbb{R})$, and that g is Lipschitz continuous and satisfies the no-blow up condition* (5.49)*. Define a sequence of functions $\{u_{\Delta t}\}$ by* (5.48)*, where a monotone difference method is used to solve the conservation law, and Euler's method is used to solve the ordinary differential equation. Then the sequence $\{u_{\Delta t}\}$ converges to the unique weak entropy solution of* (5.44)*. If $u_0 \in L^1(\mathbb{R}) \cap \text{BV}(\mathbb{R})$, then the convergence rate in Δt is $1/2$.*

Source splitting in practice. In this section we shall test how the splitting of the zeroth-order (source) term works in practice. As a test example we always use the scalar balance equation

$$u_t + \frac{1}{2}\left(u^2\right)_x = \kappa u(1-u)\left(u - \frac{1}{2}\right). \tag{5.85}$$

Here κ is a positive constant.

Example 5.20. *As initial data for* (5.85) *we chose*

$$u_0(x) = \begin{cases} \frac{e^{\kappa(x+1)}}{e^{\kappa(x+1)}+1}, & \text{for } x \in [-1,0), \\ \frac{e^{\kappa(x-1)}}{e^{\kappa(x-1)}+1}, & \text{for } x \in [0,1), \end{cases} \tag{5.86}$$

and extend this periodically for $x \notin [-1,1)$. The exact solution to this problem is $u(x,t) = u_0(x-t/2)$, and it is therefore easy to compute errors, as $u(x,4) = u(x,0)$. In Figure 5.7 we show the initial function for $\kappa = 5$ and the fronts for $t \in [0,4]$. Here we have used $\delta = \Delta x = 1/50$, and a CFL-number $\nu = 10$. At the bottom we show the resulting piecewise constant approximate solution found by using front tracking and Euler's method. From Figure 5.7 we see that the location of the shock is very accurately represented.

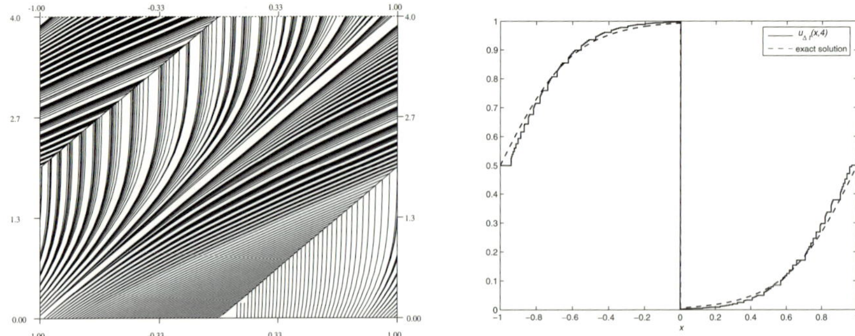

Figure 5.7. (Left) The fronts in the front-tracking approximation for $t \in [0,4]$. (Right) The initial function (5.86), and $u_{\Delta t}(x,4)$ found by front tracking and Euler's method.

Table 5.6. Convergence of the operator-splitting solution to (5.86) for various Δx and ν. Here we used $\delta = \Delta x$, and $\kappa = 5$.

Δx	1/16	1/32	1/64	1/128	1/256	1/512
error ($\nu = 50$)	3.4e-1	1.8e-1	4.7e-2	3.0e-2	1.9e-2	8.6e-3
rate	—	0.9	2.1	0.7	0.7	1.4
error ($\nu = 10$)	1.3e-1	4.2e-2	2.2e-2	9.8e-3	4.8e-3	2.2e-3
rate	—	1.6	0.9	1.2	1.0	1.1
error ($\nu = 2$)	9.0e-2	3.5e-2	1.5e-2	6.5e-3	2.6e-3	1.2e-3
rate	—	1.4	1.2	1.2	1.5	1.1

The theory in this case guarantees a convergence rate of 1/2 for this problem. Table 5.6 shows the L^1 errors at $t = 4$ for various Δx, and for $\nu = 50$ and $\nu = 10$ and $\nu = 2$. The convergence rate is actually much closer to 1 than to $\frac{1}{2}$, and this indicates that our analysis is not optimal for this problem. The time spent to compute a solution increases with decreasing time-steps, and we note that the error decreases only slightly. To investigate this, we computed the errors and the (approximate) time used for this problem for $\nu = 1, 2, 4, \ldots, 64$ with $\Delta x = 1/256$. The results are shown in Figure 5.8. From Figure 5.8 we see that doubling the effort (i.e., the CPU time), results in only a slightly smaller error.

For this specific example, with initial data given by (5.86), the characteristic speed is positive everywhere. Therefore, the simplest example of a monotone method is the upwind method, in this connection given by setting $F_i^n = (u_i^n)^2/2$ in (5.76) resulting in

$$u_i^{n+1} = u_i^n - \frac{\Delta t}{2\Delta x}\left(\left(u_i^n\right)^2 - \left(u_{i-1}^n\right)^2\right). \tag{5.87}$$

This gives a monotone method if $\frac{\Delta x}{\Delta t} \max |u_i| < 1$. Table 5.7 shows the convergence

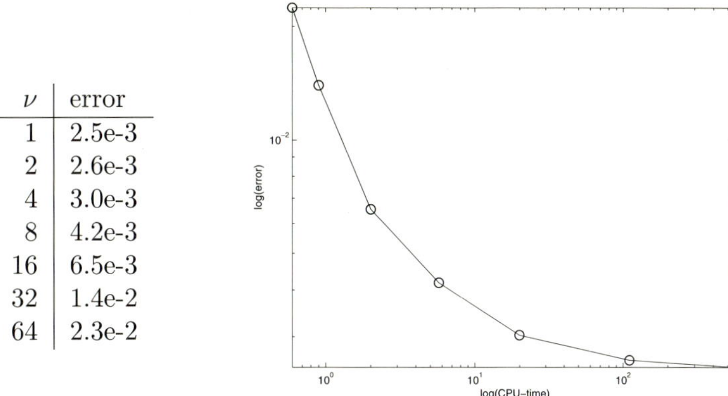

ν	error
1	2.5e-3
2	2.6e-3
4	3.0e-3
8	4.2e-3
16	6.5e-3
32	1.4e-2
64	2.3e-2

Figure 5.8. (Left) The error vs. ν for $\Delta x = 1/256$. (Right) A log-log plot of the error vs. the CPU time.

Table 5.7. Convergence of the operator-splitting solution to (5.86) using the upwind method and Euler's method, the upwind method and Heun's method, and front-tracking and Heun's method for $\nu = 10$.

Δx	1/16	1/32	1/64	1/128	1/256	1/512
error (upwind+Euler)	4.1e-2	3.6e-2	1.1e-2	6.2e-3	3.2e-3	1.7e-3
rate	—	0.2	1.7	0.8	1.0	0.9
error(upwind+Heun)	4.1e-2	2.6e-2	1.1e-2	6.2e-3	3.2e-3	1.7e-3
rate	—	0.7	1.3	0.8	1.0	0.9
error(f.t.+Heun)	1.3e-1	4.3e-2	2.1e-2	9.8e-3	4.8e-3	2.2e-3
rate	—	1.6	1.0	1.1	1.0	1.1

of the upwind method coupled with Euler's method for this example. For all the examples in this section we use $\Delta t/\Delta x = 0.9$.

It is interesting to investigate whether we can improve the splitting method in a simple way by using a second-order method for the ordinary differential equation. This is not warranted by the theory, but since we observe that the convergence rate is higher than expected anyway, it seems worth trying. Thus for the operator $\mathcal{O}_t$ we use Heun's method, i.e.,

$$u^{n+1} = u^n + \frac{\Delta t}{2} \left(g\left(u^n\right) + g\left(u^n + \Delta t g\left(u^n\right)\right)\right).$$

The results are reported in Table 5.7. Comparing Tables 5.7 and 5.6 we observe that a second order method for the ordinary differential equation does not give much better results, and certainly not better convergence rates.

6

Operator Splitting for Systems of Equations

The convergence theory presented later in this book is restricted to scalar equations or weakly coupled systems. However, in this chapter we will also illustrate operator splitting for examples not covered by the theory to show that operator splitting is a viable numeric strategy for more general problems. In order to be consistent with the spirit of the rest of the exposition, and as a reflection of our own previous research activity, we have only included examples where the hyperbolic part of the equation plays an important part. Omission of other examples and of different methods for solving them does not in any way reflect the importance of the omitted problems and techniques.

6.1 Operator splitting in porous media flow

In the two previous chapters we have studied two special classes of operator splittings: viscous splitting and dimensional splitting. Two of the examples we visited were motivated from applications in porous media flow: Examples 4.10 and 4.11 demonstrated the applicability of viscous splitting to the polymer system modeling a one-dimensional tertiary oil recovery process. In this section, we will demonstrate how one can develop efficient large time-step methods for solving the saturation equation that arises as part of standard models for flow in porous media. To this end, we will use a combination of front tracking, dimensional splitting, and viscous splitting as discussed in Sections 4.2 and 5.1. Then at the end of the section, we show how operator splitting and front tracking can be used in a Lagrangian framework to give highly efficient streamline methods that are currently the industry standard with respect to fast simulation. However, the splitting techniques applied within streamline simulation are not covered by the analysis in the second half of the book and the discussion is therefore only included to complement the overall presentation.

For the rudiments of the modeling of porous media flow processes we refer to [8, 54, 204]. We also recommend the recent introductory texts by Aarnes et al. [1, 2], which include some simplified, yet efficient, MATLAB routines for reservoir simulation on Cartesian shoe-box models in three dimensions.

Two-phase, incompressible flow (of oil and water) in porous media can be described by a system of partial differential equations consisting of an elliptic equation for the global fluid pressure $p(x,t)$,

$$\nabla \cdot (\lambda_T \nabla P) = q_1(x,t) + \nabla \cdot \left[(\lambda_w \rho_w + \lambda_o \rho_o)\, g \nabla h \right], \tag{6.1}$$

and a parabolic equation for the saturation $s(x,t)$ of the non-wetting phase (here water)

$$\phi(x) \frac{\partial s}{\partial t} + \nabla \cdot \vec{F}(s,x,t) - \varepsilon \nabla \cdot \big(d(s,x,t) \nabla s \big) = q_2(x,t), \tag{6.2}$$

coupled through a semi-empirical law called Darcy's law, which relates total fluid velocity $v(x,t)$ to pressure gradients

$$\vec{v}(x,t) = -K(x) \big[\lambda_T(s) \nabla p - \big(\lambda_w \rho_w + \lambda_o \rho_o \big) g \nabla h \big]. \tag{6.3}$$

Here ϕ is porosity, K is absolute permeability, g is gravity, $\Delta\rho = \rho_w - \rho_o$ is density difference, h is depth, and q_1 and q_2 are source terms. The global pressure is defined as

$$p = \frac{1}{2}(p_w + p_o) + \frac{1}{2} \int_{s_c}^{s} \Big(\frac{\lambda_o - \lambda_w}{\lambda} \frac{\partial p_c}{\partial s} \Big)\, ds,$$

where $p_c = p_o - p_w$ is the capillary pressure. The relative mobilities $\lambda^r(s)$ model that each phase will move slower through the rock in the presence of the other phase and are given by $k^r_\alpha(s)/\mu_\alpha$, where $k^r_\alpha(s)$ is relative permeability and μ_α is viscosity of phase α. The total mobility is given as $\lambda_T = \lambda_o + \lambda_w$. The convective flux $F(s,x)$ and the diffusive flux $d(s,x)$ in the saturation equation are given by

$$\begin{aligned} \vec{F}(s,x,t) &= f(s) \big[\vec{v}(x,t) + \lambda^r_w(s) K(x) g \Delta\rho \nabla h(x) \big], \\ d(s,x,t) &= -K(x) \frac{\lambda^r_w(s) \lambda^r_o(s)}{\lambda^r_w(s) + \lambda^r_o(s)} \frac{\partial p_c}{\partial s}, \end{aligned} \tag{6.4}$$

where $f(s)$ is the fractional flow given by $\lambda^r_w(s)/(\lambda^r_w(s) + \lambda^r_o(s))$.

To solve the coupled system (6.1)–(6.3) we will use a sequential operator splitting as outlined in Algorithm 6.1.1, which is often referred to as the IMPES (implicit pressure, explicit saturation) algorithm or sequential splitting algorithm.

Algorithm 6.1.1 Sequential splitting for porous media flow

Set initial water saturation $s^0(x) = s(x,0)$
Set $t = 0$ and $\Delta t = T/N$
For $n = 0 : N-1$
    Use $s^n(x)$ to evaluate the s-dependent coefficients in (6.1) and (6.3)
    Fix coefficients in (6.1) and (6.3) and solve for $p(x, n\Delta t)$ and $\vec{v}(x, n\Delta t)$
    Fix $\vec{v}^n(x) = \vec{v}(x, n\Delta t)$, compute $s(x, n\Delta t)$ from (6.2) with initial data $s^n(x)$
    Set $s^n(x) = s(x, n\Delta t)$ and $t = t + \Delta t$

6.1.1 Dimensional splitting. For simplicity, we start by neglecting gravity and capillary forces and set the porosity equal to unity, in which case the two-phase model reads

$$\nabla \cdot \vec{v} = 0, \qquad \vec{v} = -K(x)\lambda(s)\nabla p, \qquad s_t + \vec{v} \cdot \nabla f(s) = 0. \tag{6.5}$$

To solve the saturation equation we use dimensional splitting to obtain a sequence of one-dimensional hyperbolic problems of the form

$$s_t + v_i(x) f(s)_{x_i} = 0, \qquad s(x,0) = s_0(x). \tag{6.6}$$

One can compute approximate solutions of this equation by using the extended front-tracking method described in [181]. The dimensional-splitting method from Algorithm 5.1.1 can therefore be applied almost directly; the only change is the one-dimensional solution operators which are now given by (6.6).

Example 6.1. *In the first example we consider a quarter five-spot test case. The setup consists of a pattern of squares with an injection well in the centre and production wells at the corners* $(\pm 1, \pm 1)$*, which is repeated to infinity in all directions. By symmetry arguments, the computational domain can be reduced to the first quadrant, with an injection well at the origin, a production well at (1,1), and no-flow boundaries. The wells are modeled as point sources/sinks. Initially the reservoir is filled with oil and the permeability realization follows a log-normal distribution. The viscosity ratio of water and oil equals 0.2, which means that the water will tend to finger through the oil and give undesired early water breakthrough, that is, the sudden presence of water in the production well.*

Due to the unfavorable viscosity ratio, we consider two different production strategies: a secondary recovery in which water is injected to displace the oil and a tertiary recovery where polymer is added to the injection water to increase its viscosity. The tertiary recovery is modeled by the polymer system introduced in Example 4.10 on page 84,

$$\begin{bmatrix} s \\ sc + a(c) \end{bmatrix}_t + \vec{v} \cdot \nabla \begin{bmatrix} f(s,c) \\ cf(s,c) \end{bmatrix} = 0. \tag{6.7}$$

Here s *denotes water saturation,* c *polymer concentration, and* $a(c)$ *is the adsorption function. The model for the secondary recovery is a special case of* (6.7) *with* $c \equiv 0$*. For the adsorption function and the fractional flow function we use*

$$a(c) = \frac{0.2c}{1+c}, \qquad f(s,c) = \frac{s^2}{s^2 + 0.2(1-s)^2(1+50c)}.$$

Figure 6.1 shows a plot of the oil and water production for two different production scenarios where we either inject pure water or we inject water with a polymer concentration of 0.1. In the first case, the injected water fingers through the oil because of the unfavorable viscosity ratio and gives an early water breakthrough

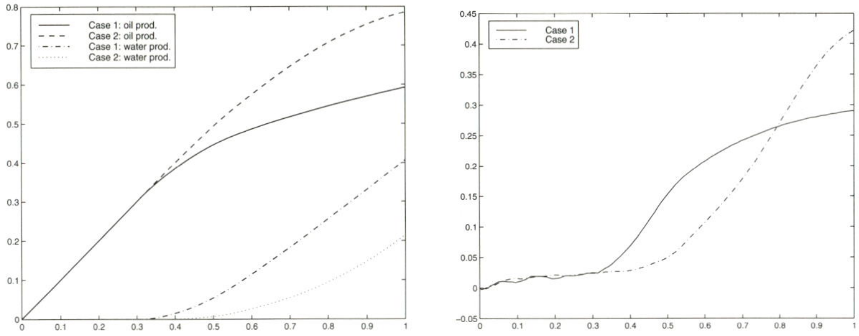

Figure 6.1. (Left) Oil and water production for secondary (Case 1) and tertiary recovery (Case 2). (Right) Percentage mass-balance error for water in the two cases.

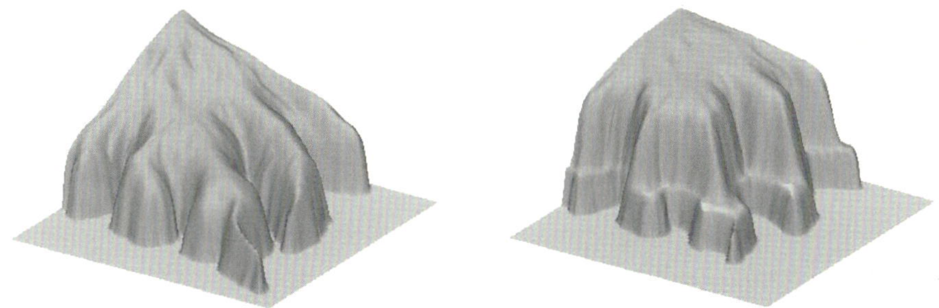

Figure 6.2. Surface plot of the water saturation for secondary recovery (left) and tertiary recovery (right).

(time $t = 0.314$*) in the production well. By adding polymer to the injected fluid in the second case, we improve the mobility ratio and obtain a later breakthrough (time* $t = 0.387$*). Moreover, the areal sweep of the injection increases, giving a much higher oil production. Figure 6.2 shows surface plots of the water saturation at time* $t = 0.3$ *in both cases.*

A major concern for reservoir engineers is conservation of mass. The front-tracking method is not strictly mass conservative due to the discretization of Riemann problems in terms of step functions or if an approximate Riemann solver is used, but this mass error is in general negligible. Dimensional splitting, on the other hand, may introduce large errors in mass conservation. A large number of experiments show that the mass error is negligible before water breakthrough, also for large splitting steps with CFL numbers an order of magnitude above unity. Since the wells are modeled by a point source, the pressure and velocity will be singular at the wells. When the water front reaches the near-well region, the mass error increases unless the splitting step has a CFL number around unity. This sug-

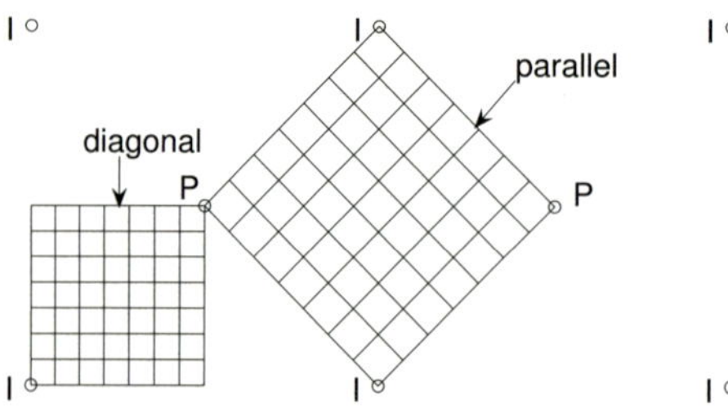

Figure 6.3. Diagonal and parallel grid of a five-spot repeated well pattern. The production and injection wells are denoted by P and I, respectively.

gests an adaptive time-stepping strategy where the splitting steps are large when the water front is in the middle of the reservoir and bounded by a CFL condition around unity after water breakthrough. See [114, 120] for a thorough discussion.

Adaptive time-stepping was used in the simulation of the two recovery strategies. Figure 6.1 *shows that the corresponding mass error never exceeds 0.45%.*

Whereas dimensional splitting (and front tracking) is a robust and efficient method for scalar transport equations in 2-D, this is unfortunately not always the case for the nonstrictly hyperbolic polymer system. As we will see in the next example (from [114]), dimensional splitting with front tracking is sensitive to the grid orientation for unstable displacements: anisotropy introduced by the splitting may lead to unphysical fingers and this effect is accentuated as the grid is refined.

Example 6.2. In this example we consider the injection of pure water into a reservoir that is initially filled with water and polymer with unit concentration. The conservation equation for water is trivial in this case since it is constant for all times. Moreover, we neglect the adsorption to obtain a linear polymer equation. This is a simplified conceptual model designed to study numerical difficulties arising for the adverse mobility displacement ($M > 1$) occurring after the injection of a polymer slug, where M is the mobility ratio between the viscosities of the displaced fluid and the injected fluid. In the following we will use $M = 10$, which implies that the total mobility is $\lambda(c) = 1/(1 + 9c)$. This will create an unstable displacement, in which the injected water tends to finger through the polymer.

The reservoir is homogeneous with a repeated five-spot well pattern, and we use two different grid orientations, see Figure 6.3, such that the grid is diagonal and parallel to the main flow directions, respectively. Figure 6.4 shows the dimensional-splitting solutions computed on a diagonal and a parallel 65×65 grid; both solutions are plotted on the diagonal grid for comparison. The solutions are clearly totally different. On the parallel grid, the water chooses preferred flow directions along the grid axes and gives an unstable finger of pure water extending from the injector and towards the producer. Similarly, on the diagonal grid, the fluid must first move in the axial directions, which make a 45 degree with the

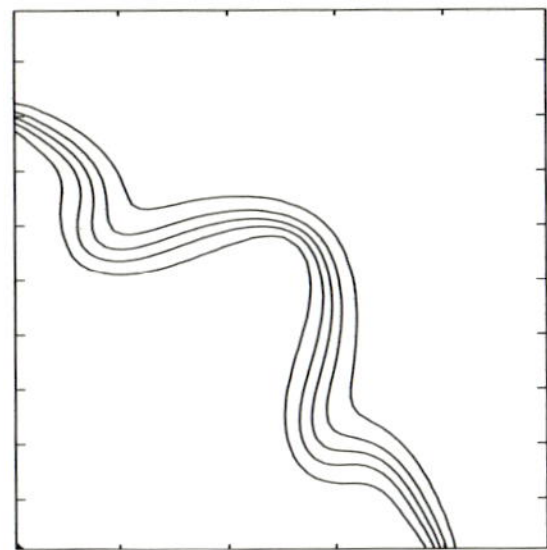
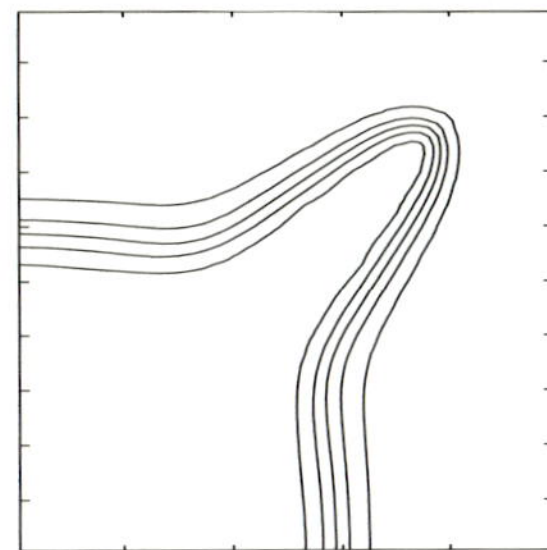

Figure 6.4. Concentration contours after 0.4 pore volumes injected on the diagonal (left) and the parallel grid (right) with $N_x = N_y = 65$ and CFL number 1.0. The velocity field was updated at each time-step.

Table 6.1. Discrepancy in the discrete L^1 norm between the numerical solutions on a diagonal and parallel grid after 0.4 pore volumes injected for different spatial discretizations. All simulations used CFL number 1.0 and pressure updates at each time-step.

$N_x = N_y$	17	33	65	129	257
difference	0.027	0.034	0.057	0.085	0.106
rate	—	-0.36	-0.73	-0.58	-0.32

main flow direction (along the diagonal). This initial movement of the injected fluid triggers unphysical fingers, and the solution on the diagonal grid therefore contains fingers of water flowing towards the injectors in addition to the finger towards the producer: see [114, 242] for a more thorough discussion.

The unphysical behavior on the diagonal grid is not eliminated by using a finer grid. On the contrary, the discrepancy of the two solutions *increases* when the spatial discretization parameters decreases, see Table 6.1. This is caused by earlier breakthrough on the parallel grid and larger unphysical fingers on the diagonal grid as the grid is refined.

6.1.2 Dimensional splitting combined with viscous splitting. Let us now also include capillary forces in the saturation equation. This will lead to a nonlinear second-order derivative term in the saturation equation, as can be seen from (6.4). For simplicity only, we once again neglect gravity, and the saturation equation can be written in the conservative form

$$s_t + \vec{v} \cdot \nabla f(s) = \varepsilon \nabla \cdot \big(K(x) \nabla D(s)\big), \qquad s(x,0) = s_0(x). \tag{6.8}$$

The diffusion function $D(s)$ is an increasing function of s since $\partial p_c/\partial s$ generally is negative. However, since the mobility of one fluid will be zero when the other fluid fills the pore space, $D'(s)$ is zero at the endpoints $s=0$ and $s=1$. Equation (6.8) will therefore degenerate to a hyperbolic equation at the endpoints in s.

To solve the equation by operator splitting, we have two choices. The first alternative is to use dimensional splitting, giving one-dimensional sub-equations of the form,

$$s_t + v_i(x,t) f_i(s)_{x_i} = \varepsilon\big(K(x)D(s)_{x_i}\big)_{x_i}. \tag{6.9}$$

These sub-equations can then be solved by viscous splitting (4.2), for instance using front tracking in combination with a standard finite-difference method as in (4.12). Alternatively, we can first use a viscous splitting to obtain a hyperbolic and a parabolic equation

$$s_t + \vec{v}\cdot\nabla f(s) = 0, \qquad s_t = \varepsilon\nabla\cdot\big(K(x)\nabla D(s)\big). \tag{6.10}$$

Dimensional splitting can then be applied to one or both of the sub-equations if necessary. Here we choose to use dimensional splitting with front tracking for the hyperbolic sub-equation and a straightforward multi-dimensional generalization of the finite-difference method (4.11) for the parabolic sub-equation. This approach avoids dimensional splitting of diffusive forces and is therefore likely to have smaller splitting errors. On the other hand, the advantage of the first approach is that it readily opens up for the use of the corrected operator-splitting method in Algorithm 4.4.1 from Section 4.4. Moreover, the parabolic substep will be faster since an implicit scheme gives a tridiagonal systems and an explicit scheme allows for larger time-steps than a corresponding method in multi-dimensions.

Altogether, we have now introduced three different splitting methods. The OS method

$$s(x,n\Delta t) \approx \big[\mathcal{H}_{\Delta t}\,\pi\mathcal{S}^m_{\Delta t}\cdots\pi\mathcal{S}^1_{\Delta t}\big]^n\pi s_0, \tag{6.11}$$

the OS^{ds} method

$$s(x,n\Delta t) \approx \big[\mathcal{H}^m_{\Delta t}\,\pi\mathcal{S}^m_{\Delta t}\cdots\mathcal{H}^1_{\Delta t}\,\pi\mathcal{S}^1_{\Delta t}\big]^n\pi s_0, \tag{6.12}$$

and the COS method

$$s(x,n\Delta t) \approx \big[\mathcal{H}^{c,m}_{\Delta t}\,\pi\mathcal{S}^m_{\Delta t}\cdots\mathcal{H}^{c,1}_{\Delta t}\,\pi\mathcal{S}^1_{\Delta t}\big]^n\pi s_0. \tag{6.13}$$

The OS splitting method is summarized in Algorithm 6.1.2 for two spatial dimensions.

Example 6.3. *Consider now a two-phase flow in the quarter five-spot configuration from Example* 6.1. *For simplicity we will assume that the diffusion function is linear $D(s)=s$. For the mobilities, we use the analytical expressions $\lambda_w(s)=s^2$ and $\lambda_o(s)=(1-s)^2$.*

A natural question is to ask how fast the algorithm converges. Since the analytical solution is unknown, an indication of the convergence rate can be obtained

Algorithm 6.1.2 OS algorithm in 2-D

Define a uniform grid $[x_{i-1/2}, x_{i+1/2}] \times [y_{j-1/2}, y_{j+1/2}]$.
Construct a piecewise constant initial function $s^0(x,y) = \pi s_0(x,y)$
Set $t = 0$ and $\Delta t = T/N$
For $n = 0 : N-1$
  For each row $j = 1, \dots, N_y$
    Use front tracking to compute solution $s(x, \Delta t)$ of
      $s_t + v_1(x; y_j) f(s)_x = 0, \quad s(x,0) = s^n(x, y_j)$
    Project solution back onto grid:
      $s^{n+1/3}(x,y) = \pi_x^j s(x,\Delta t), \quad y_{j-1/2} < y < y_{j+1/2}$
  For each column $i = 1, \dots, N_x$
    Use front tracking to compute solution $s(y, \Delta t)$ of
      $s_t + v_2(y; x_i) f(s)_y = 0, \quad s(y,0) = s^n(x_i, y)$
    Project solution back onto grid:
      $s^{n+2/3}(x,y) = \pi_y^i s(y,\Delta t), \quad x_{i-1/2} < x < x_{i+1/2}$
  Solve parabolic problem by finite differences on 2-D grid up to $t = \Delta t$
    $s_t = \varepsilon \nabla\big(K(x,y)\nabla D(s)\big), \quad s(x,y,0) = s^{n+2/3}(x,y)$
Set $s(x,y,T) = s^N(x,y)$.

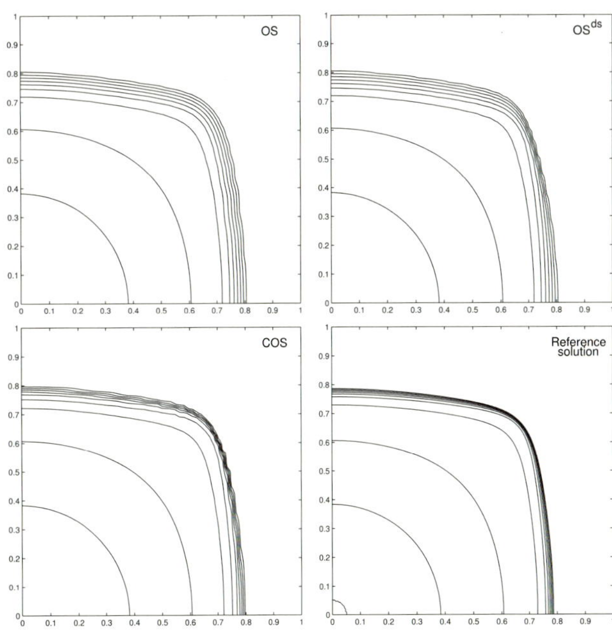

Figure 6.5. Contour plots of water saturation at time $t = 0.45$ in a quarter five-spot calculated by OS, OS^{ds}, COS and on a fine grid.

Table 6.2. Measured errors for solution on an $N \times N$ grid relative to a $2N \times 2N$ grid at time $t = 0.45$ for four different values of ε. The upper half is for OS with explicit finite differences. The lower half is for OS^{ds} with implicit differences.

N	$\varepsilon = 5 \cdot 10^{-2}$		$\varepsilon = 5 \cdot 10^{-3}$		$\varepsilon = 5 \cdot 10^{-4}$		$\varepsilon = 5 \cdot 10^{-5}$	
64	9.60e-03	—	6.05e-03	—	5.83e-03	—	5.85e-03	—
128	4.78e-03	1.01	3.38e-03	0.84	2.94e-03	0.99	3.32e-03	0.82
256	2.50e-03	0.94	1.88e-03	0.85	1.47e-03	1.00	1.54e-03	1.11
64	5.97e-03	—	5.81e-03	—	5.77e-03	—	5.84e-03	—
128	3.91e-03	0.61	3.08e-03	0.92	2.88e-03	1.00	3.32e-03	0.81
256	2.17e-03	0.85	2.01e-03	0.62	1.47e-03	0.97	1.52e-03	1.13

by considering the self-convergence of the operator splitting. That is, we define a sequence of meshes $M_{\Delta x}, M_{\Delta x/2}, \ldots$ and measure the L^1 error on mesh $M_{\Delta x}$ relative to $M_{\Delta x/2}$. Table 6.2 gives the errors and corresponding convergence rates measured by refining a 64×64 grid three times for four different values of ε. In all runs we used only one pressure update step and CFL number 8.0. If large time-steps are used, the resolution of shock-layers can be improved by using the COS splitting. Figure 6.5 shows plots of approximate solutions at time $t = 0.45$ for $\varepsilon = 0.005$ obtained on a 65×65 grid by OS, OS^{ds}, and COS with CFL number 16.0. For comparison, we include a fine grid reference solution (OS on a 257×257 grid with CFL number 2.0). Whereas the shock layer is smeared by both OS and OS^{ds}, the COS splitting seems to resolve it (almost) correctly.

Having indicated convergence in the previous example, we immediately turn to another example with a touch of realism.

Example 6.4. *In the final example we consider a heterogeneous reservoir with a synthetic geomodel consisting of uniform porosity field and a Gaussian permeability field with low-permeable blocks that are barriers to the flow. The permeability field is given on a 257×257 grid and has values varying between 13mD and 4.2D, see Figure 6.7. The reservoir has five injection and five production wells, which are all modeled as simple point sources with equal rates. The viscosity ratio $\mu_o : \mu_w$ equals $8 : 1$ and ε is set to 0.01. Figures 6.6 and 6.7 show the result of a simulation with twenty pressure updates to reach dimensionless time 1.0, which corresponds to injection of a pore volume of water. The splitting step given by a CFL number 8.0 before water breakthrough and 1.0 after. For the operator splitting we used the OS^{ds} method. At time $t = 0.1$ all the five production wells are producing oil only. The water breaks through around time $t = 0.12$ in well 10, a little later in wells 7 to 9, and around time $t = 0.3$ in well 6. At time $t = 0.5$ all wells are producing water; for well 10 the accumulated water production almost equals the oil production.*

Concerning the quality of the simulation, we observe that the mass-balance error stays below 0.02% throughout the whole simulation. Moreover, the simulation

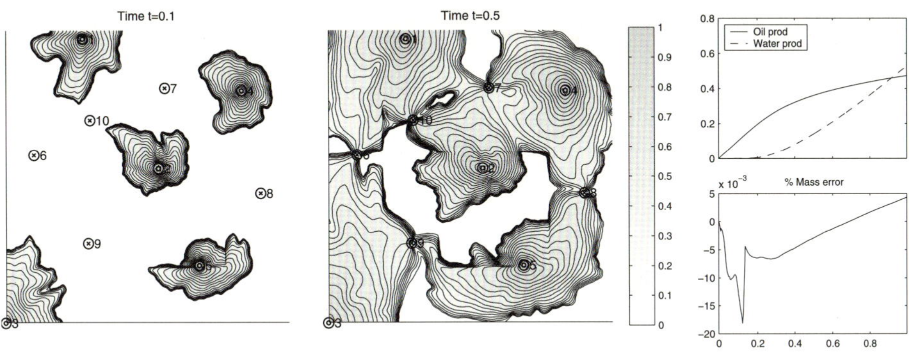

Figure 6.6. Simulation of the synthetic reservoir showing twenty equally spaced saturation contours, accumulated field production, and mass-balance error.

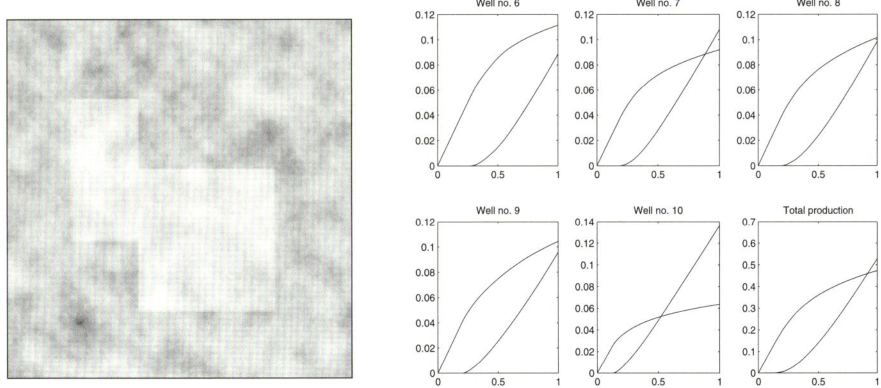

Figure 6.7. (Left) Permeability plotted on a logarithmic gray scale with dark colour for large values and light colour for small values. (Right) Accumulated production for the five production wells.

correctly reproduces sharper gradients in the low-permeability regions.

An overview of various operator-splitting methods with applications to porous media is given in [93]. Reference [120] contains more examples and a thorough discussion of the operator splittings used in the current section; that paper discusses errors in mass conservation, adaptive splitting steps, grid orientation effects, and computational efficiency for the various splittings.

In Section 5.1 we saw how dimensional splitting with front tracking is a highly efficient method for scalar problems with simple dynamics. In many cases one can use CFL numbers as high as 10–50 without affecting the accuracy significantly. Unfortunately, this situation changes when sources/sinks are included in the forcing velocity field, as is typically the case for porous media flow. Extensive numerical tests show that the method propagates fluid fronts very accurately inside the reservoir and away from the wells. Once breakthrough occurs, substantial material balance errors are introduced if too large time-steps are used. By reducing the splitting steps to a CFL number about unity, acceptable material balance errors are obtained. Still, for Cartesian grid models in 2-D, the above operator-splitting methods are surprisingly robust and are able to resolve a wide range of balances between convective and capillary forces up to breakthrough in wells.

Unfortunately, the above operator-splitting methods are not easy to extend to more realistic models in an efficient manner. First of all, real reservoir models are seldom given as Cartesian grids. The industry standard is to use so-called corner-point grids, which consist of a set of hexahedral cells that are aligned in a logical Cartesian fashion. In physical space, the rows and columns of the grid follow geological layers and these are seldom aligned with the coordinate directions. Secondly, the magnitude of the forcing velocity field will typically have large variations throughout space due to heterogeneities in the permeability field and the near-singular behavior close to injection and production wells. This may introduce severe restrictions on the global splitting steps, as observed above at water breakthrough. Furthermore, the number of calls to one-dimensional solvers increases dramatically for dimensional splitting in 3-D, and this makes dimensional splitting less attractive in terms of computational efficiency.

The above arguments explain why dimensional splitting is seldom used for industry-standard reservoir simulation. However, this does not mean that the underlying ideas do not have their merit. In the next section, we will consider the alternative setting of streamline simulation where front tracking and operator splitting is used to its full potential in a leading commercial reservoir simulator. Streamline methods [80, 258] are by many regarded as the most efficient way of simulating large and complex reservoir models where the flow is dictated by rock properties, well positions and rates, fluid mobility, etc.

6.1.3 Streamline methods. Two-phase models can be solved very efficiently by using modern streamline methods, and these methods are gaining in popularity among reservoir engineers due to their capability of solving large geological models. In the following we will discuss the operator splittings that underlie modern streamline methods, assuming a simplified flow model given by (6.5).

A streamline associated with a velocity field $\vec{v}$ is defined as the line that is everywhere tangential to the velocity field: that is, given by the relation $d\vec{x}/dr = \vec{v}/|\vec{v}|$. For a typical reservoir that produces oil by waterflooding, the fluids will flow along streamlines that start at injection wells and terminate at production wells. Rather than using the arc-length to parametrize streamlines, it is common to use the so-called *time-of-flight* τ, which is given by

$$\tau(r) = \int_0^r \frac{\phi(\zeta)}{\vec{v}(\zeta)} d\zeta,$$

or, alternatively, by the differential relation

$$\vec{v} \cdot \nabla\tau = \phi. \tag{6.14}$$

Compared with the arc-length r, the time-of-flight takes into account the relative voidage volume (i.e., the porosity) and can therefore be interpreted as the time it takes to reach a given point in the reservoir for a neutral and massless particle traveling with unit phase velocity. Next we introduce the bistream-functions ψ and χ given by $\vec{v} = \nabla\psi \times \nabla\chi$, which together with τ form a complete spatial coordinate system. Then, we can perform a formal spatial coordinate transformation $(x, y, z) \mapsto (\tau, \psi, \chi)$. The gradient operator can be expressed in the new coordinates as

$$\nabla = (\nabla\tau)\frac{\partial}{\partial\tau} + (\nabla\psi)\frac{\partial}{\partial\psi} + (\nabla\chi)\frac{\partial}{\partial\chi}.$$

Because $\vec{v}$ is orthogonal to $\nabla\psi$ and $\nabla\xi$, it follows that $\vec{v}\cdot\nabla = \phi\frac{\partial}{\partial\tau}$. Combining this with the equation for conservation of mass, $\nabla \cdot \vec{v} = 0$, we can rewrite the three-dimensional saturation equation (6.5) as a family of one-dimensional transport equations along streamlines,

$$s_t + \partial_\tau f(s) = 0, \qquad s(\tau, 0) = s_0(\tau). \tag{6.15}$$

Now we can use a dimensional splitting in the transformed coordinates to define the streamline method. That is, the solution of the saturation equation in (6.5) can be written as

$$s(x, n\Delta t) = \left[\mathcal{S}_{\Delta t}^{(\psi,\chi)}\, \mathcal{S}_{\Delta t}^{\tau}\right]^n s_0, \tag{6.16}$$

where $\mathcal{S}^\tau$ denotes the one-dimensional solution operator associated with (6.15) and the operator $\mathcal{S}^{(\psi,\chi)}$ is a trivial identity operator since $\vec{v}$ is orthogonal to the gradient of the two bistream-functions. For a constant (in time) velocity field, we now have a simple method of computing the fluid flow. In the unsteady case the velocity field is time-dependent and the coordinate transform will also depend on time. However, if we use the sequential splitting from Algorithm 6.1.1, we avoid this problem. For each step in the algorithm we thus impose a steady velocity field on the saturation equation, a velocity field that is defined by the initial velocity $\vec{v}^n(x)$ of the time-step. The streamlines will therefore be obtained from

$$\vec{v}^n \cdot \nabla\tau_n = \phi,$$

instead of from the true velocity $\vec{v}(x, t)$ as in (6.14). Revisiting the operator-splitting formula (6.16), we can now give a new meaning to the two operators

$$\begin{aligned} \mathcal{S}^\tau: &\qquad s_t + \partial_{\tau_n} f(s) = 0, \\ \mathcal{S}^{(\psi,\chi)}: &\qquad s_t + \nabla \cdot \big((\vec{v} - \vec{v}^n) f(s)\big) = 0. \end{aligned}$$

The operator $\mathcal{S}^{(\psi,\chi)}$ accounts for the transversal flux that arises because of temporal changes in the velocity field; see [218] for a more thorough discussion. In practice, the effect of $\mathcal{S}^{(\psi,\chi)}$ is often insignificant, and this (corrective) operator is therefore neglected. Instead, [218] proposes to use the CFL number ν^n defined by the velocity residual $\vec{v} - \vec{v}^n$ as an indicator of the accuracy of each step in the sequential splitting algorithm. Numerical experiments in [218] show that the splitting is stable if $\nu \leq 1$, whereas unstable effects can be observed if $\nu > 1$.

Practical implementations of the streamline method use an underlying grid in physical space (i.e., using a grid in (x, y, z)). Streamlines are computed such that each grid block contains at least one streamline. In the saturation solver, saturation values are mapped from the underlying grid in physical space and onto streamlines. This gives piecewise constant initial data defined on an irregular grid in τ for each streamline. The solution along the streamline is computed using an efficient one-dimensional method resulting in a new set of piecewise constant saturation values. These saturation values are then projected onto the underlying grid in physical space.

In general one cannot neglect gravity as was done above, and thus the saturation equation assumes a more complex form

$$s_t + \nabla \cdot \big(\vec{v} f(s) + \vec{g}\, h(s, x)\big) = 0. \tag{6.17}$$

For this equation it is natural to define a new operator splitting

$$s(x, n\Delta t) = \big[\mathcal{S}^g_{\Delta t}\, \mathcal{S}^\tau_{\Delta t}\big]^n s_0,$$

where the operator $\mathcal{S}^g$ is defined by

$$s_t + \nabla \cdot \big(\vec{g}\, h(s, x)\big) = 0$$

and acts along *gravity lines* given by $d\vec{x}/dr = \vec{g}/|\vec{g}|$. This operator splitting was first introduced by Gmelig Meyling [105, 106] and later made popular by Bratvedt et al. [38]. In a practical implementation of the algorithm, one must also include projections back to an underlying grid in physical space (as discussed above). Thus the numerical algorithm can be written

$$s(x, n\Delta t) = \big[\pi_g\, \mathcal{S}^g_{\Delta t}\, \pi_\tau\, \mathcal{S}^\tau_{\Delta t}\big]^n s_0,$$

where π_g and π_τ denote projection operators from gravity lines and streamlines to the grid in physical space. This operator splitting is used in current commercial streamline simulators.

Example 6.5. *Figures* 6.8 *and* 6.9 *show computations of a reservoir model describing a* 1000 × 500 × 100*m reservoir with five horizontal layers with different permeabilities. Within each layer the permeability is constant. The system is discretized using 100 grid blocks in the x and z directions and a single block in the y direction. A reference solution was also generated using* 500 *blocks in x and z directions. The pressure equation is discretized by a standard seven-point finite-difference approximation, see, e.g., [1]. The transport equations along*

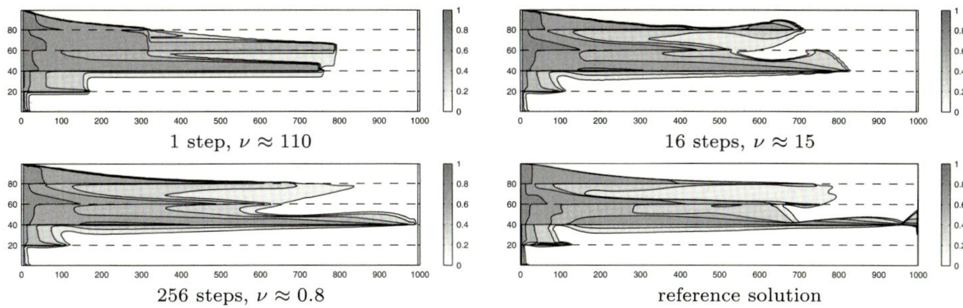

Figure 6.8. Water saturation in the (x, z) plane after 1024 days computed with different number of sequential splitting steps.

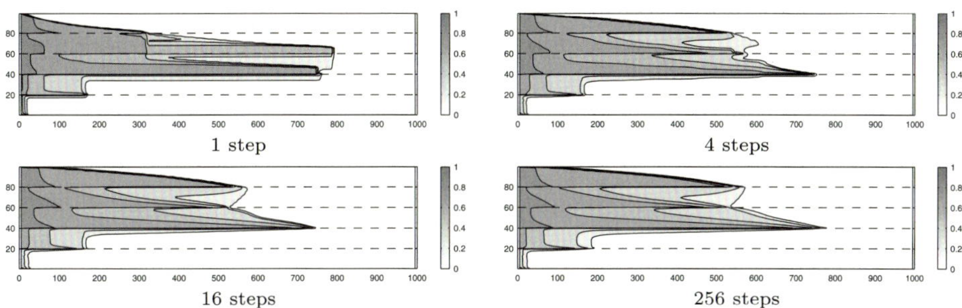

Figure 6.9. Water saturation in the (x, z) plane after 1024 days computed with a single pressure update and varying number of gravity steps.

streamlines are solved using the front-tracking method, which we have seen is capable of tracking piecewise constant saturations on an arbitrary irregular grid. For this particular case, the gravity lines are aligned with the vertical direction and this simplifies the gravity step.

Figure 6.8 *shows the resulting water saturations after simulating the above model for 1024 days using different number of sequential steps in Algorithm* 6.1.1. *Clearly the correct overall behavior, with breakthrough occurring in the third layer, is only recovered for the stable time-step with* $\nu < 1$. *The actual correction step is not performed as the contributions to the grid block saturations are very small, but the associated CFL number is nonetheless useful for classifying stable time-steps.*

Mapping streamline saturations to the grid significantly smears the solution, as we also observed for the dimensional splittings in Section 5.1. *Hence, the effect of refining the gravity splitting will often be dominated by this smearing. However, if the sequential steps are large, it is often possible to observe a significant improvement when increasing the number of gravity steps, as shown in Figure* 6.9.

Having included gravity in the saturation solver, the final question is how to include capillary forces. Referring to (6.2), the saturation equation can be written in the general form

$$s_t + \nabla \cdot \big(\vec{v} f(s) + \vec{g}\, h(s,x)\big) = \varepsilon \nabla \cdot \big(K(x) \nabla D(s)\big).$$

The obvious choice is to use a viscous splitting as in Section 6.1.1,

$$s_t + \nabla \cdot \big(\vec{v} f(s) + \vec{g}\, h(s,x)\big) = 0, \qquad s_t = \varepsilon \nabla \cdot \big(K(x) \nabla D(s)\big),$$

and solve the parabolic sub-equation on the underlying grid by some standard method. The corresponding operator-splitting method reads

$$s(x, n\Delta t) = \big[\mathcal{H}_{\Delta t}\, \pi_g\, \mathcal{S}^g_{\Delta t}\, \pi_\tau\, \mathcal{S}^\tau_{\Delta t}\big]^n s_0.$$

Unfortunately, such an operator-splitting method would suffer from the shortcomings discussed in Section 4.3, and would in general require a splitting step proportional to ε in order to resolve shock layers correctly. Still, it may prove useful in particular cases and has been investigated by several authors, e.g., [25, 227].

At the time of writing it is an open question how to extend the idea of flux corrections to this kind of splitting. Conceptually, one may apply the streamline-bistream transformation also to the diffusion operator and introduce a splitting of the form,

$$\begin{aligned} s_t + f(s)_\tau &= \varepsilon \partial_\tau \big(K \partial_\tau D(s)\big), \\ s_t &= \varepsilon \nabla_{(\psi,\chi)} \big(K \nabla_{(\psi,\chi)} D(s)\big), \\ s_t + \nabla \cdot \big(\vec{g}\, h(s,x)\big) &= 0, \end{aligned}$$

where the first operator is then approximated by a one-dimensional COS approximation. In 2-D, this may be conceivable since the bistream directions can be found by direct integration of $d\vec{x}/d\psi = \vec{v}^{\perp}/|\vec{v}|$. Whether this actually will be efficient or not is an open question.

6.2 Dimensional splitting for systems of conservation laws

The general convergence theory presented in the second half of the book only applies to scalar equations and systems of equations that are weakly coupled through source terms. Operator splitting, however, is generally applicable to any system. In this and the next section we will therefore show the application of various operator splittings to systems of conservation and balance laws, for which our convergence theory is not applicable. We will consider dimensional splitting, splitting of geometrical source terms and source terms modeling chemical reactions. By doing so, we wish to further demonstrate the general applicability of operator splitting, but also show some of the potential shortcomings and limitations one may observe for systems of conservation laws (as we already have done for the non-strictly hyperbolic polymer system in the previous section).

In the first examples, we study the Euler equations for an ideal, polytropic gas which is *the* canonical example of a hyperbolic system of conservation laws. In two spatial dimensions the equations read

$$\begin{bmatrix} \rho \\ \rho u \\ \rho v \\ E \end{bmatrix}_t + \begin{bmatrix} \rho u \\ \rho u^2 + p \\ \rho uv \\ u(E+p) \end{bmatrix}_x + \begin{bmatrix} \rho v \\ \rho uv \\ \rho v^2 + p \\ v(E+p) \end{bmatrix}_y = 0. \tag{6.18}$$

Here ρ denotes the density, u and v the velocity in the x and y directions, p the pressure, and E the total energy (kinetic plus internal energy) given by $E = \rho(u^2+v^2)/2 + p/(\gamma-1)$. In all computations we use $\gamma = 1.4$.

To solve the Euler equations we will employ two operator-splitting methods: a method based upon front tracking (FTds), as described in [124], and the Nessyahu–Tadmor scheme used componentwise (called NTds). For comparison with an unsplit scheme we will include the two-dimensional extension of the Nessyahu–Tadmor scheme (NT2d), see (A.28).

Example 6.6. *Consider the Euler equations* (6.18) *with initial data*

$$(\rho, u, v, p) = (r(x,y), \cos\theta, \sin\theta, 1),$$

where $r(x,y)$ equals one of two functions

$$r_1(x,y) = 1 + 0.1\cos(2\pi x)\cos(2\pi y),$$
$$r_2(x,y) = \begin{cases} 1.0, & \text{if } |x-0.5| < 0.1 \text{ and } |y-0.5| < 0.1, \\ 0.5, & \text{otherwise,} \end{cases}$$

defined over the unit square with periodic boundary data. For constant θ, this initial data gives a linear advection of the initial profile along a vector $\mathbf{v} = (\cos\theta, \sin\theta)$.

Table 6.3 shows estimated L^1-errors and convergence rates obtained through a grid-refinement study with FTds, NTds, and NT2d on a series of grids with $N_x \times N_x$ cells. The time-steps for the central schemes were restricted by a CFL condition of 0.3. For the smooth solution we used central differences to construct the slopes in the reconstruction (i.e., no limiter) and for the discontinuous profile we used the MM_2 limiter, which reduces to central differences if

$$1/3 \le |u_i - u_{i-1}|/|u_{i+1} - u_i| \le 3.$$

For FTds, we used $N_t/8$ splitting steps for the discontinuous case and a CFL condition of 8.0 in the smooth case.

Table 6.3. Estimated L^1-errors, convergence rates and runtimes in seconds for the solution of Euler equations at time $t = 0.5$ with initial density variation $r_1(x, y)$ and $\theta = \pi/10$ (upper part) and $r_2(x, y)$ and $\theta = \pi/4$ (lower part).

	NT2d			NTds			FTds		
N_x	error	rate	time	error	rate	time	error	rate	time
16	4.236e-03	—	0.06	2.307e-03	—	0.07	6.771e-03	—	0.01
32	8.910e-04	2.25	0.40	3.178e-04	2.86	0.47	3.406e-03	0.99	0.05
64	2.134e-04	2.06	4.24	6.216e-05	2.35	3.72	1.704e-03	1.00	0.39
128	5.320e-05	2.00	28.50	1.564e-05	1.99	28.70	8.713e-04	0.97	2.81
256	1.335e-05	1.99	211.00	4.074e-06	1.94	225.00	4.328e-04	1.01	21.20
512	3.348e-06	2.00	1820.00	1.047e-06	1.96	1680.00	2.165e-04	1.00	168.00
16	3.181e-02	—	0.09	3.140e-02	—	0.10	1.239e-02	—	0.01
32	1.549e-02	1.04	0.57	1.510e-02	1.06	0.64	7.514e-03	0.72	0.01
64	8.977e-03	0.79	5.36	8.792e-03	0.78	4.76	5.617e-03	0.42	0.04
128	5.543e-03	0.70	36.90	5.465e-03	0.69	37.50	3.733e-03	0.59	0.18
256	3.361e-03	0.72	281.00	3.323e-03	0.72	295.00	2.662e-03	0.49	1.42
512	1.974e-03	0.77	2450.00	1.959e-03	0.76	2280.00	1.860e-03	0.52	9.45

The solution for r_1 is smooth and we observe the expected first-order convergence for FTds and second-order convergence for the central schemes. Somewhat surprising, we note that NTds generally has lower error than NT2d. A comparison of error versus runtime shows that the two second-order schemes perform almost equally and are superior to the first-order front-tracking method.

The solution for r_2 is discontinuous and therefore the convergence rates are reduced to well below 1.0. Comparing error versus runtime shows that the front-tracking method is far superior with a factor up to 250 in runtime on the finer grids. The reason for this high efficiency is that there are only a small number of waves that needs to be resolve, as opposed to the smooth case, which is approximated in terms of a large number of small waves.

Although the above example was quite simple, it demonstrated important characteristics of the methods considered. The second-order central-difference scheme gives a fairly good resolution of smooth waves, but behaves poorly on linear discontinuities (contacts and shear waves). The front-tracking scheme, on the other hand, gives excellent resolution of discontinuities, but is only first-order accurate on smooth waves.

In the next example we will study the Riemann problem in two spatial dimensions, which has a richer wave structure. The 2-D Riemann problem consists of four uniform states, one in each quadrant. The corresponding wave structures can be categorized into 19 different patterns [238] (see also [165, 165, 173]) when the constant states are chosen such that the four one-dimensional Riemann problems between the quadrants lead to a single wave that is either a shock, a rarefaction or a contact/slip. Here we choose one of these configurations (number five from [173]), where the similarity solution consists of four contact discontinuities/slip-lines.

Example 6.7. *Consider the Euler equations* (6.18) *on the unit square* $[-0.5, 0.5] \times [-0.5, 0.5]$ *with constant initial data in each quadrant. In primitive variables*

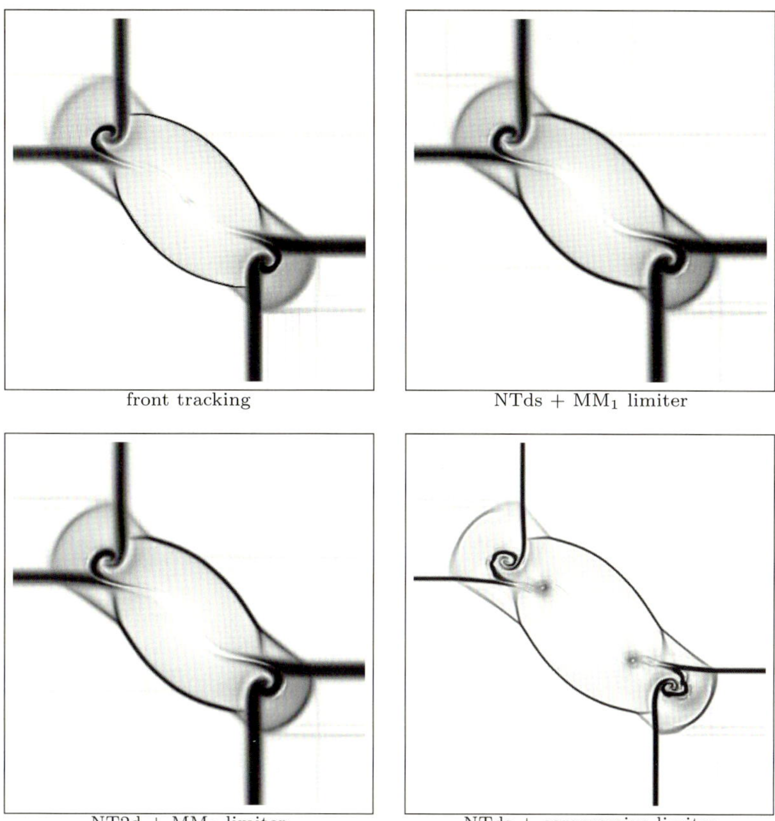

Figure 6.10. Emulated Schlieren images of the 2-D Riemann problem, configuration number 5 computed on a 400×400 grid. The plots show the magnitude of the density gradient depicted on a nonlinear grayscale.

$v_i = (p, \rho, u, v)$ *these data are*

$$v_2 = (1.0, 2.0, -0.75, 0.5), \quad v_1 = (1.0, 1.0, -0.75, -0, 5),$$
$$v_3 = (1.0, 1.0, 0.75, 0.5), \quad v_4 = (1.0, 3.0, 0.75, -0.5).$$

Figure 6.10 *shows the solutions computed by front tracking, the NTds, and a variant of the NT2d scheme where we have included improved quadrature rules for the flux computations to reduce grid orientation effects [183]. The runtimes are respectively 118, 694, and 1210 CPU seconds. For the NTds and NT2d schemes we used CFL number 0.475 (giving 264 time-steps) and for the front-tracking method we used 100 equally spaced time-steps.*

The solutions are plotted as emulated Schlieren images by depicting the norm of the density gradient in a nonlinear graymap, that is, show $(1 - \|\nabla\rho\|)^p$ *for* $p \sim$ *10–15. This visualization technique is particularly useful for depicting gradients in the density, but is also excellent for enhancing small-scale variances, for instance, the small numerical artifacts seen as shaded vertical/horizontal lines. In the visual norm, there is no apparent quality difference in the computations: the curved shocks are resolved accurately and the contacts are too diffusive. To improve the resolution of the contacts for the NTds scheme (and also for the NT2d scheme), we can employ a more compressive (but also potentially more dangerous) limiter. The lower-right plot in Figure* 6.10 *shows a computation with the SBM limiter [184] with* $\theta = 2.0$ *and* $\tau = 0.25$. *For the front-tracking method, the contacts can be sharpened by increasing the splitting step. However, this will result in the creation of more spurious small-scale oscillations in the region bounded by the curved shocks.*

To further assess the grid-orientation effects introduced by the dimensional-splitting approach we can study a problem with circular symmetry on a Cartesian grid. To be specific, we study a radially symmetric version of Sod's test problem.

Example 6.8. *Consider the Euler equations subject to the initial conditions*

$$(\rho, u, v, p)(x, y, 0) = \begin{cases} (1.0, 0, 0, 1.0), & |x|^2 + |y|^2 \le 0.16, \\ (0.0125, 0, 0, 0.1), & \textit{otherwise}. \end{cases}$$

The solution consists of a circular shock wave propagating outwards from the origin, followed by a contact discontinuity, and a rarefaction wave travelling toward the origin.

Figure 6.11 *shows a scatter plot of the solution at time* $t = 0.2$ *computed by front tracking, the NTds and the NT2d scheme (with improved quadrature rules). A scatter plot of a quantity is a plot of the value in each grid cell versus the distance of the cell centre from the origin. In this way we can present the spread in the data, as a solution with perfect radial symmetry would consist of points lying on a single line. In Figure* 6.11 *the solid line is a reference solution computed by a splitting method with shock tracking for the reduced one-dimensional, inhomogeneous system describing the radial flow (see Example* 6.10 *below).*

The NTds scheme shows small grid-orientation effects and is comparable with the multi-dimensional scheme NT2d. The front-tracking scheme, on the other hand, clearly exhibits splitting errors, but has sharper resolution of the shock and produces fewer overshoots in the post-shock region. The result of a grid-refinement study is reported in Table 6.4. *For the central schemes we used a CFL condition of 0.475 and the* MM_2 *limiter, and for the front-tracking method we used splitting steps corresponding to CFL numbers around 1.2. We see that the accuracy and efficiency of the three schemes are quite similar, with a slight preference to the central schemes over the front-tracking method, for which the analytical Riemann solver [109] applied is relatively costly compared with the Riemann solver used in the scalar case.*

In the final example we will study a dambreak problem described by the shallow

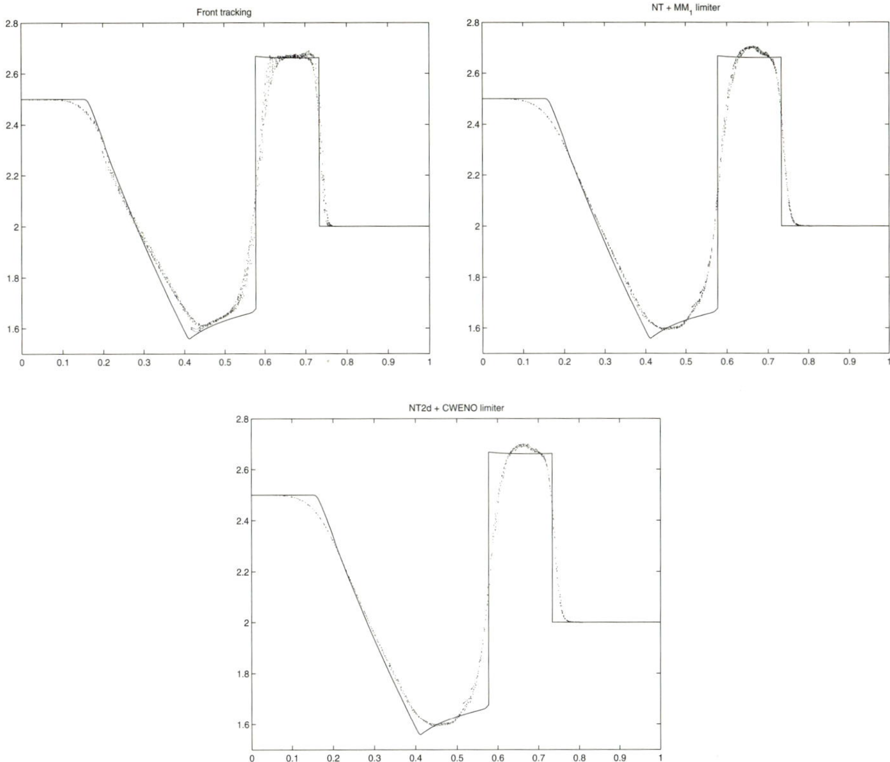

Figure 6.11. Solution of the radially symmetric Sod's problem computed on a 101×101 grid using 10 time-steps for front tracking and CFL number 0.475 (giving 23 time-steps) for the central schemes.

Table 6.4. Estimated L^1-errors, convergence rates and runtimes in seconds for the solution of the radial Sod's problem at time $t = 0.2$.

N	NT2d			NTds			FTds		
16	9.17e-02	—	0.01	9.06e-02	—	0.02	1.10e-01	—	0.01
32	7.28e-02	0.33	0.06	7.13e-02	0.35	0.07	6.57e-02	0.74	0.11
64	3.66e-02	0.99	0.60	3.74e-02	0.93	0.55	3.87e-02	0.76	0.53
128	2.00e-02	0.87	4.05	1.88e-02	0.99	4.13	2.54e-02	0.61	3.18
256	1.08e-02	0.89	30.20	9.71e-03	0.95	32.30	1.53e-02	0.73	21.10
512	6.06e-03	0.83	240.00	5.31e-03	0.87	261.00	9.31e-03	0.72	154.00

water equations, which are obtained by considering a depth-integrated flow,

$$\begin{bmatrix} h \\ hu \\ hv \end{bmatrix}_t + \begin{bmatrix} hu \\ hu^2 + \frac{1}{2}gh^2 \\ huv \end{bmatrix}_x + \begin{bmatrix} hv \\ huv \\ hv^2 + \frac{1}{2}gh^2 \end{bmatrix}_y = 0.$$

Here h denotes water depth, u and v are depth-averaged velocities in the x and y directions, respectively, and g is the acceleration of gravity (set to 9.8 $\mathrm{m/s^2}$ in the next example).

Example 6.9. *Consider a 200 m long and 200 m wide water reservoir with two different constant levels of water separated by a dam. The dam is 10 m thick and extends in the y-direction, starting at $x = 95$ m. The dam breaks at time $t = 0.0$ and the breach is 75 m wide and starts at $y = 95$ m. The initial conditions are given by*

$$h(x, y, 0) = \begin{cases} 10\,\mathrm{m}, & x \le 100\,\mathrm{m}, \\ 5\,\mathrm{m}, & \textit{otherwise}, \end{cases}$$

$$u(x, y, 0) = v(x, y, 0) = 0\,\mathrm{m/s}.$$

Figure 6.12 shows the surface elevation at times after 3.6, 7.2, and 10.8 seconds. The solution is computed by the front-tracking algorithm on a grid with 40×40 cells using a Godunov splitting with a fixed splitting step corresponding to a CFL number of 1.5. Initially the solution consists of a rarefaction wave propagating upstream in the high-water region and a bore propagating into the downstream side of the reservoir. In the second plot the bore has been reflected from the upper side wall and in the third plot it has reached the wall in the downstream direction. The front-tracking scheme resolves the bore very well, using only two points to represent the shock (after the projection) and compares favorably with other results reported in the literature; see [116] for a more thorough discussion.

As has been studied previously, the accuracy of the front-tracking method is determined by two factors: splitting errors and numerical diffusion caused by the projections onto a regular grid. The number of time-steps required to resolve the dynamics on the coarse 40×40 grid corresponds to a CFL number 1.5. On a finer grid, we expect to be able to use larger time-steps. In Figure 6.13 we show the effect of increasing the time-step on a 200×200 grid. By increasing the CFL number of the splitting step from 1.0 to 1.5, we increase the resolution of the leading bore, and by increasing the time-steps to CFL number 4.0 we also observe improved resolution of the shock wave reflected from the upper wall. This means that the projection error is dominant for low CFL numbers, as was observed in the scalar case. When the splitting step is increased further, a new factor enters the picture. In general, the projection (of shock waves) will introduce waves in the other passive families. These weak waves are resolved and tracked in the next step by the front-tracking solution, leading to a large number of weak wave interactions. As the splitting step increases, these weak waves become more and more dominant and the quality of the solution deteriorates. This effect was not observed in the scalar case, since scalar equations only have one characteristic family.

In this section we have observed that dimensional splitting is a simple, but efficient way to extend one-dimensional high-resolution methods to multi-dimensions.

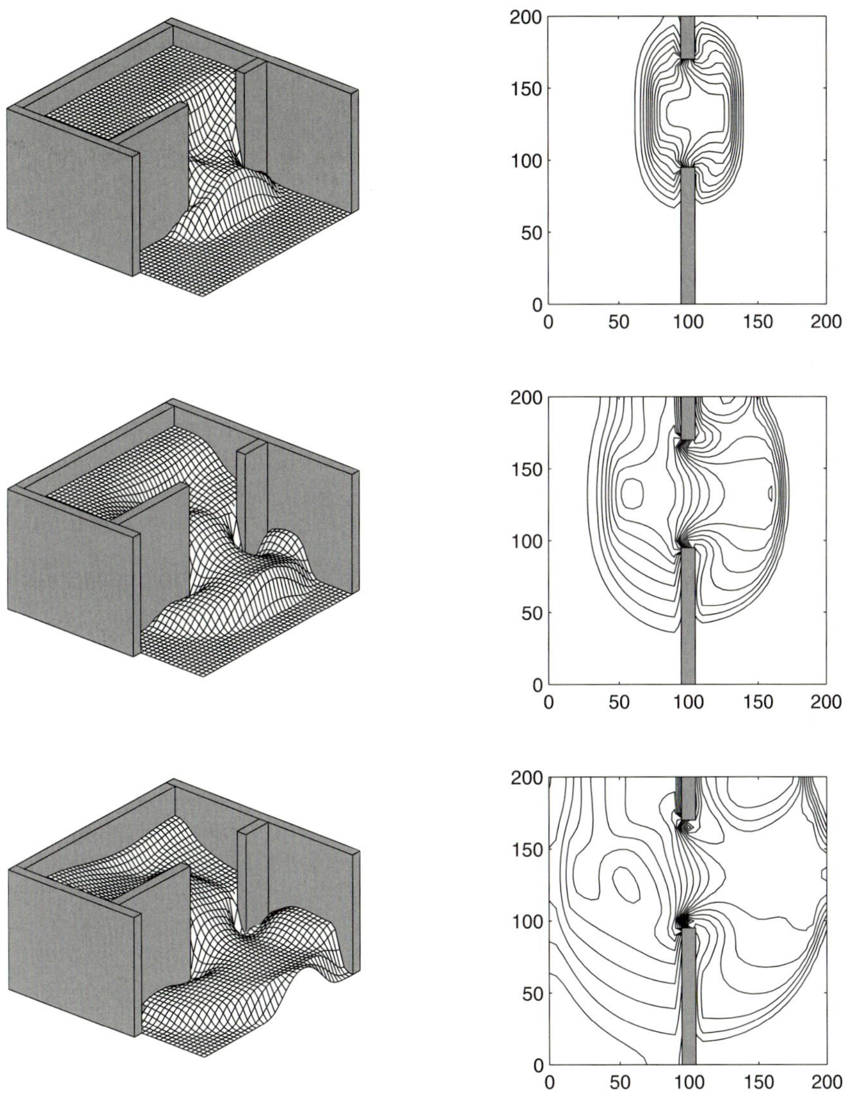

Figure 6.12. Water surface elevation for the dambreak problem after 3.6, 7.2 and 10.8 seconds.

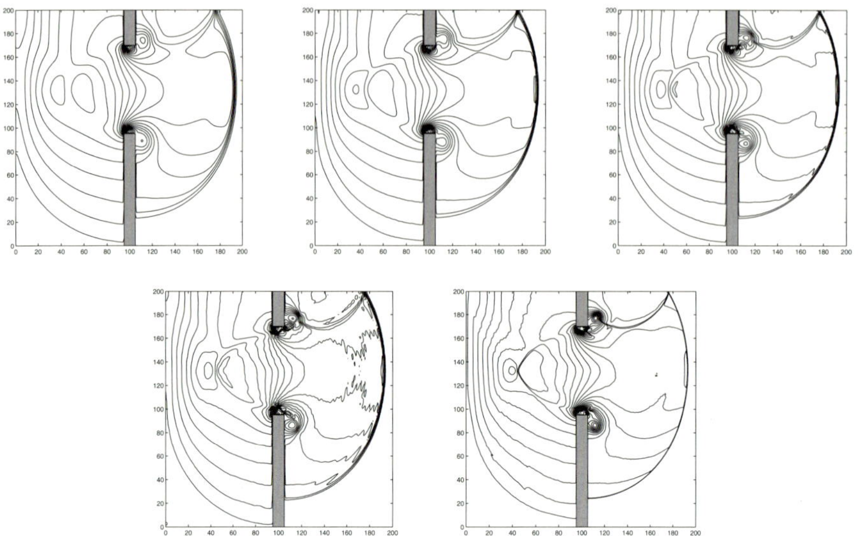

Figure 6.13. Contour plots of the water elevation at time $t = 10.0$ computed on a 200×200 grid with CFL numbers 1.0, 1.5, 4.0, and 6.0. The lower right plot is a simulation on a 800×800 grid with CFL number 2.0.

Used in this setting, the front-tracking method can be viewed as a large time-step, finite-volume scheme that reduces to the first-order Godunov method for CFL numbers below 0.5.

For (conventional) finite-volume schemes, the size of the splitting step is naturally determined by the underlying CFL condition that restricts the time-step in each one-dimensional sweep. Front tracking, on the other hand, is unconditionally stable and should in principle be ideal for application as a large-step method in which the size of the splitting step is determined by the underlying dynamics in the equation. This is indeed possible for scalar equations, as has been demonstrated by several examples. For systems of equations, the range of feasible splitting steps only extends to CFL numbers moderately larger than unity due to small-scale oscillations coming from weak waves in the passive families introduced by the projection operator. Dimensional splitting with front tracking will therefore be most efficient in cases that are dominated by strong discontinuities. For cases with smooth transitions and/or interaction of many small waves, the efficiency will be greatly reduced unless one can use a highly efficient and simplified Riemann solver, see the discussion in [174]. If smooth parts of the wave patterns are important, one is generally better off by using a high-order finite-volume method.

6.3 Operator splitting for balance laws

We will now consider splitting applied to inhomogeneous (systems of) conservation laws

$$u_t + \nabla \cdot f(u) = h(u), \qquad u(x,0) = u_0(x), \qquad u \in \mathbb{R}^m, \tag{6.19}$$

where the corresponding homogeneous conservation law $u_t + \nabla \cdot f(u) = 0$ is assumed to be hyperbolic. The function $h(u)$ is referred to as a *source term* and Equation (6.19) is often called a *balance law* rather than a conservation law.

A fractional steps method consists of splitting the inhomogeneous equation (6.19) into a homogeneous hyperbolic problem $v_t + \nabla f(v) = 0$ and an ordinary differential equation $v_t = h(v)$. Let $\mathcal{S}_t$ and $\mathcal{R}_t$ denote the corresponding solution operators. Then the fractional steps approximation to (6.19) is obtained by using either the first-order Godunov splitting

$$u(x, n\Delta t) \approx \left[\mathcal{R}_{\Delta t}\mathcal{S}_{\Delta t}\right]^n u_0,$$

or the second-order Strang splitting

$$u(x, n\Delta t) \approx \left[\mathcal{R}_{\Delta t/2}\mathcal{S}_{\Delta t}\mathcal{R}_{\Delta t/2}\right]^n u_0.$$

An alternative splitting is obtained by replacing the ODE $v_t = h(v)$ by the differential equation $\nabla f(v) = h(v)$. This is particularly useful for quasi-stationary flows for which $u_t \approx 0$.

High-resolution methods for balance laws is an important topic and there is an extensive literature in this field. An excellent overview of high-resolution methods for balance laws, with or without splitting, is given in [176, Chap. 17]. Here we will try to illustrate the use of operator splittings for balance laws by a few examples. Unlike in the preceding sections, we will not discuss numerical convergence of the splitting strategies. Instead, the examples aim to illustrate (once again) that operator splitting is a useful strategy, but also that it may have certain potential pitfalls. Rigorous convergence results and error estimates for such operator-splitting methods have been obtained in [170, 253, 254] for the scalar case, and will also be discussed in Chapters 3 and 5.

The examples will be divided into three different categories: (i) examples of geometric source terms, where we discuss radially symmetric flows and water waves over a bottom topography; (ii) examples of reacting flows, where we discuss the reactive Euler equations; and (iii) examples of flows with external forces, where we discuss Rayleigh–Taylor instabilities induced by gravity when a layer of heavy fluid is placed on top of a light fluid.

6.3.1 Geometric source terms. A physical problem that occurs in three spatial dimensions can often be described by a mathematical model in one or two spatial dimensions by using symmetries or special features of the problem. When going from a full three-dimensional model to some lower quasi-dimensional model,

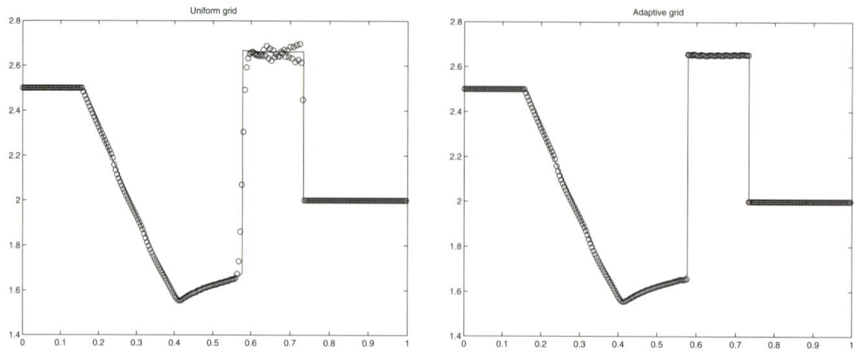

Figure 6.14. Internal energy in the radially symmetric solution computed on a grid with 250 uniform cells using 20 equal splitting steps.

source terms are often added to the equations to model the reduced dimensionality. Similarly, if the problem is posed internal or external to some given geometry, such as for flow in a duct with variable width, extra source terms will arise to model the geometry in the reduced model.

As an example let us consider the multi-dimensional Euler equations of gas dynamics. For problems with cylindrical or spherical symmetry, the flow can be described by the quasi-one-dimensional inhomogeneous equation

$$\begin{bmatrix} \rho \\ \rho u \\ E \end{bmatrix}_t + \begin{bmatrix} \rho u \\ \rho u^2 + p \\ u(E+p) \end{bmatrix}_r = -\frac{\alpha}{r} \begin{bmatrix} \rho u \\ \rho u^2 \\ u(E+p) \end{bmatrix}. \tag{6.20}$$

Here r denotes the radial direction, ρ the density, u the radial velocity, p the pressure, and E the total energy. We assume that the gas is ideal and polytropic, i.e., the energy is given by $E = \rho u^2/2 + p/(\gamma - 1)$, where the constant γ is set to 1.4. For α equal 1, (6.18) is equivalent to the two-dimensional equations with cylindrical symmetry. For three-dimensional systems with spherical symmetry, α equals 2.

Example 6.10. *Consider now an explosion problem, as in Example 6.8 in the previous section. The initial data are constant in two regions separated by a circle of radius 0.4 centred at the origin. Inside the circle* $\rho_{\mathrm{in}} = p_{\mathrm{in}} = 1.0$*, and outside* $\rho_{\mathrm{out}} = 0.125$, $p_{\mathrm{out}} = 0.1$. *The velocity is zero everywhere. The left plot in Figure 6.14 shows the radial solution (internal energy) computed using a straightforward approach with front tracking for the homogeneous conservation law and the forward Euler method for the ODE. After each hyperbolic step, the front-tracking solution is projected onto a uniform grid. Unfortunately, this projection introduces small waves in the passive wave families near strong shocks (as discussed briefly in the previous section). These small waves pollute the solution* *and lead to the oscillations observed in the post-shock zone. To overcome this problem we can introduce an adaptive grid where extra grid nodes are added at the position of all waves exceeding a prescribed threshold. In the right-hand plot in Figure* 6.14 *we have recomputed the solution using the adaptive grid. In the adaptive grid, only two extra grid points were added, one at the shock and one at the contact discontinuity. Still, the improvement in the quality of the solution is striking. The oscillations in the region between the shock and the contact discontinuity have almost disappeared and the latter wave is not smeared, as it is for the non-adaptive (fixed-grid) method.*

The depth-integrated shallow water equations for one-dimensional flows over a variable bottom topography reads

$$\begin{bmatrix} h \\ hu \end{bmatrix}_t + \begin{bmatrix} hu \\ hu^2 + \frac{1}{2}gh^2 \end{bmatrix}_x = \begin{bmatrix} 0 \\ -(z_b)_x gh \end{bmatrix}. \tag{6.21}$$

Here h denotes water depth, u is depth-averaged velocity, and g is the acceleration of gravity. The source term on the right-hand side is a geometric source term, where z_b denotes the bottom topography.

Example 6.11. *Consider a current of water flowing over an obstacle. An example can be described by* (6.21) *with initial conditions*

$$h(x,0) = 1.0 - z_b(x), \qquad u(x,0) = 1.0,$$

absorbing boundary conditions, and bottom topography

$$z_b(x) = \begin{cases} 0.2\left(1 - \frac{x^2}{4}\right), & -2 \le x \le 2, \\ 0, & \textit{otherwise}. \end{cases}$$

The spatial domain is $x \in [-10, 10]$, and we rescale the equations by setting $g = 1$. To solve this equation we use front tracking in combination with first-order Godunov splitting for the source term. As in the previous example we project the front-tracking solution onto a grid before applying the ODE solver for the source term. This is done in two ways, either by assuming a fixed grid or by adding extra grid cells at strong shocks. Figure 6.15 *shows solutions at times $t = 5$ and 10 computed with 100 and 2000 uniform grid cells by front tracking and by the composite scheme LWLF4 [190], which consists of four dispersive Lax–Wendroff steps followed by a diffusive Lax–Friedrichs step. We use a fixed CFL number 1.75 for the two front-tracking methods and 0.8 for LWLF4. In the adaptive-grid method, only two extra grid cells were added. Although front tracking with a fixed grid gives better resolution than LWLF4 on the same grid, it can by no means compete with front tracking with two extra grid cells added at the discontinuities. This method resolves both shocks within one grid cell.*

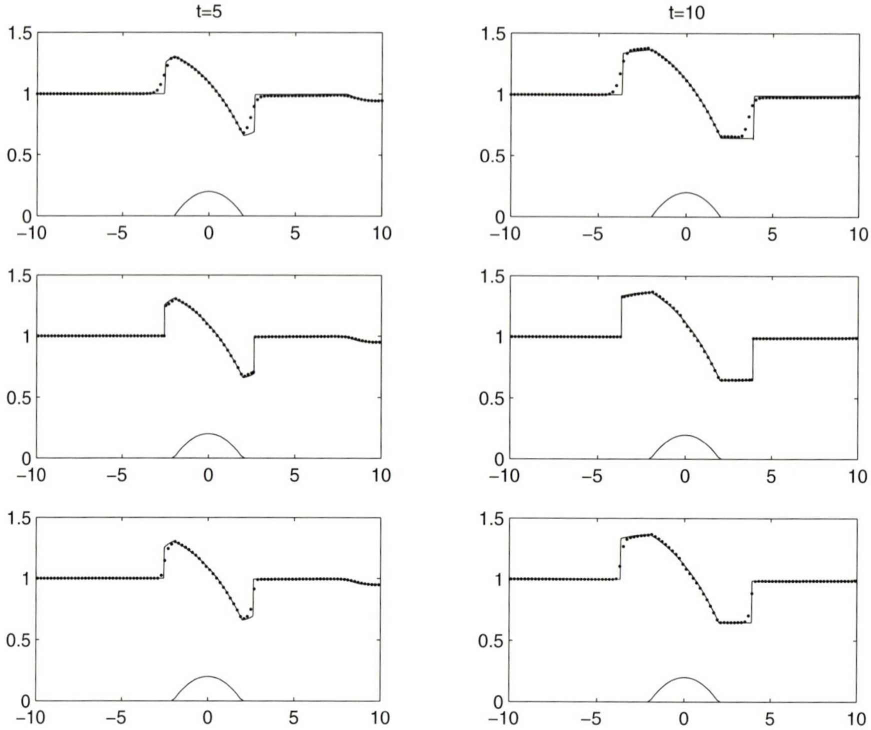

Figure 6.15. Shallow water flow over a topography in one dimension computed by the composite scheme LWLF4 (top), by front tracking with adaptive grid (middle), and by front tracking on a fixed grid (bottom).

6.3.2 Equilibrium states. In the two previous examples we have considered flows where the primary objects of interest were relatively strong waves: a pressure wave resulting from an explosion and the swell created by an underwater mountain. On the other hand, many real-life applications are merely perturbations of a steady state or some kind of underlying physical equilibrium. Resolving such phenomena turns out to be surprisingly difficult, as we will see in the next example.

Example 6.12. *The equilibrium states of the depth-integrated shallow-water equations* (6.21) *are given by*

$$hu \equiv \text{const} \quad \text{and} \quad \frac{1}{2}u^2 + gH \equiv \text{const}$$

where $H := h + z_b$ *is the water height relative to some reference level. In this example we will consider a particular equilibrium state, the lake-at-rest given by*

$$u \equiv 0 \quad \text{and} \quad H \equiv \text{const}.$$

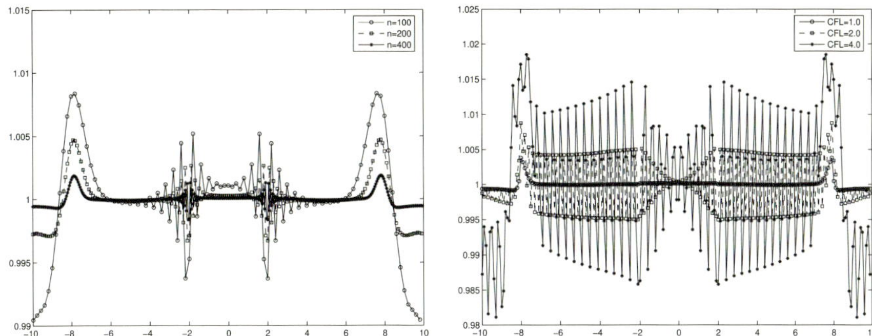

Figure 6.16. Approximate solution of the lake-at-rest computed using 100, 200, and 400 grid cells for CFL number 1.75 (left), and for CFL numbers 1.0, 2.0, and 4.0 for 200 cells (right).

To study this problem, we use the same configuration as in Example 6.11*, except that we set the water velocity to zero and assume reflective boundaries. Figure* 6.16 *shows approximate solutions computed using 100, 200, and 400 grid cells for CFL number 1.75 and CFL numbers 1.0, 2.0, and 4.0 for 200 cells. For all parameters, the approximate solution contains a 'numerical storm' that has been created due to splitting errors. As can be seen from the figure, the maximal wave height decreases as the grid is refined. Similarly, the wave height increases with the size of the time-step, as does the runtime due to a dramatic increase both in the number of wave interactions and the number of fronts needed to represent the solution.*

Similar errors are in fact observed also for unsplit schemes. This observation has led to the development of so-called well-balanced schemes that are designed such that the discrete flux and source terms balance exactly at the steady state, see, e.g., [7, 110, 214, 277].

6.3.3 Reactive flows. Many physical models involve chemical reactions. Reaction among the species in a flowing fluid will generally affect the dynamics of the fluid motion through a strong coupling between the fluxes and the source terms. In particular, if the chemical reactions take place on a much smaller time scale than the wave speeds in the fluid, the problem is said to be *stiff*. Problems with stiff source terms are generally hard to solve.

In the next two examples we will study operator-splitting methods applied to two reactive flows; one with nonstiff and one with stiff source terms. The physical problem consists of a simplified chemically reacting flow with two chemical species: "burnt gas" and "unburnt gas". The unburnt gas (the reactant) is converted to burnt gas (the product) through a one-step irreversible process where the reaction rate $K(T)$ depends only on temperature. We ignore the effects of viscosity, radiation, and heat conduction.

Then the combustion process is described by the reactive Euler equations

$$\begin{bmatrix} \rho \\ \rho u \\ E \\ \rho Z \end{bmatrix}_t + \begin{bmatrix} \rho u \\ \rho u^2 + p \\ u(E+p) \\ \rho u Z \end{bmatrix}_x = \begin{bmatrix} 0 \\ 0 \\ 0 \\ -K(T)\rho Z \end{bmatrix}.$$

Here Z denotes the fraction of unburnt gas, K is the reaction rate of the burning process and depends upon the temperature $T = p/\rho R$, where R is the universal gas constant. The unburnt gas contains chemical energy and the total energy reads

$$E = \frac{p}{\gamma - 1} + \frac{1}{2}\rho u^2 + q_0 \rho Z,$$

where q_0 is the heat released when the gas burns. The inhomogeneous system can be converted into an equivalent form,

$$\begin{bmatrix} \rho \\ \rho u \\ \hat{E} \\ \rho Z \end{bmatrix}_t + \begin{bmatrix} \rho u \\ \rho u^2 + p \\ u(\hat{E}+p) \\ \rho u Z \end{bmatrix}_x = \begin{bmatrix} 0 \\ 0 \\ q_0 K(T)\rho Z \\ -K(T)\rho Z \end{bmatrix},$$

where $\hat{E} = \rho u^2/2 + p/(\gamma - 1)$. For this equation, the homogeneous part is the standard Euler equations for a single ideal gas.

Example 6.13. *Consider first the following simple model for the reaction rate K,*

$$K(T) = K_0 e^{-E^+/T},$$

where K_0 is the reaction rate multiplier and E^+ is the activation energy. In the computations we will use the following constants: $\gamma = 1.2$, $R = 1.0$, $q_0 = 50$, and $K_0 = E^+ = 10.0$. As initial data, we use the Riemann problem with left state $(p_L, \rho_L, u_L) = (15, 1, 3)$ and right state $(p_R, \rho_R, u_R) = (1, 1, 0)$. With this initial state, the gas will start to burn immediately and a detonation wave will form. The detonation wave consists of an ordinary gas dynamical shock, in which the pressure and density increase. This will heat the gas so that it burns in a thin reaction zone behind the shock. Behind the reaction zone, there is no unburnt gas left. Figure 6.17 shows the pressure, density and reactant mass fraction at time $t = 3.0$ computed by the operator-splitting method with adaptive spatial grid. The computation is in very good agreement with the reference solution. Accurate representation of the leading shock is crucial in order to properly compute the detonation wave, and the good resolution on this relatively coarse grid can be ascribed to the adaptive grid that tracks the shock. Without grid adaptation, the operator-splitting method gives qualitatively correct results only on more refined grids, see Figure 6.18.

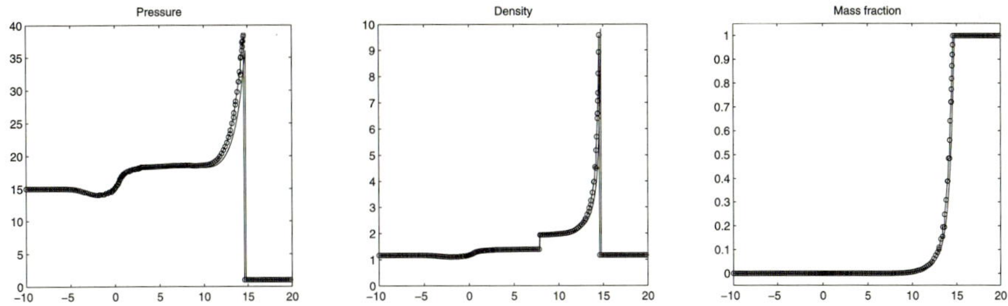

Figure 6.17. The propagating combustion front at time $t = 3.0$ computed on a 150 grid with 75 time-steps. The thin line gives a reference solution computed on a 19 200 grid.

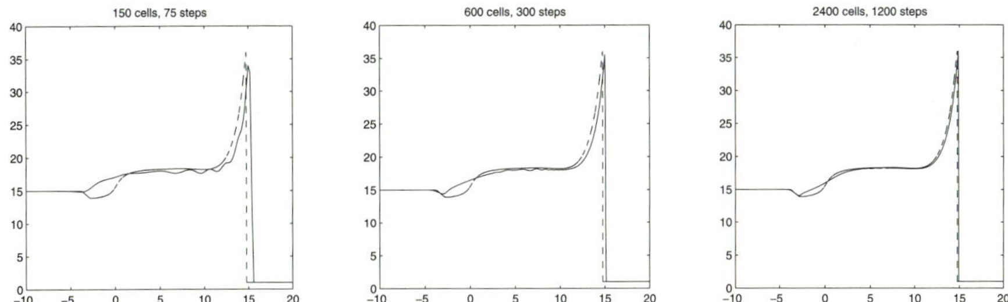

Figure 6.18. Convergence of the operator-splitting method *without* adaptive grid. The dashed line gives the reference solution.

In the above example, the chemical reaction occurs on a time scale that is almost of the same magnitude as that of the waves in the fluid flow. The operator splitting method was therefore able to resolve the dynamics of the flow. In the next example we will consider another reaction model where the reaction takes place almost instantaneously. This gives a stiff problem for which the operator-splitting strategy *fails* to compute the correct solution.

Example 6.14. *Consider as above the reactive Euler equations, but now with a simpler model for the reaction rate K,*

$$K(T) = \begin{cases} 1/\tau, & T \geq T_0, \\ 0, & T < T_0. \end{cases}$$

Here T_0 is the ignition temperature and τ is a (small) time scale of the reaction. The model reflects that reactions are negligible for temperatures below the ignition temperature and almost instantaneous above.

This system is stiff, since the reaction occurs on a very small time scale τ inside a reaction zone with spatial width proportional to τ. In practise, τ is much smaller than both the time scale of the front propagation and the possible sizes of grid cells.

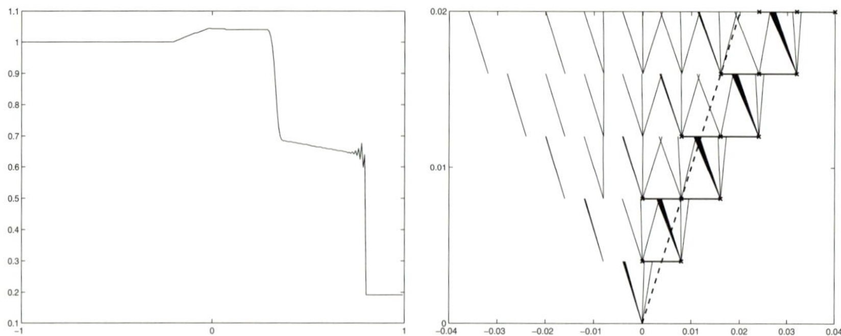

Figure 6.19. (Left) Pressure profile of the detonation front at time $t = 0.4$ computed on a 250 grid with 100 time-steps. (Right) Solution in the (x,t) plane for the first five time-steps. Solid lines give fronts in front-tracking algorithm, the dashed line gives the path of the physically correct shock front, and the horizontal lines indicate grid-cells in which the gas is burned in each time-step.

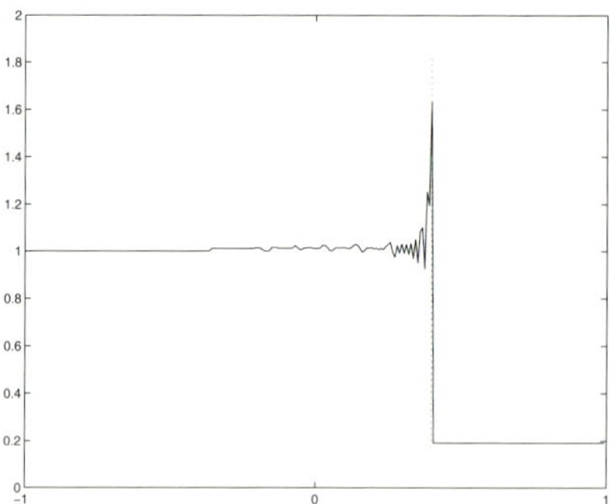

Figure 6.20. Pressure profile of the detonation front at time $t = 0.4$ computed on a 250 grid with 45 time-steps.

To integrate the ODE we therefore use an explicit method with time-step $k = \tau/5$.

In the computations we use $q_0 = R = 1$, $T_0 = 0.22$, *and* $\tau = 10^{-4}$. *Figure 6.19 shows the pressure at time* $t = 0.4$ *computed by our operating splitting method on a uniform grid with* 250 *cells. The solution is computed using* 100 *equally spaced time-steps. This corresponds to a CFL number varying between 0.56 and 0.74, that is, the front-tracking method in the hyperbolic step is similar to Godunov's method. Although the plots seem reasonable, the results are nonphysical. The exact solution consists of a shock front propagating at speed* $s = 1$, *followed by a deflagration wave. This profile is obtained if we* decrease the number of *splitting steps to 45, giving a CFL number between 1.25 and 1.58, see Figure 6.20. The unphysical waves are eliminated and the pressure peak in the so-called ZND structure is captured. On the other hand, the quality of the solution is quite poor with unstable oscillations behind the deflagration wave and oscillations of the pressure peak in time. The extent and structure of these oscillations are strongly dependent upon the splitting step (and the grid size) and the method can therefore not be considered to be stable.*

To explain what goes wrong in the computation above, we can look at the solution in the (x, t) plane. The right-hand plot in Figure 6.19 shows the solution up to time 0.02. In the first hyperbolic step, the solution consists of a shock followed by a contact discontinuity, both propagating with positive speeds, and a rarefaction wave propagating with negative speed. The passing shock heats the gas such that reaction takes place in the zone between the shock and the contact discontinuity and burns the gas almost instantaneously. The width of this reaction zone is a mere fraction of the width of the cells in the uniform grid. However, in our numerical approach, we project the solution from the hyperbolic step back onto the uniform grid before we allow the chemical reaction. Thus, the whole grid cell, and not just the true reaction zone, is heated above the ignition temperature. The zones 'burned' in the ODE steps are represented by horizontal lines in the right-hand plot of Figure 6.19. Since the time-step is much larger than τ, the gas burns instantaneously. Hence the interface between burnt and unburnt gas moves faster than it should (the true wave speed of the reaction front is indicated by a dashed line). This shortcoming is not cured by changing the order of the operators in the splitting algorithm. The problem of unphysical wave speeds was first discussed by Colella, Majda, and Roytburd [68].

To overcome these problems, various more advanced methods have been devised by several authors. A simple method called the *random projection method* was proposed by Bao and Jin [9], in which the ignition temperature T_0 is replaced by a random variable in order to obtain a shock propagation that is correct in the averaged sense.

6.3.4 External forces. Source terms may also arise when external forces are applied to a system. Physically this means that the rate of change of a conserved quantity within a volume is balanced by the flux over the edges and the external forces. As an example, we can include gravity in the Euler equations of gas dynamics

$$\begin{bmatrix} \rho \\ \rho u \\ \rho v \\ \rho w \\ E \end{bmatrix}_t + \begin{bmatrix} \rho u \\ \rho u^2 + p \\ \rho u v \\ \rho u w \\ u(E+p) \end{bmatrix}_x + \begin{bmatrix} \rho v \\ \rho u v \\ \rho v^2 + p \\ \rho v w \\ v(E+p) \end{bmatrix}_y + \begin{bmatrix} \rho v \\ \rho u w \\ \rho v w \\ \rho w^2 + p \\ w(E+p) \end{bmatrix}_z = \begin{bmatrix} 0 \\ 0 \\ 0 \\ g\rho \\ g\rho w \end{bmatrix}.$$

Here the initial momentum will not be conserved since the gravity will induce an acceleration in the vertical direction and likewise for the energy.

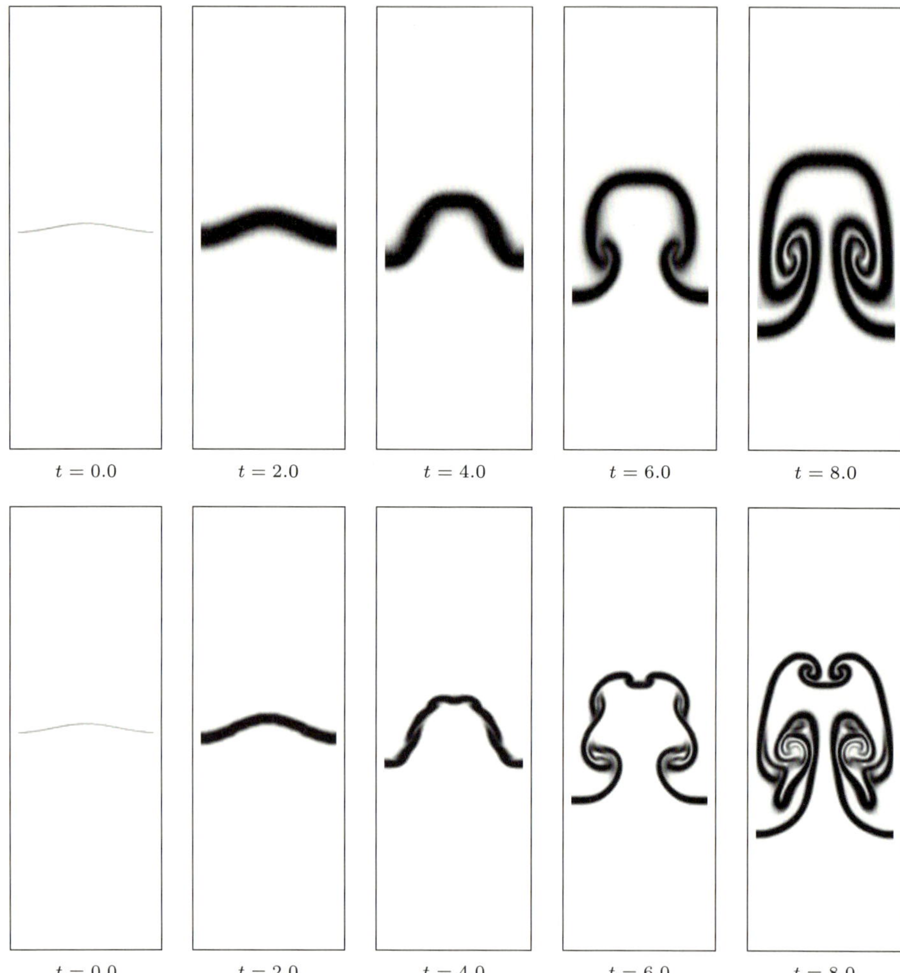

Figure 6.21. Evolution of the Rayleigh–Taylor instability depicted as emulated Schlieren images. The numerical approximation was computed by the NT2d scheme on a 266×800 grid with CFL number 0.45: MM_1 limiter (upper row) and van Leer limiter (lower row).

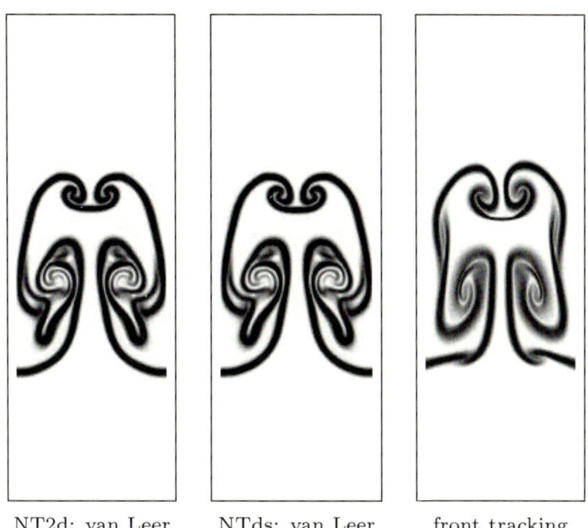

Figure 6.22. The Rayleigh–Taylor instability at time $t = 8.0$ computed with NT2d and the two dimensional splitting schemes on a 266×800 grid. The central schemes used a CFL number 0.45 and the front-tracking scheme 8500 equally spaced time-steps (giving a CFL number about 1.5).

Example 6.15. *We consider the simulation of a Rayleigh–Taylor instability. This physical instability arises when a layer of heavier fluid is placed on top of a lighter fluid and the heavier fluid is accelerated downwards by gravity. Similar phenomena occur more generally when a light fluid is accelerated towards a heavy fluid. The instability can be described by the two-dimensional Euler equations with a gravitational term. The physical domain is* $[-1/6, 1/6] \times [0, 1]$ *with a gravitational acceleration* $g = -0.1$ *in the* y*-direction. The lower fluid has unit density and the density of the upper fluid is twice as high. The interface of the fluids is* $y = 1/2 + 0.01\cos(6\pi x)$*. The fluids are initially at rest and the pressure is hydrostatic. The boundary conditions are periodic in the* x*-direction and reflective in the* y*-direction. Figure* 6.21 *shows the evolution of the Rayleigh–Taylor instability computed by a splitting method with the NT2d scheme from [183] for the homogeneous Euler equations. The interface between the light and heavy fluid is unstable. The dissipative* MM_1 *limiter suppresses the instability, whereas the interface breaks up for the van Leer limiter due to less numerical dissipation.*

Two additional schemes can be obtained by expanding the two-dimensional splitting schemes introduced in Example 4 in the previous section. Plots of the corresponding splitting solutions are given in Figure 6.22*. The numerical solutions obtained by NT2d and NTds are almost indistinguishable, while the front-tracking computation shows a clear lack of symmetry. Of the three schemes, the NTds scheme is by far the fastest; for the computation in Figure* 6.22 *the ratios of the elapsed runtimes for NTds, NT2d, and the front-tracking scheme were* $1 : 1.7 : 2.5$.

To achieve a sufficient resolution in the front-tracking computation, we eliminated only very small waves (with strength below 10^{-5}*). Thus a very large number of fronts were tracked and about* $1.4 \cdot 10^{10}$ *front collisions were resolved, resulting in the highest runtime.*

Example 6.16. *Similarly, we can extend the previous Rayleigh–Taylor simulation to three spatial dimensions. The physical domain is* $[-1/6, 1/6] \times [-1/6, 1/6] \times [0.2, 0.8]$ *with gravitational acceleration* $g = -0.1$ *in the vertical* z *direction. The upper fluid has density 2 and the lower fluid 1. The interface of the fluids is at* $z = 1/2 + 0.01\cos(6\pi \min(\sqrt{x^2 + y^2}, 1/6))$*. The fluids are initially at rest and the pressure is hydrostatic. The boundary conditions are periodic in the horizontal direction and reflective in the vertical direction. To compute the instability we use a Godunov splitting as above with a three-dimensional extension of the NT scheme (NT3d) for the homogeneous Euler equations. Figure* 6.23 *shows the evolution of the instability.*

6.4 Final remarks

We conclude this book by reiterating its scope, namely to provide the reader with an overview of contemporary mathematical and numerical techniques based on operator splitting for a class of nonlinear partial differential equations possessing solutions with limited regularity. We use "operator splitting" as a collective term to describe different techniques that share the common property that a complex operator is written as a sum of simpler operators. The simpler problems are solved, and this yields an approximate solution of the original problem. This idea is simple and has been studied and refined by several researchers for half a century.

Compared to the existing literature on splitting methods, an original aspect of our presentation lies in the fact that we are systematically utilizing numerical methods and mathematical theory associated with hyperbolic problems to construct fully discrete splitting methods, some of which are specifically designed to allow for large time (splitting) steps. This "hyperbolic" approach also enables us to provide a unifying convergence theory that applies to solutions exhibiting complex behavior such as discontinuities (shock waves), thereby covering a large class of mixed hyperbolic-parabolic evolution equations as well as weakly coupled systems of such equations.

Since the solutions of these equations lack regularity, efficient and accurate numerical computations are complicated. The operator splitting approach offers a simple solution method: Take your favorite numerical methods for the individual, simpler equations, apply operator splitting to combine the simple solutions to obtain an approximate solution for the original problem. With this set-up you can combine various techniques to produce easily implementable methods for your complicated problem. As a result it provides an excellent tool for swift numerical

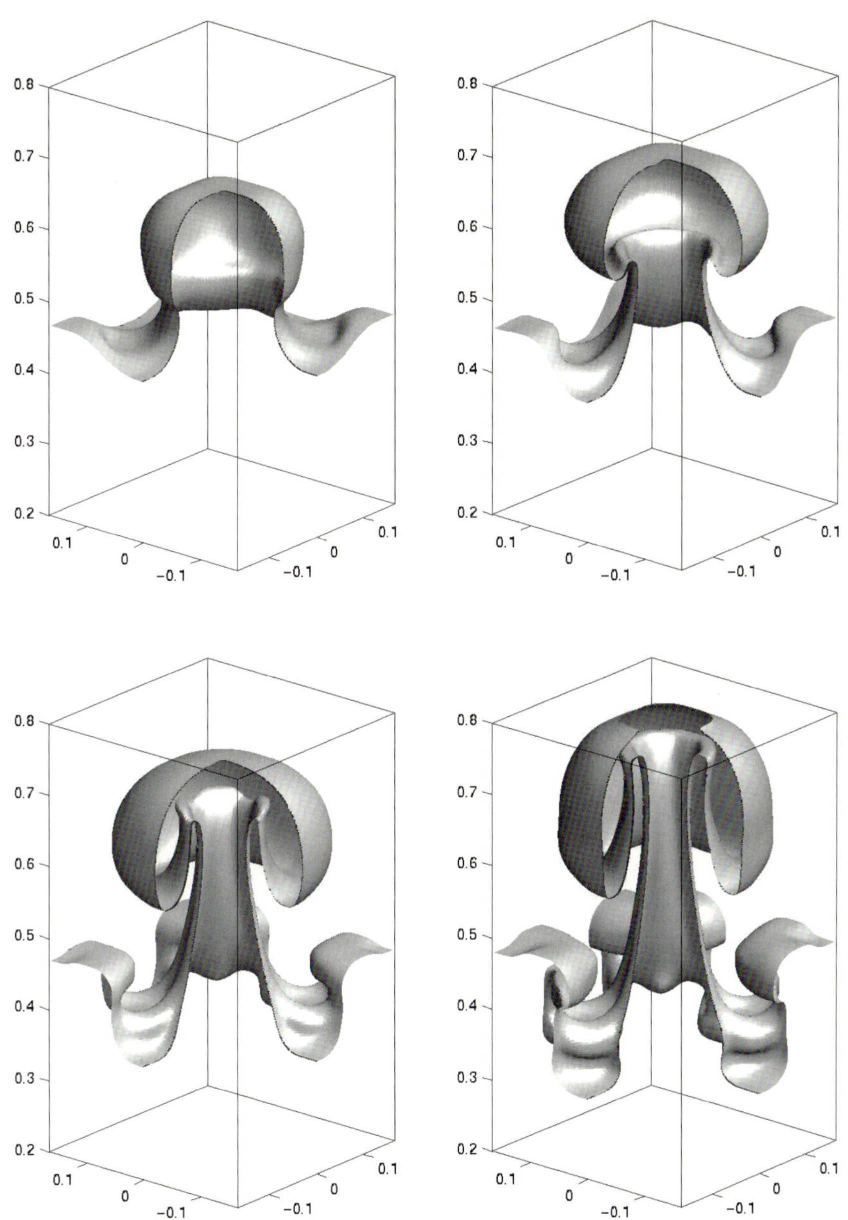

Figure 6.23. The Rayleigh–Taylor instability at times $t = 0.5$, 0.6, 0.7, and 0.8 computed with NT3d on a $100 \times 100 \times 182$ grid with the $\mathrm{MM}_{1.3}$ limiter.

investigations. Although our theoretical convergence framework only applies to scalar and weakly coupled systems of equations, operator splitting offers a general set-up for constructing methods for systems of equations, with good results as the numerical examples show. For higher-dimensional problems it might be the only applicable model. But beware! Operator splitting is not a cure-it-all method, but nor is it snake oil.

A

A Crash Course in Numerical Methods for Conservation Laws

In the previous chapter we discussed examples of operator splitting applied to fairly simple equations. In most cases, we were able to derive analytical expressions for the elementary operators and could obtain semi-analytical operator splitting approximations. This is seldom possible, and one generally has to employ some numerical method and use a fully discrete operator splitting method. Analysis of such operator splittings is the topic for this book, starting with a review of some examples in Chapter 6. As stated repeatedly throughout the book, our main emphasis is on convection-dominated problems, where the convective part of the equation plays a decisive role. This naturally leads to splitting methods where a conservation law is the primary elementary equation used to build approximate solutions. This chapter is therefore devoted to the numerical solution of such equations, giving a crash course in the most important mathematical and numerical concepts used to build computational methods for hyperbolic equations. For a broad-ranging introduction to contemporary numerical methods for hyperbolic conservation laws, we refer to [15, 67, 101, 104, 107, 108, 115, 159, 175, 176, 260, 262].

A.1 Hyperbolic conservation laws

The term hyperbolic conservation laws usually denotes a first-order, quasilinear partial differential equation of the following form (in one spatial dimension)

$$u_t + f(u)_x = 0. \tag{A.1}$$

Here u is some conserved quantity (scalar or vector) and $f(u)$ is a flux function. A conservation law usually arises from a more fundamental physical law on integral form. In one spatial dimension, this law typically reads

$$\frac{d}{dt}\int_{x_1}^{x_2} u(x,t)\,dx = f\big(u(x_1,t)\big)-f\big(u(x_2,t)\big). \tag{A.2}$$

The physical law states that the rate of change of quantity u within $[x_1, x_2]$ equals the flux across the boundaries $x = x_1$ and $x = x_2$. The partial differential equation (A.1) then follows under additional regularity assumptions on u.

Several of the elementary operators we encountered in Chapter 2 were special cases of the Cauchy problem for (A.1); that is, of the form

$$u_t + f(u)_x = 0, \qquad u(x,0) = u_0(x). \tag{A.3}$$

For a nonlinear flux function f, solutions of this equation may develop discontinuities in finite time, even for smooth initial data. This means that the solution of (A.3) is understood in the weak sense,

$$\int_0^\infty \int_{\mathbb{R}} (u\phi_t + f(u)\phi_x)\, dtdx = \int_{\mathbb{R}} u_0(x)\phi(x,0)\, dx. \tag{A.4}$$

Here $\phi(x,t)$ is a smooth test function possessing all the necessary derivatives. The function $\phi(x,t)$ is also assumed to have *compact support*, meaning that it vanishes outside a bounded region in the (x,t)-plane.

Solutions defined by the weak form (A.4) are not necessarily unique. The solution concept must therefore be extended to include additional admissibility conditions to single out the correct solution among several possible candidates satisfying the weak form. A classical method to obtain uniqueness is to add a regularizing second-order term to (A.1), giving a parabolic equation

$$u_t^\varepsilon + f(u^\varepsilon)_x = \varepsilon u_{xx}^\varepsilon,$$

that has smooth solutions. Then the unique solution of (A.1) is defined as the limit of $u^\varepsilon(x,t)$ as ε tends to zero. In models from fluid dynamics, such a second-order term can be proportional to the viscosity of the fluid, and the method is therefore called the *vanishing viscosity method.* Since the solutions $u^\varepsilon(x,t)$ are smooth, classical analysis coupled with careful limit arguments can be used to show existence, uniqueness, and stability of the solution of (A.1). This was first done by Kružkov [161], whose seminal work on 'doubling of the variables' has had a tremendous impact on the development of modern theory for nonlinear partial differential equations [41, 73, 126, 201, 221, 239, 240, 243].

However, using the viscosity limit as an admissibility condition is impractical. If u^ε is a solution of the equation with viscosity, then

$$u_t^\varepsilon + f'(u^\varepsilon)\, u_x^\varepsilon = \varepsilon u_{xx}^\varepsilon,$$

and multiplying this with a convex differentiable function $\eta'(u^\varepsilon)$ yields

$$\eta(u^\varepsilon)_t + \eta'(u^\varepsilon) f'(u^\varepsilon)\, u_x^\varepsilon = \varepsilon\left(\eta'(u^\varepsilon)\, u_x^\varepsilon\right)_x - \varepsilon\eta''(u^\varepsilon)\left(u_x^\varepsilon\right)^2 \le \varepsilon\left(\eta'(u^\varepsilon)\, u_x^\varepsilon\right)_x.$$

Now define the *entropy flux* corresponding to η by

$$q'(u) = \eta'(u) f'(u). \tag{A.5}$$

If u is to be the limit of u^ε then we can let ε to zero and obtain

$$\eta(u)_t + q(u)_x \le 0, \tag{A.6}$$

which must be interpreted in the weak sense as

$$\int_0^\infty \int_{\mathbb{R}} (\eta(u)\phi_t + q(u)\phi_x)\, dtdx + \int_{\mathbb{R}} \eta(u_0(x))\phi(x,0)\, dx \geq 0, \tag{A.7}$$

for all nonnegative test functions ϕ. We call (η, q) an entropy/entropy flux pair. This name originates from thermodynamics.

Now if u is a vector with K components, $u = (u_1, \dots, u_K)$, and flux function $f = (f^1(u), \dots, f^K(u))$, (A.5) is a system of K equations for the partial derivatives of q,

$$\frac{\partial q}{\partial u_i} = \sum_{k=1}^{K} \frac{\partial f^k}{\partial u_i} \frac{\partial \eta}{\partial u_k} \qquad i = 1, \dots, K.$$

A continuous solution q can be found if and only if the mixed partial derivatives of q are equal, which again implies that

$$\sum_k \frac{\partial f^k}{\partial u_i} \frac{\partial^2 \eta}{\partial u_k \partial u_j} = \sum_k \frac{\partial f^k}{\partial u_j} \frac{\partial^2 \eta}{\partial u_k \partial u_i} \qquad \text{for all } 1 \leq i, j \leq K.$$

Therefore, if $K > 1$, the existence of such entropy pairs is not obvious for a general system of conservation laws, but such pairs exist and have a clear physical interpretation for several important systems of equations, for instance in gas dynamics.

For scalar equations it can be shown that all entropy pairs with convex η are equivalent. A common choice is the so-called Kružkov entropy pair,

$$\eta(u) = |u - k|, \quad q(u) = \operatorname{sign}(u - k)(f(u) - f(k)), \tag{A.8}$$

where k is *any* constant. We say that $u(x,t)$ is an *entropy weak solution* of (A.1) if it satisfies (A.7) with the Kružkov entropy/entropy flux pair for all real numbers k and nonnegative test functions ϕ.

We refer to Chapter 3 for a detailed discussion of the mathematical theory for scalar conservation laws in the context of the more general degenerate parabolic equations.

A.2 Finite-volume methods

Finite-difference methods use discrete differences to approximate the derivatives in a partial differential equation. This gives discrete evolution equations for a set of point values approximating the true solution of the PDE. Once a discontinuity arises in the hyperbolic conservation law, the differential equation will cease to be pointwise valid in the classical sense. Hence, it is also to be expected that classical finite-difference approximations will break down at discontinuities, causing severe problems for standard finite-difference methods. To overcome this computational problem, it turns out that instead of seeking *pointwise* solutions to (A.1), one should look for solutions of the more fundamental integral form (A.2). To this end, we break the domain $[x_1, x_2]$ into a set of subdomains—which we call finite volumes or grid cells—and seek approximations to the *global* solution u in terms of a discrete set of *cell averages* defined over each grid cell; that is, we seek approximations to $\frac{1}{\Delta x}\int u(x,t)\,dx$ over each grid cell of size Δx. Such methods are usually called finite-volume methods.

There is a close relation between finite-difference and finite-volume methods since the formula of a specific finite-volume method in some cases may be interpreted directly as a finite-difference approximation to the underlying differential equation. However, the underlying principles are fundamentally different. Finite difference methods evolve a discrete set of point values by approximating (A.1). Finite-volume methods evolve *globally defined solutions* as given by (A.2) and *realize* them in terms of a discrete set of cell averages. The evolution of globally defined solutions is the key to the success of modern methods for hyperbolic conservation laws. There are many good books describing such methods. We can recommend the books [104, 107, 108, 115, 159, 175, 176, 260, 262], see also [15, 67, 101].

A.3 Conservative methods

The starting point for a finite-volume method for (A.1) is the cell-average defined by

$$u_i^n = \frac{1}{\Delta x_i}\int_{x_{i-1/2}}^{x_{i+1/2}} u(x,t_n)\,dx.$$

These cell averages are usually evolved in time by an explicit time-marching method, obtained by integrating (A.2) in time,

$$u_i^{n+1} - u_i^n = \frac{1}{\Delta x}\int_{t_n}^{t_{n+1}} f(u(x_{i-1/2},t))\,dt - \frac{1}{\Delta x}\int_{t_n}^{t_{n+1}} f(u(x_{i+1/2},t))\,dt. \tag{A.9}$$

Generally, we will not be able to compute the flux integrals exactly, since the point values $u(x_{i\pm1/2},t)$ vary with time and are in general unknown. However, the equation suggests that the numerical method should be of the form

$$u_i^{n+1} = u_i^n - \lambda\big(F_{i+1/2}^n - F_{i-1/2}^n\big), \tag{A.10}$$

where $\lambda = \Delta t/\Delta x$ and $F_{i\pm1/2}^n$ is some approximation to the average flux over each cell interface,

$$F_{i\pm1/2}^n \approx \frac{1}{\Delta t}\int_{t_n}^{t_{n+1}} f\big(u(x_{i\pm1/2},t)\big)\,dt.$$

Any numerical method of this form will generally be *conservative*. To see this, we can sum the equation over all i. The flux terms will cancel in pairs, and we are left with

$$\sum_{i=-M}^{M} u_i^{n+1} = \sum_{i=-M}^{M} u_i^n - \lambda\big(F_{M+1/2}^n - F_{-M-1/2}^n\big).$$

The two flux terms vanish if we assume either periodic boundary conditions or that $u(x,t)$ approaches the same constant value as $x \to \pm\infty$. Thus, the numerical method conserves the quantity u, i.e.,

$$\int u^n(x)\,dx = \int u_0(x)\,dx.$$

Schemes of the form (A.10) are denoted *conservative* schemes.

Hyperbolic conservation laws have finite speed of propagation, unless they degenerate in some form. It is therefore natural to assume that the average fluxes are given in terms of their neighboring cell averages; that is,

$$F_{i+1/2}^n = F\big(u_{i-p}^n, \dots, u_{i+q}^n\big).$$

The function F is called the *numerical flux* and will be referred to by the abbreviation $F(u^n; i+1/2)$.

It is often convenient to extend the numerical approximation u_i^n to a function defined on all of space and time. Thus we define

$$u_{\Delta x}(x,t) = \sum_{n,i} u_i^n \chi_{[x_{i-1/2}, x_{i+1/2})}(x)\chi_{[t_n, t_{n+1})}(t) \tag{A.11}$$

where χ_I denotes the characteristic function of the set I.

A.4 A few classical schemes

We have now gone through the basic underlying principles for the design of schemes for conservation laws. It is therefore time to show some examples of such schemes.

The simplest example is the upwind scheme. If $f'(u) \geq 0$, the scheme has numerical flux $F(u^n; i+1/2) = f(u_i^n)$ and reads (with $\lambda = \Delta t/\Delta x$)

$$u_i^{n+1} = u_i^n - \lambda\big[f(u_i^n) - f(u_{i-1}^n)\big]. \tag{A.12}$$

Similarly, if $f'(u) \leq 0$, the upwind scheme takes the form

$$u_i^{n+1} = u_i^n - \lambda\big[f(u_{i+1}^n) - f(u_i^n)\big].$$

In either case, the upwind scheme is a two-point scheme based upon one-sided differences in the so-called *upwind* direction, i.e., in the direction where the information flows from. The idea of upwind-differencing is the underlying design principle behind a large number of schemes of the Godunov-upwind type, which we will return to below.

Another classical scheme is the three-point Lax–Friedrichs scheme,

$$u_i^{n+1} = \tfrac{1}{2}\big(u_{i-1}^n + u_{i+1}^n\big) - \tfrac{1}{2}\lambda\big[f(u_{i+1}^n) - f(u_{i-1}^n)\big]. \tag{A.13}$$

The Lax–Friedrichs scheme is based upon central differencing and is a very stable, all-purpose scheme that will always converge, although sometimes painstakingly slowly. The scheme can be written in conservation form by introducing the numerical flux

$$F(u^n; i+1/2) = \frac{1}{2\lambda}\big(u_i^n - u_{i+1}^n\big) + \frac{1}{2}\big[f(u_i^n) + f(u_{i+1}^n)\big].$$

The upwind and the Lax–Friedrichs schemes are both examples of schemes that are formally first-order in the sense that their truncation error is of order 2, see Section A.5. Hence the schemes will converge with order 1 for *smooth* solutions; i.e., the error is $\mathcal{O}(\Delta x)$ as $\Delta x \to 0$.

Better accuracy can be obtained if we make a better approximation to the integral in the definition of the average flux. Instead of evaluating the integral at the endpoint t_n, we can evaluate it at the midpoint $t_{n+1/2} = t_n + \frac{1}{2}\Delta t$. This gives a classical second-order method called the Richtmeyer two-step Lax–Wendroff method

$$\begin{aligned} u_{i+1/2}^{n+1/2} &= \tfrac{1}{2}\big(u_i^n + u_{i+1}^n\big) - \lambda\big[f(u_{i\pm1}^n) - f(u_i^n)\big], \\ u_i^{n+1} &= u_i - \lambda\big[f(u_{i+1/2}^{n+1/2}) - f(u_{i-1/2}^{n+1/2})\big]. \end{aligned} \tag{A.14}$$

The corresponding numerical flux reads

$$F(u^n; i+1/2) = f\Big(\tfrac{1}{2}(u_i^n + u_{i+1}^n) - \tfrac{1}{2}\lambda\big[f(u_{i+1}^n) - f(u_i^n)\big]\Big).$$

Another popular variant is *MacCormack's method*

$$\begin{aligned} u_i^* &= u_i^n - \lambda\big[f(u_{i+1}^n) - f(u_i^n)\big], \\ u_i^{**} &= u_i^* - \lambda\big[f(u_i^*) - f(u_{i-1}^*)\big], \\ u_i^{n+1} &= \tfrac{1}{2}\big(u_i^n + u_i^{**}\big), \end{aligned} \tag{A.15}$$

which has the numerical flux

$$F(u^n; i+1/2) = \tfrac{1}{2}f(u_{i+1}^n) + \tfrac{1}{2}f\Big(u_i^n - \lambda\big[f(u_{i+1}^n) - f(u_i^n)\big]\Big).$$

To ensure stability one in general has to impose a restriction on the time-step through a *CFL condition*, named after Courant, Friedrichs, and Lewy, who wrote one of the first papers on finite difference methods in 1928 [70]. The CFL condition states that the true domain of dependence for the PDE (A.1) should be contained in the domain of dependence for (A.10). All the above schemes are stable under the CFL restriction

$$\lambda \max_u |f'(u)| \le 1.$$

Let us now apply the four schemes to two examples to gain some insight into their behavior.

Example A.1. *We first consider the linear advection equation with periodic boundary data*

$$u_t + u_x = 0, \qquad u(x,0) = u_0(x), \qquad u(0,t) = u(1,t). \tag{A.16}$$

As initial data $u_0(x)$ we choose a combination of a smooth, squared cosine wave and a double step function,

$$u_0(x) = \begin{cases} \cos^2\left(\pi \frac{10}{3}(x - \frac{1}{4})\right), & 0.1 \le x \le 0.4, \\ 1, & 0.6 \le x \le 0.9, \\ 0, & \textit{otherwise}. \end{cases}$$

Figure A.1 *shows approximate solutions at time $t = 10.0$ computed by the four schemes introduced above on a grid with 200 nodes using a time-step restriction $\Delta t = 0.9\Delta x$. We see that the two first-order schemes smear both the smooth part and the discontinuous path of the advected profile. The second-order schemes, on the other hand, preserve the smooth profile quite accurately, but introduce spurious oscillations around the two discontinuities.*

Example A.2. *In the next example we apply the same schemes to Burgers' equation with discontinuous initial data*

$$u_t + \left(\tfrac{1}{2}u^2\right)_x = 0, \qquad u(x,0) = \begin{cases} 1, & x \le 0.1, \\ 0, & x > 0.1. \end{cases} \tag{A.17}$$

Burgers' equation is the archetypical example of a nonlinear equation possessing discontinuous solutions (shock waves), forming even for smooth initial data.

Figure A.2 *shows approximate solutions at time $t = 0.5$ computed by all four schemes on a grid with 50 uniform grid cells and a time-step restriction $\lambda = 0.6$. Comparing the two first-order schemes, we see that the upwind scheme resolves the discontinuity quite sharply, whereas the Lax–Friedrichs smears it out over several grid cells. Both second-order schemes resolve the discontinuity sharply, but produce spurious oscillations upstream.*

Although the two examples above were fairly simple, neither of the schemes were able to compute approximate solutions with a satisfactory resolution (except for the upwind scheme in Example A.2). The first-order methods lack the resolution to prevent smooth linear waves from decaying and discontinuities to

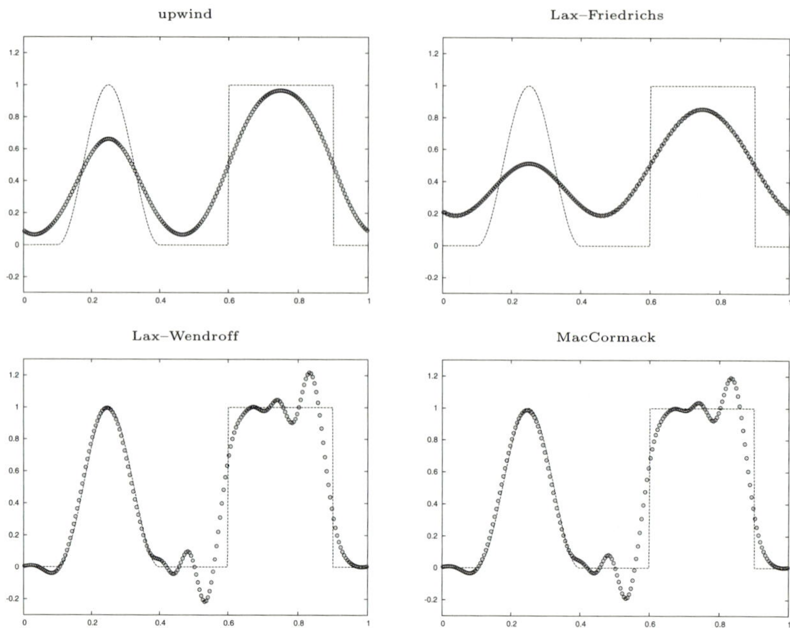

Figure A.1. Approximate solutions at time $t = 10$ for the linear advection equation (A.16) computed by four classical schemes.

being smeared, whereas the second-order methods introduce nonphysical oscillations near discontinuities. Conceptually, one could imagine a possible marriage of the two types of methods in which we try to retain the best features of each method. The resulting scheme would then have second-order (or higher) accuracy in smooth regions of the solution and at the same time have the stability of a first-order scheme where the solution is not smooth. This is a key concept underlying so-called *high-resolution schemes*. Assume now that θ_i^n is a quantity measuring the smoothness of the solution at grid cell i at time t_n such that θ_i^n is close to unity if the solution is smooth and θ_i^n is close to zero if the solution is discontinuous. Then a hybrid method with numerical flux

$$F(u^n; i + 1/2) = \big(1 - \theta_i^n\big) F_L(u^n; i + 1/2) + \theta_i^n F_H(u^n; i + 1/2)$$

would give the desired properties. Here $F_L(u^n; i)$ denotes a low-order flux like the upwind or the Lax–Friedrichs flux and $F_H(u^n; i)$ is a high-order flux like the Lax–Wendroff or the MacCormack flux. The quantity $\theta_i^n = \theta(u^n; i)$ is called a *limiter* and the method is called a *flux limiter* method. A large number of successful high-resolution methods have been developed based on the flux limiting approach. A review of such methods is outside the scope of this book. We will instead retrace our steps in Section A.6 and review a more geometrical framework for developing

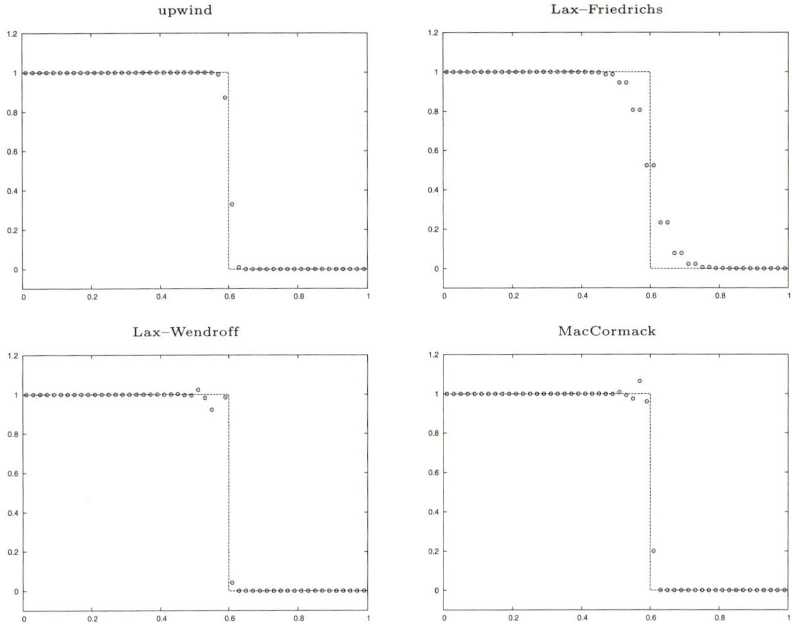

Figure A.2. Approximate solutions at time $t = 0.5$ of the Burgers' equation (A.17).

high-resolution methods. However, before we do that, let us take a careful look at the schemes introduced so far.

A.5 Convergence of conservative methods

So far, we have established an appropriate framework for designing numerical schemes for the conservation law (A.1) and shown a few examples of such schemes. But how can we ensure that these schemes will compute correct approximations to the equation? Moreover, how can we be certain that a scheme will *converge* to the true solution as the discretization parameters tend to zero?

To discuss this, we must first define what we mean by convergence. The classical way of analyzing convergence is to consider the truncation error $L_{\Delta t}$ and show that this error tends to zero with the discretisation parameters; that is, $L_{\Delta t} = \mathcal{O}(\Delta t^r)$ for $r > 0$, where r is said to be the order of the scheme. The truncation error for a scheme in the form (A.10) is defined as

$$L_{\Delta t} = \frac{1}{\Delta t}\Big\{u(x,t+\Delta t) - \Big(u(x,t) - \lambda\big[F\big(u(x,t);i\big) - F\big(u(x,t);i{-}1\big)\big]\Big)\Big\}. \quad \text{(A.18)}$$

We have seen above that the differential equation (A.1) is not valid in a pointwise sense for discontinuous solutions. Thus, the pointwise truncation error cannot be used to establish convergence. For conservation laws, the truncation error only defines the *formal order* of a scheme, i.e., the order the scheme would converge with for smooth solutions.

Since solutions of conservation laws generally are taken in the weak sense (A.4), they are not generally unique. Pointwise errors of the form $u_{\Delta t}(x,t) - u(x,t)$ are therefore not well-defined, where $u_{\Delta t}$ is defined by an appropriate interpolation of u_i^n. Instead, we must measure the deviation in some appropriate norm. It turns out that for scalar equations, the L^1 norm is the correct norm, and we say that an approximation $u_{\Delta t}$ converges to a function u if

$$\int_0^t \|u_{\Delta t}(\cdot,t) - u(\cdot,t)\|_{L^1(\mathbb{R})}\, dt \to 0, \text{ as } \Delta t \to 0.$$

A famous theorem due to Lax and Wendroff [172] says, informally, that the limits of conservative and consistent schemes are weak solutions. A precise statement will be given below; for a proof, see [126, Thm. 3.4]. We verified above that any scheme in the form (A.24) is conservative. The scheme (A.10) is said to be *consistent* if

$$F(v,\dots,v) = f(v).$$

Moreover, one generally requires the numerical flux to be Lipschitz continuous, i.e., that there is a constant L such that

$$|F(u_{i-p},\dots,u_{i+q}) - f(u)| \le L \max\big(|u_{i-p} - u|,\dots,|u_{i+q} - u|\big).$$

Theorem A.3 (Lax–Wendroff theorem). *Let $u_{\Delta t}$ be computed from a conservative and consistent scheme. Assume that* T.V.$(u_{\Delta t})$ *is uniformly bounded in Δt. Consider a subsequence $u_{\Delta t_k}$ such that $\Delta t_k \to 0$, and assume that $u_{\Delta t}$ converges in L^1_{loc} to some u as $\Delta t_k \to 0$. Then u is a weak solution of the conservation law* (A.1).

We can write (A.18) as

$$u(x,t+\Delta t) = H_{\Delta t}(u;i) + \Delta t L_{\Delta t} \tag{A.19}$$

with $H_{\Delta t}(u;i) = u(x,t)\Delta t - \Delta x[F\big(u(x,t);i\big) - F\big(u(x,t);i{-}1\big)]$. We say the scheme is *stable* on $[0,T]$ if $\|H^n_{\Delta t}\|$ remains finite in some norm $\|\cdot\|$ for all $n \in \mathbb{N}$ such that $n\Delta t \le T$. The error is defined by

$$E_{\Delta t}(x,t) = u_{\Delta x}(x,t) - u(x,t), \tag{A.20}$$

and we say that the method converges if $\|E(\cdot,t)\| \to 0$ in some norm $\|\cdot\|$ as $\Delta x \to 0$. For linear equations, the *Lax equivalence theorem* [247] states that a consistent scheme is convergent if and only if it is stable.

The Lax equivalence theorem does not hold for nonlinear equations. However, similar results hold if we can prove that the numerical scheme is contractive in some appropriate norms. The numerical method provides a family of approximate solutions, and we want to extract a sequence that converges to the true solution. We thus have to show that the approximate family is compact. We say that a subset K of a normed space $\mathcal{S}$ is compact[1] if each sequence in K has a subsequence that converges to a point in K. If the limit exists and is in $\mathcal{S}$, but not necessarily in K, we say that K is precompact. Recall that compactness was the key point to establish convergence of the operator splittings in Examples 2.5 to 2.7 in Chapter 2.

The set of functions with bounded (total) variation is compact in L^1. The total variation of a continuous function equals (see, e.g., [126, App. A])

$$\mathrm{T.V.}\,(v) = \limsup_{\varepsilon\to 0} \frac{1}{\varepsilon}\int_{-\infty}^{\infty} |v(x+\varepsilon) - v(x)|\,dx.$$

For a piecewise constant function the definition of the total variation simplifies to

$$\mathrm{T.V.}\,(v) = \sum_i |v_i - v_{i-1}|.$$

This means that we can show that a conservative and consistent scheme will converge if we can verify that the corresponding sequence $\{u_{\Delta t}\}$ has uniformly bounded total variation. There are several ways to verify uniformly bounded total variation (see, e.g., Example 2.5). The total variation of the exact solution of a scalar conservation law is nonincreasing with time

$$\mathrm{T.V.}\,(u(\,\cdot\,,t)) \le \mathrm{T.V.}\,(u(\,\cdot\,,s))\,, \qquad t \ge s.$$

An obvious way to ensure uniformly bounded variation is therefore to require that the scheme has the same property; that is,

$$\mathrm{T.V.}\,\left(u^{n+1}\right) \le \mathrm{T.V.}\,(u^n)\,. \tag{A.21}$$

Any scheme that satisfies (A.21) is called a *total variation diminishing* method, commonly abbreviated as a TVD-method. This requirement has been a popular design principle for a large number of successful schemes, see Section A.6.

In addition there are other properties that might be attractive for the scheme to fulfill:

- A scheme is *monotonicity preserving* when it ensures that if the initial data u_i^0 is monotone then so is u_i^n for any n.
- A scheme is L^1 *contractive* if $\|u_{\Delta t}(\,\cdot\,,t)\|_1 \le \|u_{\Delta t}(\,\cdot\,,0)\|_1$. The entropy weak solution of a scalar conservation law is L^1 contractive.
- A method is *monotone* if

$$u_i^n \ge v_i^n \quad i\in\mathbb{Z}, \qquad \Longrightarrow \qquad u_i^{n+1} \ge v_i^{n+1}, \quad i \in \mathbb{Z}.$$

[1]In general topology this is called sequential compactness.

For conservative and consistent methods, these properties are related as follows:

- Any monotone method is L^1 contractive.
- Any L^1 contractive method is TVD.
- Any TVD method is monotonicity preserving.

Verifying that a method is monotone is quite easy. Unfortunately, it is known that a monotone method is at most first-order accurate. A weaker assumption like TVD is therefore used when designing modern high-order schemes.

Let us now return to the question of convergence. By combining the Lax–Wendroff theorem and a compactness argument, we can verify that a numerical scheme converges to a weak solution of the conservation law. On the other hand, the theory does not say anything about whether the limit is the correct entropy weak solution or not. However, one can show that if the scheme (A.10) satisfies a so-called *cell entropy condition* in the form

$$\eta(u_i^{n+1}) \leq \eta(u_i^n) - \lambda\big(Q_{i+1/2}^n - Q_{i-1/2}^n\big), \tag{A.22}$$

then the limiting weak solution is in fact the entropy weak solution. Here Q is a numerical entropy flux that must be consistent with the entropy flux q in the same way as we required the numerical flux F to be consistent with the flux f. A cell entropy condition is an important building block when we analyze the convergence of various fully discrete operator splitting methods.

For an in-depth introduction to the concepts and convergence theories briefly surveyed above, see for example [107, 159, 175, 176, 260].

A.6 High-resolution Godunov methods

A large number of successful high-resolution methods can be classified as Godunov methods. In the following we will therefore introduce the general setup of Godunov-type methods in some detail, thereby retracing some of the steps used to derive the schemes presented in Section A.3. For simplicity, the presentation is in one spatial dimension, but the same ideas apply also in several space dimensions.

We start by defining the sliding average $\bar{u}(x,t)$ of $u(\,\cdot\,,t)$, namely

$$\bar{u}(x,t) = \frac{1}{\Delta x}\int_{I(x)} u(\xi,t)\,d\xi, \qquad I(x) = \{\xi \mid |\xi - x| \leq \tfrac{1}{2}\Delta x\}. \tag{A.23}$$

If we now integrate (A.1) over the domain $I(x)\times[t,t+\Delta t]$, we obtain an evolution equation for the sliding average $\bar{u}(x,t)$,

$$\bar{u}(x,t+\Delta t) = \bar{u}(x,t) - \frac{1}{\Delta x}\int_t^{t+\Delta t}\Big[f\big(u(x+\tfrac{1}{2}\Delta x,s)\big) - f\big(u(x-\tfrac{1}{2}\Delta x,s)\big)\Big]\,ds. \tag{A.24}$$

This equation is the general starting point for any Godunov-type finite-volume scheme, and the careful reader will notice that (A.9) is a special case of (A.24). To make a complete numerical scheme we must now define how to compute the integrals in (A.23) and (A.24). This can generally be done through a three-step algorithm called *reconstruct–evolve–average* (REA) due to Godunov:

1. Starting from known cell-averages u_i^n in grid cell $[x_{i-1/2}, x_{i+1/2})$ at time $t = t_n$, we **reconstruct** a piecewise polynomial function $\hat{u}(x, t_n)$ defined for all points x. The simplest possible choice is to use a piecewise constant approximation such that
$$\hat{u}(x, t_n) = u_i^n, \quad x \in [x_{i-1/2}, x_{i+1/2}).$$
This will generally result in a method that is formally first order. To obtain a method of higher order, we use a piecewise polynomial interpolant $p_i(x)$ such that
$$\hat{u}(x, t_n) = \sum_i p_i(x)\chi_i(x),$$
where $\chi_i(x)$ is the characteristic function of the ith grid cell $[x_{i-1/2}, x_{i+1/2})$.

2. Then we **evolve** the hyperbolic equation (A.1) exactly (or approximately) with initial data $\hat{u}(x, t_n)$ to obtain a function $\hat{u}(x, t_n + \Delta t)$ a time Δt later.

3. Finally, we **average** the function $\hat{u}(x, t_n + \Delta t)$ over an interval I as in the definition of a sliding average (A.23).

The averaging step generally leaves us with two basically different choices, leading to two classes of methods. Choosing $x = x_i$ in (A.23) gives what is referred to as *upwind* methods and choosing $x = x_{i+1/2}$ gives *central* (difference) methods, see Figure A.3. To see the fundamental difference between these two classes of methods we consider the temporal integrals in (A.24). In the upwind class, the integral of $f(u(\cdot\,, t))$ is taken over points $x_{i\pm1/2}$, where the piecewise polynomial reconstruction $\hat{u}(x, t_n)$ is discontinuous. This means that one cannot apply a standard integration rule in combination with a standard extrapolation. Instead, one must first resolve the local wave-structure arising due to the discontinuity. This amounts to solving a so-called *Riemann problem*. We will come back to this briefly below. For the central methods, the sliding average is computed over a *staggered* grid-cell $[x_i, x_{i+1}]$, which means that the flux integral is evaluated at the points x_i and x_{i+1}, where the initial data $\hat{u}(x, t_n)$ is smooth. If the discretization parameters satisfy a CFL condition which states that λ times the maximum local wave-speed is less than one half, the solution $\hat{u}(x, t)$ will remain smooth at these points for $t \in [t_n, t_n + \Delta t]$. The flux integral can thus be computed using some standard integration scheme in combination with a straightforward extrapolation according to (A.1).

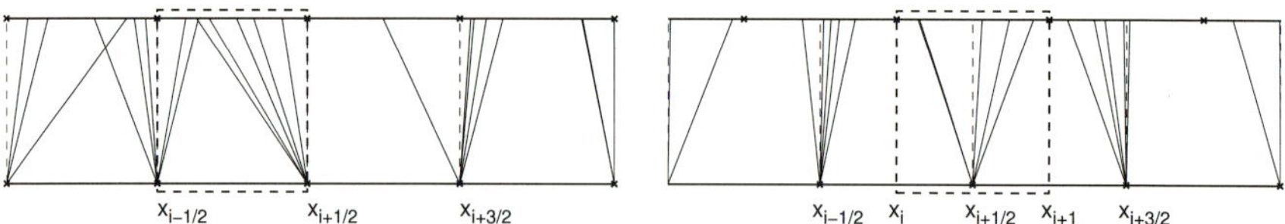

Figure A.3. Computation of sliding averages for upwind schemes (left) and central schemes (right).

A.6.1 High-resolution central schemes. The classical example of a central difference scheme is the Lax–Friedrichs scheme, as given in (A.13). The Lax–Friedrichs scheme has a *staggered* version, which can be derived within the Godunov-framework introduced above if we assume a piecewise constant reconstruction and use a one-sided quadrature rule for the flux integrals in (A.24),

$$u_{i+1/2}^{n+1} = \tfrac{1}{2}\big(u_i^n + u_{i+1}^n\big) - \lambda\big[f(u_{i+1}^n) - f(u_i^n)\big].$$

The scheme is stable under the CFL restriction $(\Delta t/\Delta x)\max_u |f'(u)| \leq 1/2$. Notice that this scheme can be converted to a *nonstaggered* scheme by averaging the staggered cell-averages over the original grid

$$u_i^{n+1} = \tfrac{1}{2}\big(u_{i-1/2}^{n+1} + u_{i+1/2}^{n+1}\big) = \tfrac{1}{4}\big(u_{i-1}^n + 2u_i^n + u_{i+1}^n\big) - \tfrac{1}{2}\lambda\big[f(u_{i+1}^n) - f(u_{i-1}^n)\big],$$

which is almost in the same form as the Lax–Friedrichs given in (A.13).

As an example of high-resolution schemes, we will now derive the second-order extension of the staggered Lax–Friedrichs scheme as introduced by Nessyahu–Tadmor [213]. This scheme, henceforth referred to as NT, is the simplest possible example of high-resolution central schemes and will be used in the next section as a building block for several discrete operator splitting schemes.

For simplicity, we first consider the scalar case. Assume a grid with uniform cell size Δx. Let u_i^n approximate the cell-average over the ith cell $[x_{i-1/2}, x_{i+1/2}]$ at time $t_n = n\Delta t$ and $u_{i+1/2}^{n+1}$ the cell-average over the staggered cell $[x_i, x_{i+1}]$ at time $t_{n+1} = (n+1)\Delta t$. Since we seek a second-order method, the scheme starts with a piecewise linear reconstruction,

$$\hat{u}_i(x, t_n) = u_i^n + (x - x_i)s_i, \qquad x \in [x_{i-1/2}, x_{i+1/2}].$$

We will come back to how to compute s_i below.

If we now insert the piecewise linear reconstruction into (A.24) and evaluate the sliding average over the staggered grid cell, we obtain

$$\begin{aligned} u_{i+1/2}^{n+1} &= \int_{x_i}^{x_{i+1}} \hat{u}(x,t)\,dx - \frac{1}{\Delta x}\int_{t_n}^{t_{n+1}} \big[f(\hat{u}(x_{i+1},t)) - f(\hat{u}(x_i,t))\big]\,dt \\ &= \frac{1}{2}\big(u_i^n + u_{i+1}^n\big) + \frac{\Delta x}{8}\big(s_i - s_{i+1}\big) \\ &\quad - \frac{1}{\Delta x}\int_{t_n}^{t_{n+1}} \big[f(\hat{u}(x_{i+1},t)) - f(\hat{u}(x_i,t))\big]\,dt. \end{aligned}$$

It turns out that it is sufficient to approximate the flux integrals by the midpoint rule to obtain second-order accuracy; that is, we set

$$\int_{t_n}^{t_{n+1}} f(\hat{u}(x_i,t))\,dt \approx \Delta t f\big(\hat{u}(x_i, t_n + \tfrac{1}{2}\Delta t)\big),$$

and similarly at point $x = x_{i+1}$. To complete the scheme, we need to determine how to compute the *point-values* $\hat{u}(x_i, t_{n+1/2})$ and $\hat{u}(x_{i+1}, t_{n+1/2})$. If we assume that the discretisation satisfies a CFL condition $(\Delta t/\Delta t)\max|f'(u)| \le 1/2$, the solution $\hat{u}(\cdot,t)$ will be continuous at the midpoints. Thus, we can use an extrapolation of (A.1) in time using a Taylor series

$$\begin{aligned} \hat{u}(x_i, t_{n+1/2}) &\approx \hat{u}(x_i,t_n) + \frac{\Delta t}{2}\,\partial_t \hat{u}(x_i,t_n) \\ &= \hat{u}(x_i,t_n) - \frac{\Delta t}{2}\,\partial_x f(\hat{u}(x_i,t_n)) \approx u_i^n - \frac{\Delta t}{2}\sigma_i. \end{aligned}$$

The flux gradient σ_i can either be computed directly as $f'(u_i^n)s_i$ or as the slope from a piecewise linear reconstruction of the fluxes of the cell averages. (The careful reader will have noticed that for a piecewise linear reconstruction, the cell averages u_i^n coincide with the point values $\hat{u}(x_i,t_n)$.)

This is almost the full story of the scheme. The only delicate point we have not touched is how to compute the slopes s_i in the piecewise linear reconstructions. A natural candidate is, of course, to use discrete differences, either one-sided or central differences. This means that the slopes s_i could be given by any of the formulas

$$s_i^- = u_i^n - u_{i-1}^n, \qquad s_i^+ = u_{i+1}^n - u_i^n, \qquad s_i^c = \frac{1}{2}\big(u_{i+1}^n - u_{i-1}^n\big).$$

Whereas the two one-sided differences are first-order approximations for smooth data, the central difference is second order and would generally be the preferred choice. However, one can show that all three choices lead to schemes that are formally second-order accurate on smooth solutions of (A.1). For discontinuous solutions, on the other hand, using any of the three approximations may lead to the formation of unphysical oscillations that spread out from a discontinuity, as seen in Examples A.1 and A.2. For scalar equations, the corresponding schemes will violate two fundamental properties of the physical solution: boundedness in L^∞ and bounded variation, see Example 2.5 in Chapter 2. To illuminate this point, let us consider the following set of cell averages

$$u_i^n = \begin{cases} 1, & i \le k, \\ 0, & i > k, \end{cases}$$

for which we have

$$s_k^- = 0, \qquad s_k^+ = -1, \qquad s_k^c = -\tfrac{1}{2}.$$

Obviously, a new maximum will be introduced in $\hat{u}(x, t_n)$ for the two candidate slopes s_k^+ and s_k^c. Similarly, if the function u_i^n is reversed from a backward to a forward step, both s_k^- and s_k^c will introduce new extrema. The formation of new extrema, and the resulting creation of unphysical oscillations, can of course be completely avoided if we use a piecewise constant approximation, but then the formal order of the scheme would be reduced to first order. Altogether, this suggests that we should try to put some more intelligence into the scheme and use the local behavior of the cell averages to determine how to compute s_i. This "intelligence" comes in the form of a nonlinear function called a *limiter*; that is,

$$s_i = \Phi\big(u_i^n - u_{i-1}^n, u_{i+1}^n - u_i^n\big), \qquad \Phi(a,b) = \phi\Big(\frac{b}{a}\Big)a. \tag{A.25}$$

This limiter has much of the same purpose as the flux limiter $\theta(u^n; i)$ introduced at the end of Section A.3.

Under certain restrictions on the function ϕ, one can show that the resulting scheme has diminishing total variation; that is,

$$\text{T.V.}\left(u_{i+1/2}^{n+1}\right) = \sum_i |u_{i+1/2}^{n+1} - u_{i-1/2}^{n+1}| \le \text{T.V.}\,(u_i^n) = \sum_i |u_i^n - u_{i-1}^n|.$$

See [184] for a more detailed discussion of the use of limiters for the Nessyahu–Tadmor scheme. A robust example of a limiter is the minmod limiter

$$\Phi^{MM}(a,b) = \begin{cases} a, & \text{if } |a| < |b| \text{ and } ab > 0, \\ b, & \text{if } |b| < |a| \text{ and } ab > 0, \\ 0, & \text{if } ab \le 0, \end{cases}$$

or alternatively

$$\phi^{MM}(r) = \max\big(0, \min(r, 1)\big).$$

The minmod limiter is 'conservative' in the sense that it uses a first-order reconstruction when the one-sided finite differences have different sign, thereby introducing an undesired clipping of local extrema. At the other extreme, the superbee limiter

$$\phi^{SB}(r;\theta) = \max\big(0, \min(\theta r, 1), \min(r, \theta)\big)$$

which yields Φ^{SB} using (A.25), generally gives a steeper (often called a more compressive) reconstruction. See [184] for an in-depth discussion of limiters for the Nessyahu–Tadmor scheme.

Summing up, we have derived a predictor-corrector scheme of the form

$$\begin{aligned} u_i^{n+1/2} &= u_i^n - \frac{\lambda}{2}\sigma_i, \\ u_{i+1/2}^{n+1} &= \frac{1}{2}\big(u_i^n + u_{i+1}^n\big) - \lambda\big(g_{i+1}^n - g_i^n\big), \\ g_i^n &= f(u_i^{n+1/2}) + \frac{1}{8\lambda}s_i, \\ s_i &= \Phi\big(u_i^n - u_{i-1}^n, u_{i+1}^n - u_i^n\big), \\ \sigma_i &= \Phi\big(f(u_i^n) - f(u_{i-1}^n), f(u_{i+1}^n) - f(u_i^n)\big). \end{aligned} \tag{A.26}$$

The scheme is formally second order. Moreover, under appropriate assumptions on the time-step and the limiter function Φ one can prove that this scheme gives solutions that are bounded by their initial data in L^∞ norm and have diminishing total variation. Thus, unlike the classical second-order schemes, the NT scheme mimics the properties of the exact scalar solution.

The major advantage of the NT scheme is that it is both compact and simple to implement, particularly since it does not require the use of any characteristic information or solution of local Riemann problems (see Section A.6.2). The only requirement is an estimate of the maximum wave-speed needed to impose a CFL restriction on the time-step.

Example A.4. *Let us now apply the NT scheme to the linear advection equation and Burgers' equation as considered in Examples* A.1 *and* A.2. *We use two different limiter functions, the dissipative minmod limiter and the compressive superbee limiter. Figure* A.4 *shows the approximate solutions computed with the same parameters as in Figures* A.1 *and* A.2, *except for the time-step which is now* $\Delta t = 0.475\Delta x$. *The improvement in the resolution is obvious. On the other hand, we see that there is some difference in the two limiter functions. The dissipative minmod limiter always chooses the lesser slopes and thus behaves more like a first-order scheme. The compressive superbee limiter picks steeper slopes and has a tendency of overcompressing smooth linear waves, as observed for the smooth cosine profile.*

If higher accuracy is wanted, one must use a spatial reconstruction of higher order and a more accurate temporal extrapolation in terms of more predictor steps such as in higher-order Runge–Kutta methods, see [28, 29, 179, 191]. Similarly, semi-discrete nonstaggered schemes have been developed [162–164], for which only the spatial derivatives are discretized, leading to a set of ordinary differential equations that can be integrated by an ODE solver. See [250] for a comprehensive overview of all schemes that have been developed within the class of nonoscillatory central schemes.

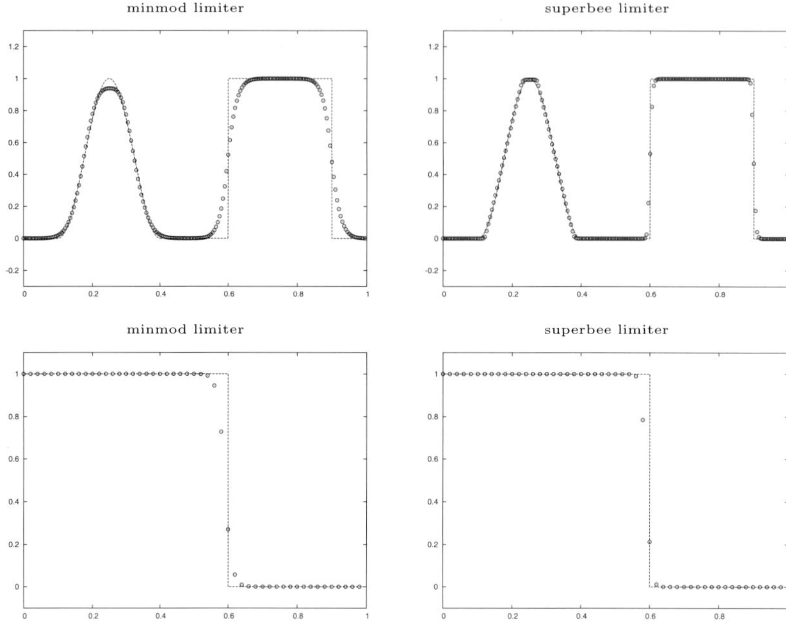

Figure A.4. Approximate solutions of the linear advection and Burgers' equation computed by the NT scheme with two different limiters.

There are different ways to extend central schemes to systems of conservation laws. The simplest method, which we will use for the NT scheme later in this book, is to apply the scheme directly to each component of the vector of unknowns [213]. This greatly simplifies the implementation of central schemes and is possible since the schemes do not use the eigenstructure of the underlying system.

To derive high-resolution schemes for conservation laws in multi-dimensions we can apply similar ideas. In two spatial dimensions, the conservation law reads

$$u_t + f(u)_x + g(u)_y = 0, \qquad u(x,y,0) = u_0(x,y). \tag{A.27}$$

As in one dimension, we introduce the sliding average

$$\bar{u}(x,y,t) = \frac{1}{\Delta x \Delta y} \int_{I(x)} \int_{J(y)} u(\xi,\eta,t)\, d\xi d\eta,$$

and integrate (A.27) over the domain $I(x) \times J(y) \times [t, t+\Delta t]$ to derive an evolution equation for the sliding average

$$
\begin{aligned}
&\bar{u}(x,t+\Delta t) \\
&\quad = \bar{u}(x,t) \\
&\qquad - \frac{1}{\Delta x \Delta y}\int_t^{t+\Delta t}\int_{J(y)}\Big[f\big(u(x+\tfrac{1}{2}\Delta x,y,s)\big) - f\big(u(x-\tfrac{1}{2}\Delta x,y,s)\big)\Big]\,dyds \\
&\qquad - \frac{1}{\Delta x \Delta y}\int_t^{t+\Delta t}\int_{I(x)}\Big[g\big(u(x,y+\tfrac{1}{2}\Delta y,s)\big) - g\big(u(x,y-\tfrac{1}{2}\Delta y,s)\big)\Big]\,dxds.
\end{aligned}
$$

Secondly, we make a piecewise linear reconstruction in each spatial direction,

$$
\begin{aligned}
\hat{u}_{ij}(x,y,t_n) = u_{ij}^n + (x-x_i)s_{ij}^x + (y-y_j)s_{ij}^y,& \\
(x,y) \in [x_{i-1/2}, x_{i+1/2}] \times [y_{j-1/2}, y_{j+1/2}].&
\end{aligned}
$$

We now evaluate the sliding average over the staggered grid cell, defined analogously as for the one-dimensional case, and use the midpoint rule to approximate the flux integrals. Altogether this gives the two-dimensional version of the Nessyahu–Tadmor scheme [134],

$$
\begin{aligned}
u_{i+1/2,j+1/2}^{n+1} &= \frac{1}{4}\big(u_{ij}^n + u_{i+1,j}^n + u_{i+1,j+1}^n + u_{i,j+1}^n\big) \\
&\quad + \frac{1}{16}\big(s_{ij}^x + s_{i,j+1}^x - s_{i+1,j}^x - s_{i+1,j+1}^x\big) \\
&\quad + \frac{1}{16}\big(s_{ij}^y - s_{i,j+1}^y + s_{i+1,j}^y - s_{i+1,j+1}^y\big) \\
&\quad - \frac{\lambda}{2}\Big[f(u_{i+1,j}^{n+1/2}) + f(u_{i+1,j+1}^{n+1/2})\Big] + \frac{\lambda}{2}\Big[f(u_{i,j}^{n+1/2}) + f(u_{i,j+1}^{n+1/2})\Big] \\
&\quad - \frac{\mu}{2}\Big[g(u_{i,j+1}^{n+1/2}) + g(u_{i+1,j+1}^{n+1/2})\Big] + \frac{\mu}{2}\Big[g(u_{i,j}^{n+1/2}) + g(u_{i+1,j}^{n+1/2})\Big], \\
u_{ij}^{n+1/2} &= u_{ij}^n - \frac{\lambda}{2}\sigma_{ij}^x - \frac{\mu}{2}\sigma_{ij}^y,
\end{aligned}
\tag{A.28}
$$

where $\lambda = \Delta t/\Delta x$, $\mu = \Delta t/\Delta y$. As for its one-dimensional counterpart, the scheme is compact and easy to implement, can be applied to systems in a componentwise fashion, and has fairly good accuracy.

High-resolution central schemes have seen a rapid development in recent years and are by now established as simple and versatile schemes for integrating conservation laws in several dimensions. We refer the reader to [249] for an introduction to central schemes. The website [250] constitutes a comprehensive source of information on central schemes, including their extensions to higher order, unstructured grids, semi-discrete versions, and applications proving the versatility of the schemes.

A.6.2 High-resolution upwind schemes. To derive upwind schemes, we return to the sliding average in (A.23). Let us for simplicity assume that the reconstructed function $\hat{u}(x,t_n)$ is piecewise constant. To evolve the solution, we see from Figure A.3 that in order to compute the integral of the flux function over the cell boundaries, we must solve a series of simple initial-value problems of the form,

$$u_t + f(u)_x = 0, \qquad u(x,0) = \begin{cases} u_L, & x<0, \\ u_R, & x>0. \end{cases}$$

This is commonly referred to as a Riemann problem, which has a self-similar solution of the form $u(x,t) = v(x/t; u_L, u_R)$ and consists of a set of constant states separated by simple waves (rarefaction, shocks and contacts). Since a hyperbolic equation has finite speed of propagation, the *global* solution $\hat{u}(x,t)$ for sufficiently small t can be constructed by piecing together the local Riemann solutions. In general it can be quite complicated to solve Riemann problems, at least for systems of conservation laws. However, to compute the flux integrals in (A.24) we only need the solution of the Riemann problem along the ray $x/t = 0$, where the solution is constant $u(0,t) = v(0; u_L, u_R)$. Thus, the general form of the upwind Godunov-methods reads

$$u_i^{n+1} = u_i^n - \lambda\Big[f\big(v(0; u_{i-1}^n, u_i^n)\big) - f\big(v(0; u_i^n, u_{i+1}^n)\big)\Big]. \tag{A.29}$$

A specific scheme is obtained by devising a method to compute the (approximate) solution of the local Riemann problems.

Example A.5. *Let us consider the scalar case with a convex flux function that satisfies $f''(u) > 0$ (the case $f''(u) < 0$ is similar). In this case the Riemann solution consists of either a single rarefaction wave or a single shock. If $f'(u)$ is either strictly positive or strictly negative, the single wave will move to one side only, and the Godunov scheme simplifies to the upwind scheme* (A.12). *If not, the solution must consist of a rarefaction wave moving in both the positive and negative direction. Inside the rarefaction wave there is a single sonic point u_s where $f'(u_s)$ is zero, and the wave is therefore called a transonic rarefaction wave. Summing up our observations, the Godunov flux reads*

$$F_{i+1/2}^n = \begin{cases} f(u_i^n), & \sigma_{i+1/2} > 0 \text{ and } u_i^n > u_s, \\ f(u_{i+1}^n), & \sigma_{i+1/2} < 0 \text{ and } u_{i+1}^n < u_s, \\ f(u_s), & u_i^n < u_s < u_{i+1}^n. \end{cases}$$

Here $\sigma_{i+1/2} = [f(u_{i+1}^n) - f(u_i^n)]/(u_{i+1}^n - u_i^n)$ is the Rankine–Hugoniot speed associated with the jump. A similar formula holds for the case when $f''(u) < 0$.

The Godunov flux derived in the above example can be written in a more compact form, which is also valid for an arbitrary nonconvex flux function $f(u)$,

$$F_{i+1/2}^n = \begin{cases} \min_{u\in[u_i^n, u_{i+1}^n]} f(u), & u_i^n \le u_{i+1}^n, \\ \max_{u\in[u_{i+1}^n, u_i^n]} f(u), & u_i^n \ge u_{i+1}^n. \end{cases} \tag{A.30}$$

If $f(u)$ is nonconvex, the flux may have several sonic points, one at each of its local critical points.

Working with the exact Godunov flux in an actual implementation is a bit cumbersome since the formula requires the computation of the minimum or maximum of $f(u)$ over an interval. It is therefore customary to replace formula (A.30) with another formula based upon an approximation to the Riemann problem. This approach is also much easier to generalise to systems of conservation laws.

Let us first assume that the solution is a continuous wave (i.e., a rarefaction wave). This gives the Engquist–Osher scheme, which is a natural extension of the upwind scheme to nonconvex flux functions. The Engquist–Osher flux function reads

$$F^n_{i+1/2} = f(0) + \int_0^{u_i^n} \max(f'(v), 0)\, dv + \int_0^{u_{i+1}^n} \min(f'(v), 0)\, dv. \tag{A.31}$$

Alternatively, we can approximate the Riemann problem by a single shock. Then the flux can be written as

$$F^n_{i+1/2} = f(u_i^n) + \sigma^-_{i+1/2}(u^n_{i+1} - u^n_i),$$

or alternatively as

$$F^n_{i+1/2} = f(u^n_{i+1}) - \sigma^+_{i+1/2}(u^n_{i+1} - u^n_i).$$

Here $s^+ = \max(s, 0)$ and $s^- = \min(s, 0)$. By averaging the equivalent expressions we obtain the numerical flux

$$F^n_{i+1/2} = \tfrac{1}{2}\big[f(u^n_i) + f(u^n_{i+1}) - |\sigma_{i+1/2}|(u^n_{i+1/2} - u^n_i)\big].$$

This can be interpreted as a central flux approximation plus a viscous correction with coefficient $|\sigma_{i+1/2}|$. The formula can quite easily be extended to systems of equations and gives what is commonly referred to as the Roe linearization of the Riemann problem. For transonic waves the coefficient $\sigma_{i+1/2}$ may vanish or be close to zero and the added dissipation is insufficient to stabilise the computation. It is therefore customary to add extra dissipation in the form of an entropy fix. For more details, consult for instance [176].

High-resolution versions of the upwind schemes can be obtained by using a higher-order reconstruction of the cell averages. This is beyond the scope of the current book. The interested reader can find the relevant details in a number of books, for example [107, 108, 115, 159, 175, 176, 260, 262].

A.7 Front tracking

In this section we will introduce a completely different method that does not fall into the class (A.10). Still, the method will prove to be particularly versatile when applied to one-dimensional equations as part of an operator splitting approach.

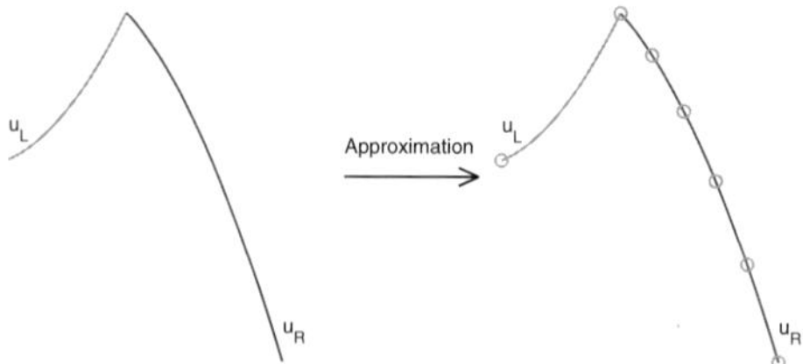

Figure A.5. Piecewise constant approximation of a Riemann solution in state space. The shock wave connecting u_L is kept, while the rarefaction wave connecting u_R is replaced by a set of (small) jump discontinuities.

The term front tracking has been used to describe a number of methods that perform some form of explicit tracking of discontinuities. Most of these methods use a regular grid to represent continuous parts of the solution and add extra degrees of freedom to represent (moving) discontinuities.

We will use the term front tracking to signify a different method, described in, e.g., [126]. For earlier papers on the method we refer to [72, 117, 118]. Front tracking is a mathematical algorithm for computing *exact* solutions of a certain subclass of equations on the form (A.1). The algorithm can also be used for numerical computations; the resulting method is unconditionally stable, has no numerical dissipation and is very fast. Due to its versatility as a computational (and mathematical) method, the front-tracking method will be used extensively in the rest of the book as a basic solution method for hyperbolic conservation laws in one spatial dimension. We will therefore introduce the method in more detail.

Consider the Cauchy problem (A.3). Let $\widetilde{u}_0$ be a piecewise constant approximation to u_0. Since the conservation law is L^1 continuous with respect to its initial data, the solution $v(x,t)$ of the Cauchy problem

$$v_t + f(v)_x = 0, \qquad v(x,0) = \widetilde{u}_0(x), \tag{A.32}$$

can be made to approximate $u(x,t)$ arbitrarily well by choosing an appropriate approximation $\widetilde{u}_0(x)$ to $u_0(x)$. Let us therefore look at the solution to (A.32) in more detail. Since the initial data is piecewise constant, the Cauchy problem initially consists of a set of local Riemann problems. The solution of each of these Riemann problems is a similarity solution consisting of a set of constant states separated by simple waves giving either smooth transitions (for rarefaction waves) or discontinuities along space-time rays (for shocks or contacts).

The key idea in the front-tracking algorithm is to approximate each Riemann solution by a step function; that is, replace the smooth rarefaction waves by a series of (small) jump discontinuities and keep shocks and contacts. For scalar equations, the approximation of the Riemann problem is obtained implicitly through a piecewise linear approximation of the flux function. For systems, the Riemann solution is discretized in the phase plane so that shocks and contacts are retained and rar-

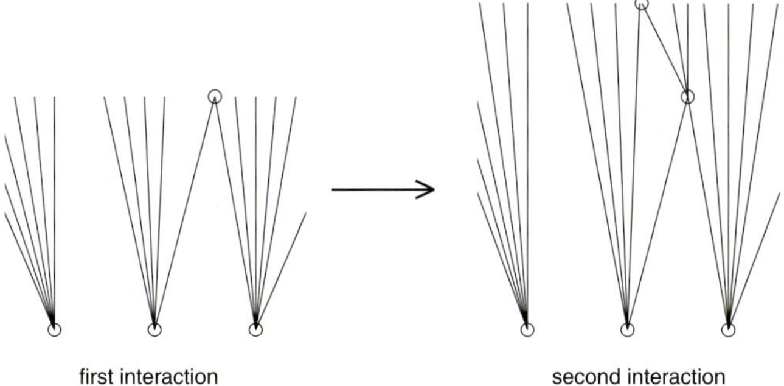

Figure A.6. Front-tracking construction of a piecewise constant solution depicted in the (x,t)-plane.

efaction waves are approximated by step functions, see Figure A.5. In either case, the result is a *discrete*, approximate Riemann fan consisting of constant states separated by space-time rays of discontinuity (fronts). The global solution is obtained by piecing together the local Riemann fans and is evolved forward in time by tracking the discontinuities. The piecewise constant solution is valid as long as the discontinuities do not interact. When two (or more) rays collide, they define a new local Riemann problem. This Riemann problem can be solved and the approximate piecewise constant wave fan can be inserted in the global solution, see Figure A.6. This construction can be continued forward to an arbitrary time.

The front-tracking algorithm is summarized in Algorithm A.7.1. In an implementation of a standard finite-volume method, the basic data-structure consists of a grid and an associated array of cell averages u_i^n. In an implementation of the front-tracking algorithm, the basic data structure is a set of *front objects* that represent the propagating discontinuities. The following data is associated with each discontinuity: left and right states u_L and u_R, point of origin (x_0, t_0), propagation speed σ, and possible collision point (x_c, t_c) with another front object. To track the fronts, we use two lists, a spatial list F, where the fronts are sorted from left to right, and a collision list C, where front collisions are sorted with respect to collision time in ascending order. A thorough discussion of the implementation of the front-tracking algorithm is given in [169].

Most of the algorithmic functions in Algorithm A.7.1 should be self-explanatory, except for the Riemann solver. We will therefore explain this function in more detail. To this end, consider a scalar Riemann problem

$$u_t + f(u)_x = 0, \qquad u(x,0) = \begin{cases} u_L, & x < 0, \\ u_R, & x > 0. \end{cases}$$

Algorithm A.7.1 The front-tracking algorithm

```
Construct a piecewise constant initial function ũ_0(x) = ũ_i, for x ∈ [x_{i-1/2}, x_{i+1/2}]
F = {∅}, C = {∅}, and t = 0
Solve all initial Riemann problems in ũ^0(x)
   {f_L, ..., f_R} = RiemannSolver(ũ_i, ũ_{i+1}; x_{i+1/2}, t)
   f_end = RightMostFront(F)
   Append(F, {f_L, ..., f_R})
   c = ComputCollision(f_end, f_L)
   C=Sort({C, c})
While (t ≤ T) and C ≠ {∅} do
     (c, x_c, t_c) = GetNextCollision(C) and remove c from C
     {f_L, ..., f_R} = FindAllFrontsInCollision(c)
     f_p = FindNeighbor(f_L,to left)
     f_n = FindNeighbor(f_R,to right)
     F = {..., f_p, f_n, ...}
     u_L = GetLeftState(f_L)
     u_R = GetRightState(f_R)
     {f_L, ..., f_R} = RiemannSolver(u_L, u_R; x_c, t_c)
     F = {..., f_p, f_L, ..., f_R, f_n, ...}
     c_L = ComputeCollision(f_p, f_L)
     c_R = ComputeCollision(f_R, f_n)
     C = Sort({C, c_L, c_R})
     set t = t_c
endwhile
```

If $u_L > u_R$ the solution of the Riemann problem is a self-similar function given by

$$u(x,t) = \begin{cases} u_L, & x/t < f_c'(u_L), \\ \left(f_c'\right)^{-1}(x/t), & f_c'(u_L) < x/t < f_c'(u_R), \\ u_L, & x/t \geq f_c'(u_R). \end{cases}$$

Here $f_c(u)$ is the upper concave envelope of f over the interval $[u_R, u_L]$ and $(f_c')^{-1}$ the inverse of its derivative. Intuitively, the concave envelope is constructed by imagining a rubber band attached at u_R and u_L and stretched above f in between. When released, the rubber band assumes the shape of the concave envelope and ensures that $f''(u) < 0$ for all $u \in [u_R, u_L]$. The case $u_L < u_R$ is treated symmetrically with f_c denoting the lower convex envelope. Assume now that the flux function $f(u)$ is continuous piecewise linear between the points $\{u_1, u_2, \dots, u_N\}$, where $u_1 = u_R$ and $u_N = u_L$. In this case, the Riemann solution takes a particu-

Algorithm A.7.2 Piecewise linear envelope function in case $u_1 < u_N$

```
i = 1, k = 1, I = {1}
while i < N
    σ_max = −∞
    for j = N, ..., i + 1
        Δ = (f(u_j) − f(u_i))/(u_j − u_i)
        If Δ > σ_max: k = i and σ_max = Δ
    end
    I = {I, k}
    i = k
endwhile
```

larly simple form

$$u(x,t) = \begin{cases} u_N = u_L, & x < x_{N-1}(t), \\ u_i, & x_i(t) < x < x_{i-1}(t), \quad i = N-1, \dots, 2, \\ u_1 = u_R, & x \geq x_1(t), \end{cases}$$

where each space-time ray $x_i(t)$ satisfies the Rankine–Hugoniot condition

$$\frac{dx_i}{dt} = \frac{f_c(u_{i+1}) - f_c(u_i)}{u_{i+1} - u_i} = \sigma_i, \quad x_i(0) = 0.$$

Having obtained an explicit formula for the Riemann solution, the only remaining problem is to determine the envelope function. Algorithm A.7.2 gives a simple algorithm for the case $u_1 < u_N$. After the algorithm, the piecewise linear envelope function is given by the points $\{u_{I(1)} = u_1, u_{I(2)}, \dots, u_{I(M)} = u_N\}$, where $M \leq N$. When used as a numerical method, front tracking is unconditionally stable and has first order convergence with respect to the approximation of the initial data and the Riemann problems. For scalar equations it can even be shown that the number of steps in the algorithm is finite. Moreover, since the front-tracking algorithm computes the exact solution of an equation within the same class of equations, the approximation will automatically satisfy the same mathematical properties as the true solution. This means, for instance, that the front-tracking approximation is by definition bounded in L^∞, has bounded variation, and satisfies a Kružkov entropy inequality.

Example A.6. *Figure A.7 shows the front-tracking construction for a scalar equation with piecewise constant initial data and nonconvex flux function $f(u) = u^3$. Initially, each Riemann problem is solved by a single discontinuity. As t increases, the solution develops two shocks as constant states with larger values of $|u|$ overtake constant states with smaller values of $|u|$.*

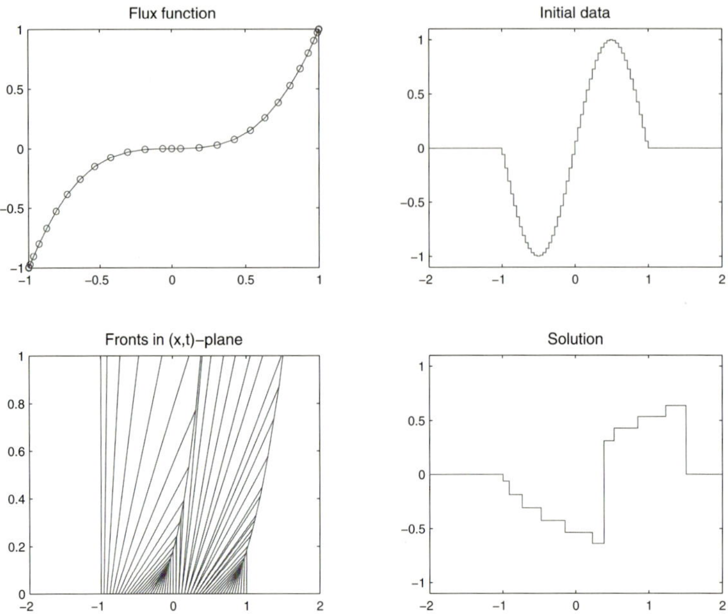

Figure A.7. Front-tracking construction of the exact analytical solution.

The front-tracking method has an obvious generalization to equations of the form

$$u_t + V(x,t) f(u)_x = 0, \qquad u(x,0) = u_0(x). \tag{A.33}$$

This equation will arise in some of the operator splitting examples studied in Chapter 6. Equation (A.33) has (almost) the same permissible discontinuities as the original conservation law $u_t + f(u)_x = 0$. The difference is that instead of following straight lines, the permissible discontinuities of (A.33) follow paths that satisfy a differential equation of the form

$$\dot{x}(t) = V(x,t)\sigma, \qquad x(t_0) = x_0, \tag{A.34}$$

where σ is the usual Rankine–Hugoniot velocity. Tracking discontinuities and computing (potential) intersections is the core of the front-tracking algorithm. For a general discontinuity path satisfying (A.34), this is a nontrivial exercise, since one would typically have to rely on some ODE solver to compute the solution of (A.34). However, if $V(x,t)$ is the tensor product of two piecewise linear functions, (A.34) can be solved explicitly, giving formulas for computing possible intersections of two shock paths in explicit form. This means that the front-tracking algorithm can be used to compute solutions to (A.33) if we make such a piecewise linear spline approximation to the velocity $V(x,t)$ in (A.33); see [181] for more details.

For a comprehensive introduction to the front-tracking method, both from a mathematical and a numerical point of view, we refer to [126]. A variant of the front-tracking method was also used by Bressan in his seminal work on the theory of systems of conservation laws, see, e.g., [41].

Lucier [195] has developed a second-order front-tracking method for scalar conservation laws. Instead of approximating the flux function by a polygon, i.e., a piecewise linear and continuous function, Lucier considered a quadratic spline approximation. Furthermore, the initial data is replaced by a continuous piecewise linear approximation rather than the piecewise constant approximation considered here. The method applies to convex flux functions.

References

[1] J. E. Aarnes, T. Gimse, and K.-A. Lie. An introduction to the numerics of flow in porous media using Matlab. In G. Hasle, K.-A. Lie, and E. Quak, editors, *Geometric Modelling, Numerical Simulation, and Optimization: Applied Mathematics at SINTEF*, pages 265–306. Springer Verlag, Berlin Heidelberg New York, 2007.

[2] J. E. Aarnes, V. Kippe, K.-A. Lie, and A. B. Rustad. Modelling of multiscale structures in flow simulations for petroleum reservoirs. In G. Hasle, K.-A. Lie, and E. Quak, editors, *Geometric Modelling, Numerical Simulation, and Optimization: Applied Mathematics at SINTEF*, pages 307–360. Springer Verlag, Berlin Heidelberg New York, 2007.

[3] H. W. Alt and S. Luckhaus. Quasilinear elliptic-parabolic differential equations. *Math. Z.*, 183(3):311–341, 1983.

[4] L. Ambrosio, N. Fusco, and D. Pallara. *Functions of bounded variation and free discontinuity problems.* Oxford Mathematical Monographs. The Clarendon Press Oxford University Press, New York, 2000.

[5] B. A. Andreianov, M. Gutnic, and P. Wittbold. Convergence of finite volume approximations for a nonlinear elliptic-parabolic problem: a "continuous" approach. *SIAM J. Numer. Anal.*, 42(1):228–251 (electronic), 2004.

[6] D. Aregba-Driollet, R. Natalini, and S. Tang. Explicit diffusive kinetic schemes for nonlinear degenerate parabolic systems. *Math. Comp.*, 73(245):63–94 (electronic), 2004.

[7] E. Audusse, F. Bouchut, M.-O. Bristeau, R. Klein, and B. Perthame. A fast and stable well-balanced scheme with hydrostatic reconstruction for shallow water flows. *SIAM J. Sci. Comput.*, 25(6):2050–2065, 2004.

[8] K. Aziz and A. Settari. *Petroleum reservoir simulation.* Applied Science Publishers, London, 1979.

[9] W. Bao and S. Jin. The random projection method for stiff detonation capturing. *SIAM J. Sci. Comput.*, 23(3):1000–1026 (electronic), 2001.

[10] W. Bao, S. Jin, and P. A. Markowich. On time-splitting spectral approximations for the Schrödinger equation in the semiclassical regime. *J. Comput. Phys.*, 175(2):487–524, 2002.

[11] W. Bao, S. Jin, and P. A. Markowich. Numerical study of time-splitting spectral discretizations of nonlinear Schrödinger equations in the semiclassical regimes. *SIAM J. Sci. Comput.*, 25(1):27–64 (electronic), 2003.

[12] G. Barles. Convergence of numerical schemes for degenerate parabolic equations arising in finance theory. In *Numerical methods in finance*, pages 1–21. Cambridge Univ. Press, Cambridge, 1997.

[13] G. Barles, C. Daher, and M. Romano. Convergence of numerical schemes for parabolic equations arising in finance theory. *Math. Models Methods Appl. Sci.*, 5(1):125–143, 1995.

[14] G. Barles and P. E. Souganidis. Convergence of approximation schemes for fully nonlinear second order equations. *Asymptotic Anal.*, 4(3):271–283, 1991.

[15] T. Barth and M. Ohlberger. Finite volume methods: foundation and analysis. In E. Stein, R. de Borst, and T. J. R. Hughes, editors, *Encyclopedia of Computational Mechanics*, volume 1, pages 439–474. John Wiley and Sons Ltd, 2004.

[16] J. T. Beale and C. Greengard. Convergence of Euler-Stokes splitting of the Navier–Stokes equations. *Comm. Pure Appl. Math.*, 47(8):1083–1115, 1994.

[17] J. T. Beale, C. Greengard, and E. Thomann. Operator splitting for Navier–Stokes and Chorin–Marsden product formula. In *Vortex flows and related numerical methods (Grenoble, 1992)*, volume 395 of *NATO Adv. Sci. Inst. Ser. C Math. Phys. Sci.*, pages 27–38. Kluwer Acad. Publ., Dordrecht, 1993.

[18] J. T. Beale and A. Majda. Rates of convergence for viscous splitting of the Navier–Stokes equations. *Math. Comp.*, 37(156):243–259, 1981.

[19] M. Bendahmane and K. H. Karlsen. Renormalized entropy solutions for quasi-linear anisotropic degenerate parabolic equations. *SIAM J. Math. Anal.*, 36(2):405–422 (electronic), 2004.

[20] M. Bendahmane and K. H. Karlsen. Uniqueness of entropy solutions for doubly nonlinear anisotropic degenerate parabolic equations. In *Nonlinear partial differential equations and related analysis*, volume 371 of *Contemp. Math.*, pages 1–27. Amer. Math. Soc., Providence, RI, 2005.

[21] P. Bénilan and H. Touré. Sur l'équation générale $u_t = \varphi(u)_{xx} - \psi(u)_x + v$. *C. R. Acad. Sci. Paris Sér. I Math.*, 299(18):919–922, 1984.

[22] P. Bénilan and H. Touré. Sur l'équation générale $u_t = a(\cdot, u, \phi(\cdot, u)_x)_x + v$ dans L^1. I. Étude du problème stationnaire. In *Evolution equations (Baton Rouge, LA, 1992)*, pages 35–62. Dekker, New York, 1995.

[23] P. Bénilan and H. Touré. Sur l'équation générale $u_t = a(\cdot, u, \phi(\cdot, u)_x)_x + v$ dans L^1. II. Le problème d'évolution. *Ann. Inst. H. Poincaré Anal. Non Linéaire*, 12(6):727–761, 1995.

[24] P. Benilan and P. Wittbold. On mild and weak solutions of elliptic-parabolic problems. *Adv. Differential Equations*, 1(6):1053–1073, 1996.

[25] R. A. Berenblyum, A. A. Shapiro, K. Jessen, E. H. Stenby, and J. Franklin M. Orr. Black oil streamline simulator with capillary effects. In *SPE Annual Technical Conference and Exhibition*, Denver, Colorado, 5-8 October 2003. SPE 84037.

[26] S. Berres, R. Bürger, K. H. Karlsen, and E. M. Tory. Strongly degenerate parabolic-hyperbolic systems modeling polydisperse sedimentation with compression. *SIAM J. Appl. Math.*, 64(1):41–80 (electronic), 2003.

[27] C. Besse, B. Bidégaray, and S. Descombes. Order estimates in time of splitting methods for the nonlinear Schrödinger equation. *SIAM J. Numer. Anal.*, 40(1):26–40 (electronic), 2002.

[28] F. Bianco, G. Puppo, and G. Russo. High order central schemes for hyperbolic systems of conservation laws. In *Hyperbolic problems: theory, numerics, applications, Vol. I (Zürich, 1998)*, volume 129 of *Internat. Ser. Numer. Math.*, pages 55–64. Birkhäuser, Basel, 1999.

[29] F. Bianco, G. Puppo, and G. Russo. High-order central schemes for hyperbolic systems of conservation laws. *SIAM J. Sci. Comput.*, 21(1):294–322 (electronic), 1999.

[30] J. J. Blum and M. C. Reed. A model for fast axonal-transport. *Cell motility and the cytoskeleton*, 5(6):507–527, 1985.

[31] F. Bouchut, C. Bourdarias, and B. Perthame. A MUSCL method satisfying all the numerical entropy inequalities. *Math. Comp.*, 65(216):1439–1461, 1996.

[32] F. Bouchut, F. R. Guarguaglini, and R. Natalini. Diffusive BGK approximations for nonlinear multidimensional parabolic equations. *Indiana Univ. Math. J.*, 49(2):723–749, 2000.

[33] F. Bouchut and B. Perthame. Kružkov's estimates for scalar conservation laws revisited. *Trans. Amer. Math. Soc.*, 350(7):2847–2870, 1998.

[34] F. Bratvedt, K. Bratvedt, C. F. Buchholz, T. Gimse, H. Holden, L. Holden, and N. H. Risebro. Front tracking for petroleum reservoirs. In *Ideas and Methods in Mathematical Analysis, Stochastics, and Applications. In Memory of Raphael Høegh-Krohn (1938-1988).* Cambridge University Press, Cambridge, 1992, pp. 409-427.

[35] F. Bratvedt, K. Bratvedt, C. F. Buchholz, T. Gimse, H. Holden, L. Holden, R. Olufsen, and N. H. Risebro. Three-dimensional reservoir simulation based on front tracking. In *North Sea Oil and Gas Reservoirs – III*, pages 247–257. Kluwer Academic Publishers, 1994.

[36] F. Bratvedt, K. Bratvedt, C. Buchholz, H. Holden, L. Holden, and N. H. Risebro. A new front-tracking method for reservoir simulation. *SPE Reservoir Engineering* 7:107–116, 1992.

[37] F. Bratvedt, K. Bratvedt, C. F. Buchholz, T. Gimse, H. Holden, L. Holden, and N. H. Risebro. FRONTLINE and FRONTSIM: two full scale, two-phase, black oil reservoir simulators based on front tracking. *Surveys Math. Indust.*, 3(3):185–215, 1993.

[38] F. Bratvedt, T. Gimse, and C. Tegnander. Streamline computations for porous media flow including gravity. *Transport in Porous Media*, 25:63–78, 1996.

[39] Y. Brenier. Averaged multivalued solutions for scalar conservation laws. *SIAM J. Numer. Anal.*, 21(6):1013–1037, 1984.

[40] S. C. Brenner and L. R. Scott. *The mathematical theory of finite element methods*, volume 15 of *Texts in Applied Mathematics*. Springer-Verlag, New York, second edition, 2002.

[41] A. Bressan. *Hyperbolic systems of conservation laws*, volume 20 of *Oxford Lecture Series in Mathematics and its Applications*. Oxford University Press, Oxford, 2000.

[42] M. Briani, C. La Chioma, and R. Natalini. Convergence of numerical schemes for viscosity solutions to integro-differential degenerate parabolic problems arising in financial theory. *Numer. Math.*, 98(4):607–646, 2004.

[43] K. Brusdal, H. K. Dahle, K. H. Karlsen, and T. Mannseth. A study of the modelling error in two operator splitting algorithms for porous media flow. *Comput. Geosci.*, 2(1):23–36, 1998.

[44] R. Bürger, A. Coronel, and M. Sepúlveda. A semi-implicit monotone difference scheme for an initial-boundary value problem of a strongly degenerate parabolic equation modeling sedimentation-consolidation processes. *Math. Comp.*, 75(253):91–112 (electronic), 2006.

[45] R. Bürger, S. Evje, and K. H. Karlsen. On strongly degenerate convection-diffusion problems modeling sedimentation-consolidation processes. *J. Math. Anal. Appl.*, 247(2):517–556, 2000.

[46] R. Bürger and K. H. Karlsen. On some upwind difference schemes for the phenomenological sedimentation-consolidation model. *J. Engrg. Math.*, 41(2-3):145–166, 2001.

[47] M. C. Bustos, F. Concha, R. Bürger, and E. M. Tory. *Sedimentation and thickening*, volume 8 of *Mathematical Modelling: Theory and Applications*. Kluwer Academic Publishers, Dordrecht, 1999.

[48] J. Carrillo. On the uniqueness of the solution of the evolution dam problem. *Nonlinear Anal.*, 22(5):573–607, 1994.

[49] J. Carrillo. Solutions entropiques de problèmes non linéaires dégénérés. *C. R. Acad. Sci. Paris Sér. I Math.*, 327(2):155–160, 1998.

[50] J. Carrillo. Entropy solutions for nonlinear degenerate problems. *Arch. Ration. Mech. Anal.*, 147(4):269–361, 1999.

[51] R. Cavazzoni. Diffusive approximations for a class of nonlinear parabolic equations. *NoDEA Nonlinear Differential Equations Appl.*, 12(3):275–293, 2005.

[52] A. Chalabi and J.-P. Vila. Operator splitting, fractional steps method and entropy condition. In *Third International Conference on Hyperbolic Problems, Vol. I, II (Uppsala, 1990)*, pages 226–240. Studentlitteratur, Lund, 1991.

[53] A. Chambolle and B. J. Lucier. Un principe du maximum pour des opérateurs monotones. *C. R. Acad. Sci. Paris Sér. I Math.*, 326(7):823–827, 1998.

[54] G. Chavent and J. Jaffré. *Mathematical models and finite elements for reservoir simulation: single phase, multiphase, and multicomponent flows through porous media*, volume v. 17. North-Holland, Amsterdam, 1986.

[55] G.-Q. Chen and E. DiBenedetto. Stability of entropy solutions to the Cauchy problem for a class of nonlinear hyperbolic-parabolic equations. *SIAM J. Math. Anal.*, 33(4):751–762 (electronic), 2001.

[56] G.-Q. Chen and K. H. Karlsen. Quasilinear anisotropic degenerate parabolic equations with time-space dependent diffusion coefficients. *Commun. Pure Appl. Anal.*, 4(2):241–266, 2005.

[57] G.-Q. Chen and K. H. Karlsen. L^1-framework for continuous dependence and error estimates for quasilinear anisotropic degenerate parabolic equations. *Trans. Amer. Math. Soc.*, 358(3):937–963 (electronic), 2006.

[58] G.-Q. Chen and B. Perthame. Well-posedness for non-isotropic degenerate parabolic-hyperbolic equations. *Ann. Inst. H. Poincaré Anal. Non Linéaire*, 20(4):645–668, 2003.

[59] G.-Q. Chen and D. Wang. The Cauchy problem for the Euler equations for compressible fluids. In *Handbook of mathematical fluid dynamics, Vol. I*, pages 421–543. North-Holland, Amsterdam, 2002.

[60] Z. Chen and G. Ji. Sharp L^1 a posteriori error analysis for nonlinear convection-diffusion problems. *Math. Comp.*, 75(253):43–71 (electronic), 2006.

[61] A. Chertock, A. Kurganov, and G. Petrova. Fast explicit operator splitting method. Application to the polymer system. In *Finite Volumes for Complex Applications IV (2005)*, pages 63–72. ISTE Publishing Company, 2006.

[62] A. Chertock, A. Kurganov, and G. Petrova. Fast explicit operator splitting method for convection-diffusion equation. *Internat. J. Numer. Methods Fluids*, 59(3): 309–332.

[63] A. J. Chorin, M. F. McCracken, T. J. R. Hughes, and J. E. Marsden. Product formulas and numerical algorithms. *Comm. Pure Appl. Math.*, 31(2):205–256, 1978.

[64] B. Cockburn, F. Coquel, and P. Le Floch. An error estimate for finite volume methods for multidimensional conservation laws. *Math. Comp.*, 63(207):77–103, 1994.

[65] B. Cockburn and P.-A. Gremaud. A priori error estimates for numerical methods for scalar conservation laws. I. The general approach. *Math. Comp.*, 65(214):533–573, 1996.

[66] B. Cockburn and G. Gripenberg. Continuous dependence on the nonlinearities of solutions of degenerate parabolic equations. *J. Differential Equations*, 151(2):231–251, 1999.

[67] B. Cockburn, C. Johnson, C.-W. Shu, and E. Tadmor. *Advanced numerical approximation of nonlinear hyperbolic equations*, volume 1697 of *Lecture Notes in Mathematics*. Springer-Verlag, Berlin, 1998.

[68] P. Colella, A. Majda, and V. Roytburd. Theoretical and numerical structure for reacting shock waves. *SIAM J. Sci. Stat. Comput.*, 7(4):1059–1080, 1986.

[69] V. Comincioli and G. Toscani. Operator splitting for the Boltzmann equation of a Maxwell gas. In *Boundary value problems for partial differential equations and applications*, volume 29 of *RMA Res. Notes Appl. Math.*, pages 345–350. Masson, Paris, 1993.

[70] R. Courant, K. Friedrichs, and H. Lewy. Über die partiellen Differenzengleichungen der mathematischen Physik. *Math. Ann.*, 100(1):32–74, 1928.

[71] M. Crandall and A. Majda. The method of fractional steps for conservation laws. *Numer. Math.*, 34(3):285–314, 1980.

[72] C. M. Dafermos. Polygonal approximations of solutions of the initial value problem for a conservation law. *J. Math. Anal. Appl.*, 38:33–41, 1972.

[73] C. M. Dafermos. *Hyperbolic conservation laws in continuum physics*, volume 325 of *Grundlehren der Mathematischen Wissenschaften*. Springer-Verlag, Berlin, third edition, 2010.

[74] H. K. Dahle. *Adaptive characteristic operator splitting techniques for convection-dominated diffusion problems in one and two space dimensions.* PhD thesis, Department of Mathematics, University of Bergen, Norway, 1988.

[75] H. K. Dahle, M. S. Espedal, and R. E. Ewing. Characteristic Petrov–Galerkin subdomain methods for convection-diffusion problems. In *Numerical simulation in oil recovery (Minneapolis, Minn., 1986)*, volume 11 of *IMA Vol. Math. Appl.*, pages 77–87. Springer, New York, 1988.

[76] H. K. Dahle, M. S. Espedal, R. E. Ewing, and O. Sævareid. Characteristic adaptive subdomain methods for reservoir flow problems. *Numer. Methods Partial Differential Equations*, 6(4):279–309, 1990.

[77] H. K. Dahle, M. S. Espedal, and O. Sævareid. Characteristic, local grid refinement techniques for reservoir flow problems. *International Journal for Numerical Methods in Enginering*, 34:1051–1069, 1992.

[78] H. K. Dahle, R. E. Ewing, and T. F. Russell. Eulerian–Lagrangian localized adjoint methods for a nonlinear advection-diffusion equation. *Comput. Methods Appl. Mech. Engrg.*, 122(3-4):223–250, 1995.

[79] P. D'Ancona and S. Spagnolo. The Cauchy problem for weakly parabolic systems. *Math. Ann.*, 309(2):307–330, 1997.

[80] A. Datta-Gupta and M. J. King. *Streamline Simulation: Theory and Practice.* Textbook Series. Society of Petroleum Engineers, 2007.

[81] C. Dawson. Godunov-mixed methods for advection-diffusion equations in multidimensions. *SIAM J. Numer. Anal.*, 30(5):1315–1332, 1993.

[82] C. Dawson. High resolution upwind-mixed finite element methods for advection-diffusion equations with variable time-stepping. *Numer. Methods Partial Differential Equations*, 11(5):525–538, 1995.

[83] C. N. Dawson. Godunov-mixed methods for advective flow problems in one space dimension. *SIAM J. Numer. Anal.*, 28(5):1282–1309, 1991.

[84] C. N. Dawson and M. F. Wheeler. Time-splitting methods for advection-diffusion-reaction equations arising in contaminant transport. In *ICIAM 91 (Washington, DC, 1991)*, pages 71–82. SIAM, Philadelphia, PA, 1992.

[85] K. Deimling. *Nonlinear functional analysis.* Springer-Verlag, Berlin, 1985.

[86] S. Descombes. Convergence of a splitting method of high order for reaction-diffusion systems. *Math. Comp.*, 70(236):1481–1501 (electronic), 2001.

[87] S. Descombes and M. Massot. Operator splitting for nonlinear reaction-diffusion systems with an entropic structure: singular perturbation and order reduction. *Numer. Math.*, 97(4):667–698, 2004.

[88] S. Descombes and M. Schatzman. On Richardson extrapolation of Strang formula for reaction-diffusion equations. In *Équations aux dérivées partielles et applications*, pages 429–452. Gauthier-Villars, Éd. Sci. Méd. Elsevier, Paris, 1998.

[89] S. Descombes and M. Schatzman. Strang's formula for holomorphic semi-groups. *J. Math. Pures Appl. (9)*, 81(1):93–114, 2002.

[90] J. Douglas, Jr. and H. H. Rachford, Jr. On the numerical solution of heat conduction problems in two and three space variables. *Trans. Amer. Math. Soc.*, 82:421–439, 1956.

[91] S. D. Èĭdel′man. *Parabolic systems.* North-Holland Publishing Co., Amsterdam, 1969.

[92] M. S. Espedal and R. E. Ewing. Characteristic Petrov–Galerkin subdomain methods for two-phase immiscible flow. *Comput. Methods Appl. Mech. Engrg.*, 64(1-3):113–135, 1987.

[93] M. S. Espedal and K. H. Karlsen. Numerical solution of reservoir flow models based on large time step operator splitting algorithms. In *Filtration in Porous Media and Industrial Applications (Cetraro, Italy, 1998)*, volume 1734 of *Lecture Notes in Mathematics*, pages 9–77. Springer, Berlin, 2000.

[94] L. C. Evans. *Partial differential equations.* American Mathematical Society, Providence, RI, 1998.

[95] L. C. Evans and R. F. Gariepy. *Measure theory and fine properties of functions.* Studies in Advanced Mathematics. CRC Press, Boca Raton, FL, 1992.

[96] S. Evje and K. H. Karlsen. Degenerate convection-diffusion equations and implicit monotone difference schemes. In *Hyperbolic problems: theory, numerics, applications, Vol. I (Zürich, 1998)*, pages 285–294. Birkhäuser, Basel, 1999.

[97] S. Evje and K. H. Karlsen. Viscous splitting approximation of mixed hyperbolic-parabolic convection-diffusion equations. *Numer. Math.*, 83(1):107–137, 1999.

[98] S. Evje and K. H. Karlsen. Discrete approximations of BV solutions to doubly nonlinear degenerate parabolic equations. *Numer. Math.*, 86(3):377–417, 2000.

[99] S. Evje and K. H. Karlsen. Monotone difference approximations of *BV* solutions to degenerate convection-diffusion equations. *SIAM J. Numer. Anal.*, 37(6):1838–1860 (electronic), 2000.

[100] S. Evje, K. H. Karlsen, K.-A. Lie, and N. H. Risebro. Front tracking and operator splitting for nonlinear degenerate convection-diffusion equations. In *Parallel solution of partial differential equations (Minneapolis, MN, 1997)*, volume 120 of *IMA Vol. Math. Appl.*, pages 209–227. Springer, New York, 2000.

[101] R. Eymard, T. Gallouët, and R. Herbin. Finite volume methods. In *Handbook of numerical analysis, Vol. VII*, Handb. Numer. Anal., VII, pages 713–1020. North-Holland, Amsterdam, 2000.

[102] R. Eymard, T. Gallouët, R. Herbin, and A. Michel. Convergence of a finite volume scheme for nonlinear degenerate parabolic equations. *Numer. Math.*, 92(1):41–82, 2002.

[103] M. Falcone, P. Lanucara, and A. Seghini. A splitting algorithm for Hamilton–Jacobi–Bellman equations. *Appl. Numer. Math.*, 15(2):207–218, 1994.

[104] M. Feistauer, J. Felcman, and I. Straskraba. *Mathematical and Computational Methods for Compressible Flow.* Oxford University Press, USA, 2003.

[105] R. H. J. Gmelig Meyling. A characteristic finite element method for solving non-linear convection-diffusion equations on locally refined grids. In D. Guerillot and O. Guillon, editors, *2nd European Conference on the Mathematics of Oil Recovery*, pages 255–262, Arles, France, Sept 11-14 1990. Editions Technip.

[106] R. H. J. Gmelig Meyling. Numerical methods for solving the nonlinear hyperbolic equations of porous media flow. In *Third International Conference on Hyperbolic Problems, Vol. I, II (Uppsala, 1990)*, pages 503–517. Studentlitteratur, Lund, 1991.

[107] E. Godlewski and P.-A. Raviart. *Hyperbolic systems of conservation laws*, volume 3/4 of *Mathématiques & Applications (Paris).* Ellipses, Paris, 1991.

[108] E. Godlewski and P.-A. Raviart. *Numerical approximation of hyperbolic systems of conservation laws*, volume 118 of *Applied Mathematical Sciences.* Springer-Verlag, New York, 1996.

[109] J. J. Gottlieb and C. P. T. Groth. Assessment of riemann solvers for unsteady one-dimensional inviscid flows of perfect gases. *J. Comput. Phys.*, 78(2):437–458, Oct 1988.

[110] J. M. Greenberg and A. Y. Leroux. A well-balanced scheme for the numerical processing of source terms in hyperbolic equations. *SIAM J. Numer. Anal.*, 33(1):1–16, 1996.

[111] F. R. Guarguaglini, V. Milišić, and A. Terracina. A discrete BGK approximation for strongly degenerate parabolic problems with boundary conditions. *J. Differential Equations*, 202(2):183–207, 2004.

[112] W. Hackbusch. *Elliptic differential equations*, volume 18 of *Springer Series in Computational Mathematics*. Springer-Verlag, Berlin, 1992.

[113] H. Hanche-Olsen and H. Holden. The Kolmogorov–Riesz compactness theorem. *Expo. Math.*, to appear.

[114] V. Haugse, K. H. Karlsen, K.-A. Lie, and J. R. Natvig. Numerical solution of the polymer system by front tracking. *Transp. Porous Media*, 44(1):63–83, 2001.

[115] C. Hirsch. *Numerical computation of internal and external flows*. John Wiley & Sons, 1990.

[116] R. Holdahl, H. Holden, and K.-A. Lie. Unconditionally stable splitting methods for the shallow water equations. *BIT*, 39(3):451–472, 1999.

[117] H. Holden and L. Holden. On scalar conservation laws in one-dimension. In S. Albeverio, J. E. Fenstad, H. Holden, and T. Lindstrøm, editors, *Ideas and Methods in Mathematics and Physics*, pages 480–509. Cambridge University Press, Cambridge, 1988.

[118] H. Holden, L. Holden, and R. Høegh-Krohn. A numerical method for first order nonlinear scalar conservation laws in one dimension. *Comput. Math. Appl.*, 15(6-8):595–602, 1988.

[119] H. Holden, K. H. Karlsen, and K.-A. Lie. Operator splitting methods for degenerate convection-diffusion equations. I. Convergence and entropy estimates. In *Stochastic processes, physics and geometry: new interplays, II (Leipzig, 1999)*, volume 29 of *CMS Conf. Proc.*, pages 293–316. Amer. Math. Soc., Providence, RI, 2000.

[120] H. Holden, K. H. Karlsen, and K.-A. Lie. Operator splitting methods for degenerate convection-diffusion equations II: numerical examples with emphasis on reservoir simulation and sedimentation. *Comput. Geosci.*, 4(4):287–322, 2000.

[121] H. Holden, K. H. Karlsen, and N. H. Risebro. Operator splitting methods for generalized Korteweg–de Vries equations. *J. Comput. Phys.*, 153(1):203–222, 1999.

[122] H. Holden, K. H. Karlsen, and N. H. Risebro. On uniqueness and existence of entropy solutions of weakly coupled systems of nonlinear degenerate parabolic equations. *Electron. J. Differential Equations*, pages No. 46, 31 pp. (electronic), 2003.

[123] H. Holden, K. H. Karlsen, N. H. Risebro, and T. Tao. Operator splitting for the KdV equation. *Math. Comp.*, to appear.

[124] H. Holden, K.-A. Lie, and N. H. Risebro. An unconditionally stable method for the Euler equations. *J. Comput. Phys.*, 150(1):76–96, 1999.

[125] H. Holden and N. H. Risebro. A method of fractional steps for scalar conservation laws without the CFL condition. *Math. Comp.*, 60(201):221–232, 1993.

[126] H. Holden and N. H. Risebro. *Front Tracking for Hyperbolic Conservation Laws.* Volume 152 of *Applied Mathematical Sciences.* Springer-Verlag, New York. Corrected 2nd printing 2007.

[127] Z. Huang, S. Jin, P. A. Markowich, C. Sparber, and C. Zheng. A time-splitting spectral scheme for the Maxwell–Dirac system. *J. Comput. Phys.*, 208(2):761–789, 2005.

[128] F. Hubert. Global existence for hyperbolic-parabolic systems with large periodic initial data. *Differential Integral Equations*, 11(1):69–83, 1998.

[129] W. Hundsdorfer and J. Verwer. *Numerical solution of time-dependent advection-diffusion-reaction equations*, volume 33 of *Springer Series in Computational Mathematics.* Springer-Verlag, Berlin, 2003.

[130] N. Igbida and J. M. Urbano. Uniqueness for nonlinear degenerate problems. *NoDEA Nonlinear Differential Equations Appl.*, 10(3):287–307, 2003.

[131] E. R. Jakobsen and K. H. Karlsen. Convergence rates for semi-discrete splitting approximations for degenerate parabolic equations with source terms. *BIT Numerical Mathematics*, 45(1):37–67, 2005.

[132] E. R. Jakobsen, K. H. Karlsen, and N. H. Risebro. On the convergence rate of operator splitting for Hamilton–Jacobi equations with source terms. *SIAM J. Numer. Anal.*, 39(2):499–518 (electronic), 2001.

[133] E. R. Jakobsen, K. H. Karlsen, and N. H. Risebro. On the convergence rate of operator splitting for weakly coupled systems of Hamilton–Jacobi equations. In *Hyperbolic problems: theory, numerics, applications, Vol. I, II (Magdeburg, 2000)*, volume 141 of *Internat. Ser. Numer. Math., 140*, pages 553–562. Birkhäuser, Basel, 2001.

[134] G.-S. Jiang and E. Tadmor. Nonoscillatory central schemes for multidimensional hyperbolic conservation laws. *SIAM J. Sci. Comput.*, 19(6):1892–1917 (electronic), 1998.

[135] S. Jin and Z. P. Xin. The relaxation schemes for systems of conservation laws in arbitrary space dimensions. *Comm. Pure Appl. Math.*, 48(3):235–276, 1995.

[136] C. Johnson. *Numerical solution of partial differential equations by the finite element method.* Cambridge University Press, Cambridge, 1987.

[137] J. Kačur. Solution of some free boundary problems by relaxation schemes. *SIAM J. Numer. Anal.*, 36(1):290–316 (electronic), 1999.

[138] J. Kačur. Solution of degenerate convection-diffusion problems by the method of characteristics. *SIAM J. Numer. Anal.*, 39(3):858–879 (electronic), 2001.

[139] K. H. Karlsen. On the accuracy of a dimensional splitting method for scalar conservation laws. Master's thesis, Department of Mathematics, University of Oslo, Oslo, Norway, 1994.

[140] K. H. Karlsen, K. Brusdal, H. K. Dahle, S. Evje, and K.-A. Lie. The corrected operator splitting approach applied to a nonlinear advection-diffusion problem. *Comput. Methods Appl. Mech. Engrg.*, 167(3-4):239–260, 1998.

[141] K. H. Karlsen and K.-A. Lie. An unconditionally stable splitting scheme for a class of nonlinear parabolic equations. *IMA J. Numer. Anal.*, 19(4):609–635, 1999.

[142] K. H. Karlsen, K.-A. Lie, J. R. Natvig, H. F. Nordhaug, and H. K. Dahle. Operator splitting methods for systems of convection-diffusion equations: nonlinear error mechanisms and correction strategies. *J. Comput. Phys.*, 173(2):636–663, 2001.

[143] K. H. Karlsen, K.-A. Lie, N. H. Risebro, and J. Frøyen. A front-tracking approach to a two-phase fluid-flow model with capillary forces. *In Situ*, 22(1):59–89, 1998.

[144] K. H. Karlsen and M. Ohlberger. A note on the uniqueness of entropy solutions of nonlinear degenerate parabolic equations. *J. Math. Anal. Appl.*, 275(1):439–458, 2002.

[145] K. H. Karlsen and N. H. Risebro. An operator splitting method for nonlinear convection-diffusion equations. *Numer. Math.*, 77(3):365–382, 1997.

[146] K. H. Karlsen and N. H. Risebro. Corrected operator splitting for nonlinear parabolic equations. *SIAM J. Numer. Anal.*, 37(3):980–1003 (electronic), 2000.

[147] K. H. Karlsen and N. H. Risebro. Convergence of finite difference schemes for viscous and inviscid conservation laws with rough coefficients. *M2AN Math. Model. Numer. Anal.*, 35(2):239–269, 2001.

[148] K. H. Karlsen and N. H. Risebro. On the uniqueness and stability of entropy solutions of nonlinear degenerate parabolic equations with rough coefficients. *Discrete Contin. Dyn. Syst.*, 9(5):1081–1104, 2003.

[149] K. H. Karlsen, N. H. Risebro, and J. D. Towers. Upwind difference approximations for degenerate parabolic convection-diffusion equations with a discontinuous coefficient. *IMA J. Numer. Anal.*, 22(4):623–664, 2002.

[150] K. H. Karlsen, N. H. Risebro, and J. D. Towers. L^1 stability for entropy solutions of nonlinear degenerate parabolic convection-diffusion equations with discontinuous coefficients. *Skr. K. Nor. Vidensk. Selsk.*, (3):1–49, 2003.

[151] K. H. Karlsen and J. D. Towers. Convergence of the Lax–Friedrichs scheme and stability for conservation laws with a discontinous space-time dependent flux. *Chinese Ann. Math. Ser. B*, 25(3):287–318, 2004.

[152] T. Kato. On the Trotter–Lie product formula. *Proc. Japan Acad.*, 50:694–698, 1974.

[153] C. Klingenberg and N. H. Risebro. Convex conservation laws with discontinuous coefficients. Existence, uniqueness and asymptotic behavior. *Comm. Partial Differential Equations*, 20(11-12):1959–1990, 1995.

[154] P. Knabner and L. Angermann. *Numerical methods for elliptic and parabolic partial differential equations*, volume 44 of *Texts in Applied Mathematics*. Springer-Verlag, New York, 2003.

[155] Y. Kobayashi. Product formula for nonlinear semigroups in Hilbert spaces. *Proc. Japan Acad. Ser. A Math. Sci.*, 58(10):425–428, 1982.

[156] Y. Kobayashi. A product formula approach to first order quasilinear equations. *Hiroshima Math. J.*, 14(3):489–509, 1985.

[157] Y. Kobayashi. Product formula for nonlinear contraction semigroups in Banach spaces. *Hiroshima Math. J.*, 17(1):129–140, 1987.

[158] H.-O. Kreiss and J. Lorenz. *Initial-boundary value problems and the Navier–Stokes equations.* Academic Press Inc., Boston, MA, 1989.

[159] D. Kröner. *Numerical schemes for conservation laws.* Wiley-Teubner Series Advances in Numerical Mathematics. John Wiley & Sons Ltd., Chichester, 1997.

[160] S. N. Kružkov. Results on the nature of the continuity of solutions of parabolic equations, and certain applications thereof. *Mat. Zametki*, 6:97–108, 1969.

[161] S. N. Kružkov. First order quasilinear equations with several independent variables. *Mat. Sb. (N.S.)*, 81 (123):228–255, 1970.

[162] A. Kurganov and D. Levy. A third-order semidiscrete central scheme for conservation laws and convection-diffusion equations. *SIAM J. Sci. Comput.*, 22(4):1461–1488 (electronic), 2000.

[163] A. Kurganov, S. Noelle, and G. Petrova. Semidiscrete central-upwind schemes for hyperbolic conservation laws and Hamilton–Jacobi equations. *SIAM J. Sci. Comput.*, 23(3):707–740 (electronic), 2001.

[164] A. Kurganov and E. Tadmor. New high-resolution central schemes for nonlinear conservation laws and convection-diffusion equations. *J. Comput. Phys.*, 160(1):241–282, 2000.

[165] A. Kurganov and E. Tadmor. Solution of two-dimensional Riemann problems for gas dynamics without Riemann problem solvers. *Numer. Methods Partial Differential Equations*, 18(5):584–608, 2002.

[166] N. N. Kuznetsov. Stable methods for the solution of a first order quasilinear equation in a class of discontinuous functions. *Soviet Math. Dokl.*, 16(6):1569–1573, 1975.

[167] N. N. Kuznetsov. The accuracy of certain approximate methods for the computation of weak solutions of a first order quasilinear equation. *U.S.S.R. Computational Math. and Math. Phys.*, 16(6):105–119, 1976.

[168] O. A. Ladyženskaja, V. A. Solonnikov, and N. N. Ural′ceva. *Linear and quasilinear equations of parabolic type.* American Mathematical Society, Providence, R.I., 1967.

[169] J. O. Langseth. On an implementation of a front tracking method for hyperbolic conservation laws. *Advances in Engineering Software*, 26(1):45–63, 1996.

[170] J. O. Langseth, A. Tveito, and R. Winther. On the convergence of operator splitting applied to conservation laws with source terms. *SIAM J. Numer. Anal.*, 33(3):843–863, 1996.

[171] D. Lanser and J. G. Verwer. Analysis of operator splitting for advection-diffusion-reaction problems from air pollution modelling. *J. Comput. Appl. Math.*, 111(1-2):201–216, 1999.

[172] P. Lax and B. Wendroff. Systems of conservation laws. *Comm. Pure Appl. Math.*, 13:217–237, 1960.

[173] P. D. Lax and X.-D. Liu. Solution of two-dimensional Riemann problems of gas dynamics by positive schemes. *SIAM J. Sci. Comput.*, 19(2):319–340 (electronic), 1998.

[174] R. J. LeVeque. A large time step generalization of Godunov's method for systems of conservation laws. *SIAM J. Numer. Anal.*, 22(6):1051–1073, 1985.

[175] R. J. LeVeque. *Numerical methods for conservation laws.* Lectures in Mathematics ETH Zürich. Birkhäuser Verlag, Basel, second edition, 1992.

[176] R. J. LeVeque. *Finite volume methods for hyperbolic problems.* Cambridge Texts in Applied Mathematics. Cambridge University Press, Cambridge, 2002.

[177] L. Lévi and F. Peyroutet. A time-fractional step method for conservation law related obstacle problems. *Adv. in Appl. Math.*, 27(4):768–789, 2001.

[178] L. Lévi, E. Rouvre, and G. Vallet. Weak entropy solutions for degenerate parabolic-hyperbolic inequalities. *Appl. Math. Lett.*, 18(5):497–504, 2005.

[179] D. Levy, G. Puppo, and G. Russo. Central WENO schemes for hyperbolic systems of conservation laws. *M2AN Math. Model. Numer. Anal.*, 33(3):547–571, 1999.

[180] K.-A. Lie. A dimensional splitting method for quasilinear hyperbolic equations with variable coefficients. *BIT*, 39(4):683–700, 1999.

[181] K.-A. Lie. Front tracking for one-dimensional quasilinear hyperbolic equations with variable coefficients. *Numer. Algorithms*, 24(3):275–298, 2000.

[182] K.-A. Lie, V. Haugse, and K. Hvistendahl Karlsen. Dimensional splitting with front tracking and adaptive grid refinement. *Numer. Methods Partial Differential Equations*, 14(5):627–648, 1998.

[183] K.-A. Lie and S. Noelle. An improved quadrature rule for the flux-computation in staggered central difference schemes in multidimensions. *J. Sci. Comput.*, 18(1):69–81, 2003.

[184] K.-A. Lie and S. Noelle. On the artificial compression method for second-order nonoscillatory central difference schemes for systems of conservation laws. *SIAM J. Sci. Comput.*, 24(4):1157–1174 (electronic), 2003.

[185] E. H. Lieb and M. Loss. *Analysis*, volume 14 of *Graduate Studies in Mathematics*. American Mathematical Society, Providence, RI, second edition, 2001.

[186] P. Lin, K. W. Morton, and E. Süli. Euler characteristic Galerkin scheme with recovery. *RAIRO Modél. Math. Anal. Numér.*, 27(7):863–894, 1993.

[187] P. Lin, K. W. Morton, and E. Süli. Characteristic Galerkin schemes for scalar conservation laws in two and three space dimensions. *SIAM J. Numer. Anal.*, 34(2):779–796, 1997.

[188] P.-L. Lions and B. Mercier. Splitting algorithms for the sum of two nonlinear operators. *SIAM J. Numer. Anal.*, 16(6):964–979, 1979.

[189] P.-L. Lions and B. Mercier. Approximation numérique des équations de Hamilton–Jacobi–Bellman. *RAIRO Anal. Numér.*, 14(4):369–393, 1980.

[190] R. Liska and B. Wendroff. Composite schemes for conservation laws. *SIAM J. Numer. Anal.*, 35(6):2250–2271 (electronic), 1998.

[191] X.-D. Liu and E. Tadmor. Third order nonoscillatory central scheme for hyperbolic conservation laws. *Numer. Math.*, 79(3):397–425, 1998.

[192] T. Lu, P. Neittaanmäki, and X.-C. Tai. A parallel splitting up method and its application to Navier–Stokes equations. *Appl. Math. Lett.*, 4(2):25–29, 1991.

[193] T. Lu, P. Neittaanmäki, and X.-C. Tai. A parallel splitting-up method for partial differential equations and its applications to Navier–Stokes equations. *RAIRO Modél. Math. Anal. Numér.*, 26(6):673–708, 1992.

[194] B. J. Lucier. Error bounds for the methods of Glimm, Godunov and LeVeque. *SIAM J. Numer. Anal.*, 22(6):1074–1081, 1985.

[195] B. J. Lucier. A moving mesh numerical method for hyperbolic conservation laws. *Math. Comp.*, 46(173):59–69, 1986.

[196] B. J. Lucier. On nonlocal monotone difference schemes for scalar conservation laws. *Math. Comp.*, 47(175):19–36, 1986.

[197] E. Maitre. Numerical analysis of nonlinear elliptic-parabolic equations. *M2AN Math. Model. Numer. Anal.*, 36(1):143–153, 2002.

[198] A. Majda. A qualitative model for dynamic combustion. *SIAM J. Appl. Math.*, 41(1):70–93, 1981.

[199] A. Majda and R. Rosales. Resonantly interacting weakly nonlinear hyperbolic waves. I. A single space variable. *Stud. Appl. Math.*, 71(2):149–179, 1984.

[200] A. J. Majda and A. L. Bertozzi. *Vorticity and incompressible flow*, volume 27 of *Cambridge Texts in Applied Mathematics*. Cambridge University Press, Cambridge, 2002.

[201] J. Málek, J. Nečas, M. Rokyta, and M. Ružička. *Weak and measure-valued solutions to evolutionary PDEs*, volume 13 of *Applied Mathematics and Mathematical Computation*. Chapman & Hall, London, 1996.

[202] M. Maliki and H. Touré. Uniqueness of entropy solutions for nonlinear degenerate parabolic problems. *J. Evol. Equ.*, 3(4):603–622, 2003.

[203] G. I. Marchuk. Splitting and alternating direction methods. In *Handbook of numerical analysis, Vol. I*, Handb. Numer. Anal., I, pages 197–462. North-Holland, Amsterdam, 1990.

[204] C. Marle. *Multiphase flow in porous media.* Institut Français du Pétrole Publications. Editions Technip, Paris, 1981.

[205] C. Mascia, A. Porretta, and A. Terracina. Nonhomogeneous Dirichlet problems for degenerate parabolic-hyperbolic equations. *Arch. Ration. Mech. Anal.*, 163(2):87–124, 2002.

[206] R. I. McLachlan and G. R. W. Quispel. Splitting methods. *Acta Numer.*, 11:341–434, 2002.

[207] A. Michel and J. Vovelle. Entropy formulation for parabolic degenerate equations with general Dirichlet boundary conditions and application to the convergence of FV methods. *SIAM J. Numer. Anal.*, 41(6):2262–2293 (electronic), 2003.

[208] C. Moler and C. Van Loan. Nineteen dubious ways to compute the exponential of a matrix, twenty-five years later. *SIAM Rev.*, 45(1):3–49 (electronic), 2003.

[209] K. W. Morton. *Numerical solution of convection-diffusion problems*, volume 12 of *Applied Mathematics and Mathematical Computation.* Chapman & Hall, London, 1996.

[210] K. Nakajo and W. Takahashi. Strong and weak convergence theorems by an improved splitting method. *Comm. Appl. Nonlinear Anal.*, 9(2):99–107, 2002.

[211] R. Natalini. Recent results on hyperbolic relaxation problems. In *Analysis of systems of conservation laws (Aachen, 1997)*, pages 128–198. Chapman & Hall/CRC, Boca Raton, FL, 1999.

[212] R. Natalini and B. Hanouzet. Weakly coupled systems of quasilinear hyperbolic equations. *Differential Integral Equations*, 9(6):1279–1292, 1996.

[213] H. Nessyahu and E. Tadmor. Nonoscillatory central differencing for hyperbolic conservation laws. *J. Comput. Phys.*, 87(2):408–463, 1990.

[214] S. Noelle, N. Pankratz, G. Puppo, and J. R. Natvig. Well-balanced finite volume schemes of arbitrary order of accuracy for shallow water flows. *J. Comput. Phys.*, 213(2):474–499, 2006.

[215] J. Norbury and A. M. Stuart. A model for porous-medium combustion. *Quart. J. Mech. Appl. Math.*, 42(1):159–178, 1989.

[216] M. Ohlberger. A posteriori error estimates for vertex centered finite volume approximations of convection-diffusion-reaction equations. *M2AN Math. Model. Numer. Anal.*, 35(2):355–387, 2001.

[217] M. Ohlberger and C. Rohde. Adaptive finite volume approximations for weakly coupled convection dominated parabolic systems. *IMA J. Numer. Anal.*, 22(2):253–280, 2002.

[218] I. Osako, A. Datta-Gupta, and M. J. King. Timestep selection during streamline simulation through transverse flux correction. *SPE Journal*, 9(4):450–464, Dec 2004.

[219] F. Otto. L^1-contraction and uniqueness for quasilinear elliptic-parabolic equations. *J. Differential Equations*, 131(1):20–38, 1996.

[220] D. W. Peaceman and H. H. Rachford, Jr. The numerical solution of parabolic and elliptic differential equations. *J. Soc. Indust. Appl. Math.*, 3:28–41, 1955.

[221] B. Perthame. *Kinetic formulation of conservation laws*, volume 21 of *Oxford Lecture Series in Mathematics and its Applications*. Oxford University Press, Oxford, 2002.

[222] B. Perthame and P. E. Souganidis. Dissipative and entropy solutions to non-isotropic degenerate parabolic balance laws. *Arch. Ration. Mech. Anal.*, 170(4):359–370, 2003.

[223] F. Peyroutet and M. Madaune-Tort. Error estimate for a splitting method applied to convection-reaction equations. *Math. Models Methods Appl. Sci.*, 11(6):1081–1100, 2001.

[224] P. A. C. Raats, P. Dewilligen, and R. G. Gerritse. Transport and fixation of phosphate in acid, homogeneous soils. 1. physico-mathematical model. *Agriculture and Environment*, 7(2):149–160, 1982.

[225] M. Reed and B. Simon. *Methods of modern mathematical physics. I. Functional analysis.* Academic Press Inc., New York, second edition, 1980.

[226] M. Renardy. A degenerate parabolic-hyperbolic system modeling the spreading of surfactants. *SIAM J. Math. Anal.*, 28(5):1048–1063, 1997.

[227] P. G. Rodriguez, M. K. Segura, and F. J. M. Moreno. Streamline methodology using an efflcient operator splitting for accurate modelling of capillarity and gravity effects. In *SPE Reservoir Simulation Symposium*, Houston, Texas, 3–5 February 2003. SPE 79693.

[228] C. Rohde. Entropy solutions for weakly coupled hyperbolic systems in several space dimensions. *Z. Angew. Math. Phys.*, 49(3):470–499, 1998.

[229] C. Rohde. Upwind finite volume schemes for weakly coupled hyperbolic systems of conservation laws in 2D. *Numer. Math.*, 81(1):85–123, 1998.

[230] H.-G. Roos, M. Stynes, and L. Tobiska. *Numerical methods for singularly perturbed differential equations*, volume 24 of *Springer Series in Computational Mathematics*. Springer-Verlag, Berlin, 1996.

[231] É. Rouvre and G. Gagneux. Solution forte entropique de lois scalaires hyperboliques-paraboliques dégénérées. *C. R. Acad. Sci. Paris Sér. I Math.*, 329(7):599–602, 1999.

[232] E. Rouvre and G. Gagneux. Formulation forte entropique de lois scalaires hyperboliques-paraboliques dégénérées. *Ann. Fac. Sci. Toulouse Math. (6)*, 10(1):163–183, 2001.

[233] A. A. Samarskii, V. A. Galaktionov, S. P. Kurdyumov, and A. P. Mikhailov. *Blow-up in quasilinear parabolic equations*, volume 19 of *de Gruyter Expositions in Mathematics*. Walter de Gruyter & Co., Berlin, 1995.

[234] R. Sanders. On convergence of monotone finite difference schemes with variable spatial differencing. *Math. Comp.*, 40(161):91–106, 1983.

[235] M. Schatzman. Numerical integration of reaction-diffusion systems. *Numer. Algorithms*, 31(1-4):247–269, 2002.

[236] E. Schneid, P. Knabner, and F. Radu. A priori error estimates for a mixed finite element discretization of the Richards' equation. *Numer. Math.*, 98(2):353–370, 2004.

[237] E. Schneid, P. Knabner, and F. Radu. A priori error estimates for a mixed finite element discretization of the Richards' equation. *Numer. Math.*, 98(2):353–370, 2004.

[238] C. W. Schulz-Rinne, J. P. Collins, and H. M. Glaz. Numerical solution of the Riemann problem for two-dimensional gas dynamics. *SIAM J. Sci. Comput.*, 14(6):1394–1414, 1993.

[239] D. Serre. *Systems of conservation laws. 1.* Cambridge University Press, Cambridge, 1999.

[240] D. Serre. *Systems of conservation laws. 2.* Cambridge University Press, Cambridge, 2000.

[241] M. Shashkov. *Conservative finite-difference methods on general grids*. Symbolic and Numeric Computation Series. CRC Press, Boca Raton, FL, 1996. With 1 IBM-PC floppy disk (3.5 inch; HD).

[242] G. R. Shubin and J. B. Bell. An analysis of the grid orientation effect in numerical simulation of miscible displacement. *Comp. Meth. Appl. Mech. Eng.*, 47(1-2):47–71, 1984.

[243] J. Smoller. *Shock waves and reaction-diffusion equations*, volume 258 of *Grundlehren der Mathematischen Wissenschaften.* Springer-Verlag, New York, second edition, 1994.

[244] P. E. Souganidis. Max-min representations and product formulas for the viscosity solutions of Hamilton–Jacobi equations with applications to differential games. *Nonlinear Anal.*, 9(3):217–257, 1985.

[245] B. Sportisse. An analysis of operator splitting techniques in the stiff case. *J. Comput. Phys.*, 161(1):140–168, 2000.

[246] G. Strang. On the construction and comparison of difference schemes. *SIAM J. Numer. Anal.*, 5:506–517, 1968.

[247] J. C. Strikwerda. *Finite difference schemes and partial differential equations.* Wadsworth & Brooks/Cole Advanced Books & Software, Pacific Grove, CA, 1989.

[248] M. Sun. Alternating direction algorithms for solving Hamilton–Jacobi–Bellman equations. *Appl. Math. Optim.*, 34(3):267–277, 1996.

[249] E. Tadmor. Approximate solutions of nonlinear conservation laws. In *Advanced numerical approximation of nonlinear hyperbolic equations (Cetraro, 1997)*, volume 1697 of *Lecture Notes in Math.*, pages 1–149. Springer, Berlin, 1998.

[250] E. Tadmor. Central Station, http://www.cscamm.umd.edu/centpack/, 2006.

[251] E. Tadmor and T. Tao. Velocity averaging, kinetic formulations and regularizing effects in quasilinear pdes. *Comm. Pure Appl. Math.*, 60(10):1488–1521, 2007.

[252] S. Takagi. Renormalized dissipative solutions for quasilinear anisotropic degenerate parabolic equations. *Commun. Appl. Anal.*, 9(3-4):481–503, 2005.

[253] T. Tang. Convergence analysis for operator-splitting methods applied to conservation laws with stiff source terms. *SIAM J. Numer. Anal.*, 35(5):1939–1968 (electronic), 1998.

[254] T. Tang and Z. H. Teng. Error bounds for fractional step methods for conservation laws with source terms. *SIAM J. Numer. Anal.*, 32(1):110–127, 1995.

[255] F. Tappert. Numerical solutions of the Korteweg–de Vries equation and its generalizations by the split-step Fourier method. In A. C. Newell, editor, *Nonlinear Wave Motion*, volume 15 of *Lectures in Applied Mathematics*, pages 215—216. American Mathematical Society, Providence, R.I., 1974.

[256] M. E. Taylor. *Partial differential equations. III*, volume 117 of *Applied Mathematical Sciences*. Springer-Verlag, New York, 1997.

[257] Z. H. Teng. On the accuracy of fractional step methods for conservation laws in two dimensions. *SIAM J. Numer. Anal.*, 31(1):43–63, 1994.

[258] M. R. Thiele. Streamline simulation. In *8th International Forum on Reservoir Simulation*, Stresa/Lago Maggiore, Italy, June 20-24 2005. `http://www.streamsim.com/papers/rss2005.pdf`.

[259] J. W. Thomas. *Numerical partial differential equations: finite difference methods*, volume 22 of *Texts in Applied Mathematics*. Springer-Verlag, New York, 1995.

[260] J. W. Thomas. *Numerical partial differential equations*, volume 33 of *Texts in Applied Mathematics*. Springer-Verlag, New York, 1999.

[261] V. Thomée. *Galerkin finite element methods for parabolic problems*, volume 25 of *Springer Series in Computational Mathematics*. Springer-Verlag, Berlin, 1997.

[262] E. F. Toro. *Riemann solvers and numerical methods for fluid dynamics*. Springer-Verlag, Berlin, 1997.

[263] A. Tourin. Splitting methods for Hamilton–Jacobi equations. *Numer. Methods Partial Differential Equations*, 22(2):381–396, 2006.

[264] A. Tourin and T. Zariphopoulou. Numerical schemes for investment models with singular transactions. *Comput. Econom.*, 7(4):287–307, 1994.

[265] A. Tourin and T. Zariphopoulou. Viscosity solutions and numerical schemes for investment/consumption models with transaction costs. In *Numerical methods in finance*, Publ. Newton Inst., pages 245–269. Cambridge Univ. Press, Cambridge, 1997.

[266] J. D. Towers. Convergence of a difference scheme for conservation laws with a discontinuous flux. *SIAM J. Numer. Anal.*, 38(2):681–698 (electronic), 2000.

[267] J. D. Towers. A difference scheme for conservation laws with a discontinuous flux: the nonconvex case. *SIAM J. Numer. Anal.*, 39(4):1197–1218 (electronic), 2001.

[268] H. F. Trotter. On the product of semi-groups of operators. *Proc. Amer. Math. Soc.*, 10:545–551, 1959.

[269] A. Tveito and R. Winther. *Introduction to partial differential equations*, volume 29 of *Texts in Applied Mathematics*. Springer-Verlag, New York, 1998.

[270] G. Vallet. Dirichlet problem for a degenerated hyperbolic-parabolic equation. *Adv. Math. Sci. Appl.*, 15(2):423–450, 2005.

[271] A. I. Vol′pert and S. I. Hudjaev. The Cauchy problem for second order quasilinear degenerate parabolic equations. *Mat. Sb. (N.S.)*, 78 (120):374–396, 1969.

[272] J. Weickert, K. Zuiderveld, B. ter Haar Romeny, and W. Niessen. Parallel implementations of aos schemes: A fast way of nonlinear diffusion filtering. *icip*, 03:396, 1997.

[273] J. A. C. Weideman and B. M. Herbst. Split-step methods for the solution of the nonlinear Schrödinger equation. *SIAM J. Numer. Anal.*, 23(3):485–507, 1986.

[274] M. F. Wheeler, W. A. Kinton, and C. N. Dawson. Time-splitting for advection-dominated parabolic problems in one space variable. *Comm. Appl. Numer. Methods*, 4(3):413–423, 1988.

[275] Z. Wu, J. Zhao, J. Yin, and H. Li. *Nonlinear diffusion equations.* World Scientific Publishing Co. Inc., River Edge, NJ, 2001.

[276] Z. Q. Wu and J. X. Yin. Some properties of functions in BV_x and their applications to the uniqueness of solutions for degenerate quasilinear parabolic equations. *Northeast. Math. J.*, 5(4):395–422, 1989.

[277] Y. Xing and C.-W. Shu. High-order well-balanced finite difference weno schemes for a class of hyperbolic systems with source terms. *J. Sci. Comput.*, 27(1-3):477–494, 2006.

[278] N. N. Yanenko. *The method of fractional steps. The solution of problems of mathematical physics in several variables.* Springer-Verlag, New York, 1971.

[279] L. Ying. Viscosity splitting method for three-dimensional Navier–Stokes equations. *Acta Math. Sinica (N.S.)*, 4(3):210–226, 1988. A Chinese summary appears in Acta Math. Sinica **32** (1989), no. 4, 575.

[280] L. A. Ying. On the viscosity splitting method for initial-boundary value problems of the Navier–Stokes equations. *Chinese Ann. Math. Ser. B*, 10(4):487–512, 1989. A Chinese summary appears in Chinese Ann. Math. Ser. A **10** (1989), no. 5, 641.

[281] L. A. Ying. Viscosity splitting method in bounded domains. *Sci. China Ser. A*, 32(8):908–921, 1989.

[282] L. A. Ying. The viscosity splitting solutions of the Navier–Stokes equations. *J. Partial Differential Equations*, 3(4):31–48, 1990.

[283] L.-A. Ying. Viscous splitting for the unbounded problem of the Navier–Stokes equations. *Math. Comp.*, 55(191):89–113, 1990.

[284] L.-A. Ying. Viscous splitting for the unbounded problem of the Navier–Stokes equations. *Math. Comp.*, 55(191):89–113, 1990.

[285] L.-A. Ying. Viscosity-splitting scheme for the Navier–Stokes equations. *Numer. Methods Partial Differential Equations*, 7(4):317–338, 1991.

[286] L.-A. Ying. Convergence of Chorin–Marsden formula for the Navier–Stokes equations on convex domains. *J. Comput. Math.*, 17(1):73–88, 1999.

[287] L.-A. Ying. Convergence study of the Chorin–Marsden formula. *Math. Comp.*, 72(241):307–333 (electronic), 2003.

[288] W. P. Ziemer. *Weakly differentiable functions*, volume 120 of *Graduate Texts in Mathematics*. Springer-Verlag, New York, 1989.

Index